# *NONLINEAR EFFECTS IN PLASMA*

# NONLINEAR EFFECTS IN PLASMA


*Vadim N. Tsytovich*

*P. N. Lebedev Institute of Physics*
*Academy of Sciences of the USSR*
*Moscow*


*Translated from Russian by James S. Wood*

*Translation edited by Sydney M. Hamberger*


*Culham Laboratory*
*United Kingdom Atomic Energy Authority*
*Abingdon, England*



*PLENUM PRESS • NEW YORK–LONDON • 1970*

*Vadim Nikolaevich Tsytovich,* born in 1929, is a Senior Scientist of the P. N. Lebedev Institute of Physics of the Academy of Sciences of the USSR and a Doctor of Physicomathematical Sciences. Prior to 1956 his scientific interests centered on quantum electrodynamics. From 1957 to 1960 he elaborated the theory proposed by Wexler for coherent methods of charged particle acceleration. Since 1959 he has studied plasma physics and astrophysics. Tsytovich has been responsible, in particular, for investigations of the properties of relativistic plasma, the physics of turbulent plasma, and nonlinear effects in plasma.

The original Russian text, first published by Nauka Press in Moscow in 1967, has been corrected by the author for this edition. The present translation is published under an agreement with Mezhdunarodnaya Kniga, the Soviet book export agency.

*Вадим Николаевич Цытович*

**Нелинейные эффекты в плазме**

NELINEINYE EFFEKTY V PLAZME

*Library of Congress Catalog Card Number 69-12545*

SBN 306-30425-2


A Division of Plenum Publishing Corporation
227 West 17th Street, New York, N.Y. 10011

United Kingdom edition published by Plenum Press, London
A Division of Plenum Publishing Company, Ltd.
Donington House, 30 Norfolk Street, London W.C.2, England



Printed in the United States of America

# Preface to English Edition

In the short time that has elapsed since publication of the book the need has not yet arisen for major additions or alterations to the material presented. Nevertheless, the rate of development of scientific research is so enormous, particularly in the study of nonlinear wave processes in plasma, that there is always reason to take at least a cursory second look at some of the problems discussed in the book and to say a few words concerning further trends and applications.

In 1966 the author gave a short series of lectures on the nonlinear interaction of waves in plasma for members of the Lebedev Institute of Physics of the Academy of Sciences of the USSR. In revising this material for the present book, however, it was found necessary to expand considerably upon many of the pertinent problems and to discuss them in greater detail. The outcome has been a book that combines a lecture and a monograph style of presentation. The author has endeavored as far as possible to make the book assimilable to a wide range of readers, undergraduate and graduate students, and others interested in the most recent problems in plasma physics. Even among physicists directly concerned with research into plasma physics there exists a kind of "potential barrier," which stands between the ability to utilize nonlinear theory and the desire or need to utilize it. This situation arises primarily from the complexity and laboriousness of the original investigations, as well as the lack of simple approximate methods.

One of the objectives of the present book has been to try an break down this barrier, to show that there is nothing overly complex about nonlinear theory and that it can be formulated in fairly general terms. Using straightforward and almost trivial arguments about induced and spontaneous processes, the author has sought to focus attention primarily on the physical problems and their applications. At the same time the general proof of the equivalence of these concepts, for example, to the commonly-used methods of averaging over a statistical ensemble proves exceedingly difficult and presents a separate problem of a mathematical, rather than a physical, character.

These and other similar problems are deliberately avoided. Many of the estimates incorporated into the lectures for illustration have been retained where possible. Many different nonlinear effects become significant under certain, more or less complex experimental conditions, and these examples should give some idea of the method for obtaining certain desired information from theory. Some detailed remarks concerning the specific results of investigations on nonlinear interactions may be found in the Preface to the Russian edition.

Since publication of the book the development of the theory of the nonlinear properties of wave processes in plasma has led to an accumulation of certain material on the probabilities for various interaction processes, particularly for plasma in external magnetic and electric fields and for inhomogeneous and unconfined plasma, as well as on the development of experimental studies. Definite progress has been made in furthering the growth of the ideas discussed in the last chapter of the book. There has also been a stronger tendency to use the results of nonlinear theory in the physics of turbulent plasma, in the theory of electromagnetic wave propagation in turbulent plasma, in astronomical problems, and in problems involving the interaction of high-intensity electromagnetic waves with plasma. Any one of these is vast enough in scope to serve as the subject of a separate treatise.

The author would like to take this opportunity to express his grattitude to Sidney Hamberger for his excellent editting of the English edition.

The author hopes that these few remarks will prove helpful to the reader, and that the English language edition of the book will contribute to increased scientific communication and exchanges of views.

# Preface

For some time now there has been an interest in the nonlinear interaction of electromagnetic waves in plasma [1,2]. But only in the last few years has the theory of nonlinear wave interaction effects undergone such vigorous development as to result in the formulation of clear physical concepts regarding the mechanisms of interaction. This development has been engendered by attempts to solve many of the plasma-physical problems accompanying the tremendous growth of experimental research [3]. The importance of nonlinear effects in modern plasma physics is discussed in detail in Chap. I. At this point we merely stress the fact that today the analysis of nonlinear effects is a practical necessity in any experiment involving plasma instabilities. We should also point out that plasma instabilities can assert themselves extensively in solids (solid state plasma) and play an important part in the study of cosmic plasma. Consequently, the problems of nonlinear wave interaction in plasma are of concern to those working in widely different areas of physics. Yet it is difficult to assimilate the results of investigations on nonlinear effects, owing to the complicated way in which the results of original research are presented. In the present book the author hopes in some measure to fill the need for a text on the physics of nonlinear effects that is accessible to a fairly general audience.

The book represents the revision of some lectures delivered by the author on nonlinear effects before members of the Lebedev Institute of Physics of the Academy of Sciences of the USSR. The purpose of the book is to familiarize the reader with the fundamental physical notions and concepts and methods of estimating various nonlinear interactions, and to shed some light on the wide range of possible applications. This aspect is given special attention (Chaps. V, VI, and IX). General theoretical considerations are discussed only insofar as they contribute to the comprehension of the physical foundations of nonlinear interactions. As a rule, a particular type of interaction is first analyzed in a simple example, and only then are possible generalizations of the results indicated. The author hopes this will help the greatest number of readers to use the results of research on nonlinear effects in various fields of study. Some of the more complicated calculations are saved for the end of the book (Appendices) to simplify the overall presentation. Also discussed in the Appendices are certain methodological aspects of the analytical treatment of nonlinear interactions, along with some important reference material.

A few remarks are in order concerning the history of research on nonlinear wave interaction in plasma. The nonlinear interaction of Langmuir waves was first expounded by Sturrock [4], and the nonlinear interaction of Langmuir and transverse waves was described by Ginzburg and Zheleznyakov [5]. The theory of weak nonlinear interactions was systematically developed in a kinematic treatment by Kadomtsev and Petviashvili [6], and was later expanded in many papers by a number of authors.

Along with the study of weak nonlinearities the problem of strongly nonlinear waves [7-10] and the related problem of collisionless shock waves were posed some time ago.* Some investigations in which the present author was involved were based on the possible application of the straightforward notions of induced processes in plasma for weak nonlinearities [11, 12]. These have been adopted as the basis of the present book. They provide a way around many of the complex mathematical calculations and permit a very simple description of the basic physical concepts. The notions of induced processes are widely used today in the physics of quantum oscillators (masers, lasers, etc.) and should be familiar to many readers.

Frequently in the book an analogy is drawn between the processes of nonlinear wave interaction in plasma and the processes which occur in so-called "negative-temperature" systems. In

* Strong nonlinearity is discussed in Chap. X in connection with the outlook for future research on nonlinear plasma effects.

particular a strong analogy is exhibited between the processes of nonlinear interaction of two wave packets and the processes in a two-level system, as well as between the reduction of Landau damping and saturation effects in a two-level system. These analogies will, it is hoped, aid in the understanding of the physics of nonlinear interactions in plasma.

The theory of nonlinear effects in plasma and the theory of interaction of optical modes have been developed in somewhat different directions. Whereas in plasma we are mainly interested in the interaction of waves in random phase (see, e.g., the survey [13]), in nonlinear optics the prime concern is the interaction of waves whose phases are not random [14, 15]. Even in the original literature, however, there is confusion regarding the limits of applicability of each approach and its potential use. In the present book, therefore, before discussing the notions of induced processes we give an analysis of the elementary nonlinear interaction of three wave packets in a cold plasma (Chap. III). This analysis makes it possible to set the limits of applicability of the notions of induced processes.

The equations derived in Chap. III by direct averaging over the phases are later obtained in Chap. IV from simple physical considerations regarding induced processes. Those interested solely in the physical aspect of the problem need not delve into the problems in Chaps. II and III concerning the limits of applicability of the physical notions, but can proceed directly from Chap. I to Chap. IV, omitting Chaps. II and III. For a grasp of the physical sense of nonlinear instabilities of a plasma (Chap. X), however, the reader is advised to scan the material of Chaps. II and III.

The author has had a direct hand in the theoretical development of many of the problems discussed in the book, a fact that could not help but reflect a bias in the presentation of the material. In particular, much of the actual material obtained by many authors on nonlinear interaction has been revised and refined, the limits of applicability of the results stated, etc. In addition, all the results have been "reduced" to the simple language of induced effects.

A great many diverse applications of nonlinear effects are described in the book. These applications affect a wide range of problems, from the origin of cosmic rays and the physics of cosmic plasma to the heating of plasma, the efficiency of interaction of beams with plasma, etc. We have limited the treatment here by and large to physical concepts and simple estimates, referring the reader to the literature for the details.

Considerable attention is also given in the book to the nonlinear theory of hydrodynamic instabilities of beams, and estimates of various nonlinear effects which can affect the development of beam instability are discussed in detail (Chaps. V, IX, and X).

It is the author's hope that the analysis of the various applications will prove useful to theoreticians in need of a more detailed formulation of the theory of nonlinear interactions applied to specific circumstances.

The effects of nonlinear interaction are illustrated mainly by examples of wave interaction in isotropic plasma (Chap. VIII). At the same time, the general theoretical considerations also apply to any other waves, as, for example, when magnetic fields are present in the plasma (this extension is not a fundamental one), and even for other homogeneous media. Consequently, the material of the book should be of interest for the analysis of nonlinear wave interactions in plasma situated in a strong external magnetic field, etc., as well as the interaction of waves in solids.

The author would like to express his utmost appreciation to V. L. Ginzburg, B. B. Kadomtsev, M. A. Leontovich, L. I. Rudakov, and Ya. B. Fainberg for reviewing the manuscript and offering a number of valuable and important suggestions, as well as to M. A. Livshits for assisting with the layout of the book.

V. N. Tsytovich

## Literature Cited

1. V. L. Ginzburg, Propagation of Electromagnetic Waves in a Plasma, Gordon and Breach, New York (1962).
2. V. L. Ginzburg and A. V. Gurevich, "Nonlinear effects in a plasma in a variable electromagnetic field," Uspekhi Fiz. Nauk, 70:201, 393 (1960).
3. B. B. Kadomtsev, Plasma Turbulence, Academic Press, New York (1965).
4. P. A. Sturrock, "Nonlinear effect in an electron plasma," Proc. Roy. Soc., A242: 277 (1957).
5. V. L. Ginzburg and V. V. Zheleznyakov, "Possible mechanisms of the sporadic radio emission of the sun (investigation in an isotropic plasma)," Astron. Zh., 35:694 (1958).

6. B. B. Kadomtsev and V. I. Petviashvili, "Weakly turbulent plasma in a magnetic field," Zh. Éksp. Teor. Fiz., 43:2234 (1962).
7. A. I. Akhiezer, G. Ya. Lyubarskii, and Ya. B. Fainberg, "Nonlinear theory of oscillations in a plasma," Uch. Zap. Khar'kovsk, Gos. Univ., Vol. 64: Trudy Fizich. Otd. Fiz.-Matem. Fak., 6:73 (1955).
8. R. Z. Sagdeev, "Collective processes and shock waves in a rarefied plasma," Problems in Plasma Theory, Vol. 4. Atomizdat (1964), p. 20.
9. N. J. Zabusky and M. D. Kruskal, "Interaction of 'solitons' in a collisionless plasma and the recurrence of initial states," Phys. Rev. Lett., 15:240 (1965).
10. V. P. Silin, "Parametric resonance in a plasma," Zh. Éksp. Teor. Fiz., 48:1679 (1965).
11. A. Gailitis and V. N. Tsytovich, "Investigation of transverse electromagnetic waves in the scattering of charged particles by plasma waves," Zh. Éksp. Teor. Fiz., 46:1726 (1964).
12. V. N. Tsytovich and A. B. Shvartsburg, "Theory of nonlinear wave interaction in a magnetoactive anisotropic plasma," Zh. Éksp. Teor. Fiz., 49:797 (1965).
13. A. A. Galeev, V. I. Karpman, and R. Z. Sagdeev, "Many-particle aspects of the theory of a turbulent plasma," Yadernyi Sintez, 5:20 (1965).
14. J. A. Armstrong, N. Bloembergen, J. Ducuing, and P. S. Pershan, "Interactions between light waves in a nonlinear dielectric," Phys. Rev., 127:1918 (1962).
15. S. Akhmanov and R. V. Khokhlov, Problems of Nonlinear Optics. VINITI, Moscow (1965).

# Contents

*Chapter I*

# Nonlinear Effects and Plasma Physics

## 1. ROLE OF NONLINEAR EFFECTS IN PLASMA PHYSICS

The last few years have been marked by a tremendous surge of activity in theoretical and experimental research on plasma physics. Plasma effects are finding ever-increasing application in astrophysics, solid state physics, etc., not to mention the physics of gas discharges and research on plasma confinement and heating. Very recently there has been a considerable advance in the study of nonlinear effects, which hold a special place in plasma physics. In order to stress the significance of these research efforts we need to say a few words about plasma as a state of matter and about some of the applications of plasma research.

It is well known that so-called collective processes play a very important part in plasma. They are associated in particular with various plasma instabilities. As a rule the development of instability is accompanied by an increase in the electric field strength, which can attain large values. Consequently, even in the absence of intense external fields, relatively strong fields can still occur spontaneously in a plasma due to the growth of instability. It is important to mention the presence of various instabilities as certainly one of the most characteristic attributes of plasma as a state of matter. Normally in the initial stage of development of instability the electromagnetic fields are weak, so that the growth rate of the instability can be described within the framework of the linear approximation, in which the current **j** stimulated by a field **E** in the plasma is linearly dependent upon **E**:

$$j_i = \sigma_{ij} E_j. \tag{1.1}$$

A simple generalization of (1.1) to the case of a slightly inhomogeneous plasma in external magnetic fields yields equations that suitably describe the majority of the most important plasma instabilities. Studies of these dispersion equations for specific cases have supplied the material for a great many papers on linear plasma theory [1-4].

However, the inadequacy of the linear approximation is shown merely by the fact that it does not answer the fundamental question, namely, are instabilities likely to play a significant role, or does their development produce only slight distortions of the plasma parameters? It is important to know, for example, what fraction of the energy goes into the development of instability. The very existence of instability indicates the inapplicability of the linear theory for the description of a plasma, because the exponential growth of the perturbations very rapidly invalidates the linear theory. But if the growth of the perturbation field is rapidly limited by nonlinear effects, thereby arresting the further growth of instability, and the plasma enters a kind of quasistationary state that differs only slightly from the initial unstable state, it is safe to say that the given instability produces only minor effects. On the other hand, in the case when the final state differs sharply, say in temperature, from the initial state, one can expect a significant increase of the energy and, for example, plasma heating. From this point of view alone the study of nonlinear effects can supply the answer to the question of the importance of any kind of plasma instability. These primarily concern the heating and confinement of plasma, both of which have important practical implications.

Moreover, considerable interest attaches to the study of the quasistationary states of a plasma that occur as the result of nonlinear effects in that these states are clearly those most often to be encountered in nature.

Let us see how the investigation of nonlinear effects can provide the answers to questions about the heating of a plasma. Suppose that the initial state of the plasma is unstable. The growth of instability leads to an increase of the electric fields, which reach saturation through nonlinear effects. Suppose that the temperature of the plasma assumes a value T. One might ask the following: in the future how will the plasma cool, and will this cooling be the same as in the absence of the increased fields or not? If the plasma is insulated from the walls, radiation will be the natural cause of energy losses from it. The question is then, to what extent does radiation from a plasma in the quasistationary nonlinear state differ from the radiation from the initial state before the growth of instability? It is obvious that the answer to this question hinges on the amplitudes of the electric fields generated during the growth of instability. For example, if longitudinal modes, which have a difficult time escaping from the plasma, are generated, then with increasing amplitude the effect of nonlinear conversion of the longitudinal modes into others capable of leaving the plasma tends to increase. Consequently, at large oscillation amplitudes it is possible to have an additional radiation from the plasma, which can affect the cooling time of the plasma and in some cases even control it.

It is also clear that, under conditions such that energy is continuously supplied to the plasma, the steady-state temperature depends on the energy balance between plasma heating and energy loss. Ordinarily in a high-temperature plasma two types of radiative losses are considered: bremsstrahlung due to the slowing down of fast thermal electrons in the electric field of heavy ions, and magnetic radiation from electrons arising from their motion in applied or self-consistent magnetic fields.

It is apparent from the above that the radiation due to nonlinear conversion of plasma oscillations into waves that are easily emitted from the plasma can provide an additional source of radiative loss. This loss can become dominant at certain amplitudes of the electric fields of the plasma oscillations (Chap. V).

Therefore not only is the problem of the amplitudes of the electric fields generated in the nonlinear stage important, but also that of the nonlinear conversion of one mode into others. It is essential to note that the radiation from a plasma is about the only information we have concerning the processes that take place in interstellar or solar plasma. This nonlinear conversion plays an important part in the interpretation of many astrophysical phenomena, such as chromospheric flares (i.e., the giant explosions on the sun's surface which affect processes in space close to Earth) or the emission from supernovae, whose investigation can yield considerable information about the origin of cosmic rays. This also applies to the investigation of the most interesting astronomical objects, namely the quasars discovered in 1962 [5]. These objects, with a mass on the order of $10^8$ times that of the sun, are clearly plasma formations, whose emission can by explained in terms of the effects of nonlinear conversion of plasma waves into transverse waves [6, 7]. And indeed the astrophysical data attest to the development in the plasma of nonlinear conversion of longitudinal into transverse waves, which are recorded on Earth. Thus, normally in solar flares plasma emission is observed at frequencies $2\omega_{oe}(\omega_{oe} = \sqrt{4\pi ne^2/m_e})$, which is interpreted as effects involving the nonlinear interaction of two plasma waves [8]. Plasma emission due to nonlinear conversion can be observed under experimental conditions, for example, by the transmission of beams of charged particles through the plasma [9, 10].

Next we see how the study of nonlinear effects can answer questions concerning the confinement of plasma. A very crude example would be the instability of a self-constricted discharge, caused by the so-called pinch effect. We know that a confined straight cylindrical discharge in a plasma is unstable with respect to bending, which distorts the straight current path into a serpentine series of kinks. The curvature can become so great as to bring the discharge into contact with the walls of the chamber. If the maximum possible amplitude of these kinks is limited by nonlinear effects, then the development of this instability would not disrupt the confinement of the plasma in a discharge chamber whose dimensions slightly exceed the maximum amplitude. More generally it is apparent that large-scale instabilities and the possibilities for their nonlinear stabilization are extremely important with respect to the confinement problem.

Intimately related to the confinement problem is that of plasma diffusion. Suppose, for example, that the plasma is confined by a magnetic field. This means that in some region the density of the plasma is quite large, decreasing with increasing magnetic field (Fig. 1). Due to collisions the plasma diffuses into the region occupied by the magnetic field. This diffusion ultimately, under

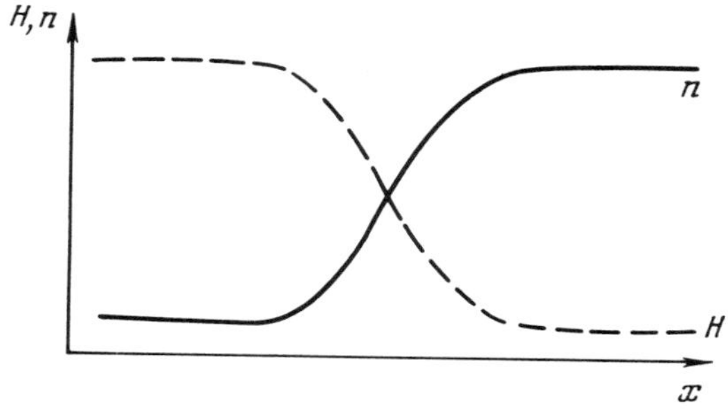


Fig. 1. Confinement of a plasma by a magnetic field.

laboratory conditions, brings the plasma to the walls of the chamber or causes energy losses from the system (heat transfer by the electrons).

We recall the familiar estimate of the diffusion coefficient in a strong magnetic field. It is known from elementary kinetic considerations that the diffusion coefficient can be estimated in terms of the mean free path $l$ and collision frequency $1/\tau$:

$$D \simeq \frac{l^2}{\tau}. \tag{1.2}$$

Each charge of the plasma moves across the strong magnetic field in a circle of radius $r \sim v_t/\omega_H$, where $v_t$ is the mean thermal velocity of the charge, and $\omega_H = eH/mc$ is the gyrofrequency. It is also clear that for every collision in a strong magnetic field the plasma particle can traverse a distance only on the order of the Larmor radius, i.e., $l \sim r$. Assuming $1/\tau \sim \nu_{ei}$ (where $\nu_{ei}$ is the collision frequency), from (1.2) we obtain the classical expression for the diffusion coefficient of a plasma across the magnetic field

$$D_\perp \sim \frac{\text{const}}{H^2}. \tag{1.3}$$

The rapid decrease in $D_\perp$ with increasing H suggests the possibility of good plasma confinement in strong magnetic fields. It must be emphasized that the diffusion (1.3) results from collisions.

One logically inquires how the collective effects so typical in plasma can distort the classical diffusion picture. This problem was posed by Bohm [11]. As mentioned, collective interactions become particularly significant during the development of instabilities. It turns out than an inhomogeneous plasma confined by a magnetic field is usually unstable [12, 13], this instability arising from the very existence of inhomogeneity, i.e., it will always occur in the confinement problem. For this reason it has come to be called the universal instability [12]. It is now fully apparent how important a role nonlinear effects can play in plasma confinement. Specifically, they can determine the amplitude of the steady-state nonlinear perturbations generated in the development of universal instability.

Let us suppose that a certain stationary situation (determined by nonlinear effects) with a certain distribution of self-consistent fields has arisen due to the development of instability. Let us examine, for example, randomly distributed domins, inside which the field is greater than outside. The individual plasma particle, on falling into such a domain, comes under the powerful influence of the fields and changes its motion. This change of motion may be interpreted as a "collision." After experiencing a great many of these "collisions," the particle diffuses over a certain distance. One might say that such diffusion is attributable to a "collision" of the particle with the collective fields produced by the development of instability. Under certain conditions this diffusion, which is called anomalous diffusion, can greatly exceed the classical diffusion due to collisions, and, consequently the confinement of the plasma deteriorates [13].

It is appropriate in general to speak of the occurrence of collective effects in a plasma whenever the action of the collective fields on the individual plasma particle exceeds the collisional effects, i.e., whenever the anomalous exceeds the classical diffusion. The interaction of the particle with the collective fields can be assessed from the amplitudes of these fields, which are governed by nonlinear effects.

In light of our discussion of the anomalous diffusion problem, it is appropriate to add one very important comment. It is apparent from (1.2) that the diffusion coefficient is greater, the larger the dimension $l$ characterizing the collective fields, i.e., the larger the scale of the instability. Large-scale instabilities, which correspond to long wavelengths in the linear approximation, are more dangerous from the point of view of plasma confinement. Small-scale instabilities affect the diffusion only moderately and can be used, for instance, for the heating of a plasma.

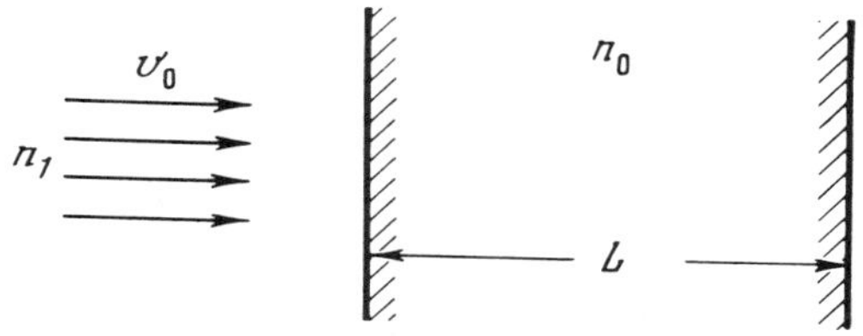


Fig. 2. Injection of a beam of charged particles into a confined plasma.

Actually the development of instability in a plasma can in general cause a change in several plasma parameters, including the temperature. The heating of plasma during the development of instabilities is often called turbulent heating. Suppose that the plasma is heated during the development of a small-scale instability, where the amplitudes of the small-scale fields attain relatively large values. A very important question arises in this connection: can small-scale instabilities generate large-scale instabilities, which then affect the diffusion of the plasma? The answer to this can be found only through nonlinear theory, which predicts the rates of nonlinear conversion of one oscillatory mode of the plasma into others (Chaps. V, IX, and X).

It is appropriate at this point to mention turbulent heating. As already pointed out, this type of heating accompanies the development of instability or is a consequence of this process. Of basic interest in this connection is the problem of what fraction of the incoming energy goes into plasma heating. As a rule the amplitude of the instability fields and the plasma temperature are interrelated, hence the investigation of nonlinear effects makes it possible in principle to assess the possibilities of adding energy to the plasma, and heating it. The problem becomes particularly simple in the case of heating by beams of charged particles, which either occur in the plasma or are injected into it. Effective heating of the plasma during the development of beam instability has clearly been observed in a number of studies [9, 10].

We illustrate the above with the example of a beam injected into a plasma (Fig. 2). Let a beam of charged particles with a velocity $v_0$ and concentration $n_1$ be injected into a plasma of finite length L and density $n_0$. The energy delivered across 1 cm$^2$ in

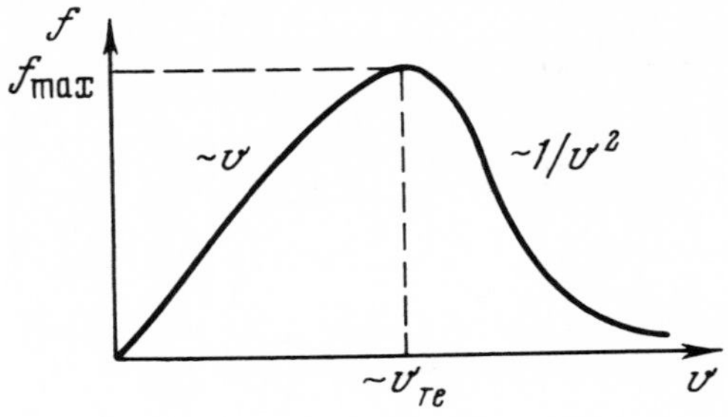


Fig. 3. Frictional force between electrons and ions as a function of the velocity of the electrons relative to the ions.

1 sec is $n_1 \frac{mv_0^2}{2} v_0$. We wish to find out what must be the beam velocity $v_0$, length of the plasma L, and the relation between the other parameters ($n_0$, $n_1$, . . . ) in order for energy on the same order as that transmitted by the beam to be dissipated in the plasma (even with 20-10% energy dissipation considerable heating would be observed). There are values of the parameters such that the heating is ineffective and others for which considerable heating is possible. Nonlinear effects must be explored to find the answer to this problem. Often in this case what is needed is not a detailed theoretical model, which may have only a remote bearing on reality, but some fairly simple physical ideas regarding the nature of the nonlinear interactions and some very simple rough estimates.

We have now arrived at the point of considering the role of nonlinear effects caused by external influences on the plasma. In the above example the plasma is acted upon by a beam of charged particles and acquires energy, which induces plasma instability in the absence of external electric fields. It is important to mention that even the influence of relatively weak external fields on the plasma can produce additional instabilities. Modern technology enables strong fields to be applied to the plasma. Here we include strong steady or quasisteady external electric fields, high-frequency and ultrahigh-frequency fields, and, finally, the powerful emissions of lasers. The application of powerful external fields in itself can produce a great many nonlinear effects which depend on the amplitude of the external fields. Usually, however, the development of instability under the influence of the external fields proves to be very substantial, resulting in the generation of intrinsic plasma fields and nonlinear interaction of these with the external fields. We illustrate this in the example of the behavior of a plasma in a strong constant (quasistationary) electric field.

Under the influence of the electric field the electrons acquire a drift velocity relative to the ions. If collisions are absent, the velocity of the electrons increases linearly with time: $v = \frac{eE}{m_e} t$. If the electron suffers collisions and completely loses its directed velocity, its mean drift velocity then is of the order

$$\bar{v} \approx \frac{eE}{2m_e \nu_{ei}}, \tag{1.4}$$

where $\nu_{ei}$ is the collision frequency. From (1.4) we readily obtain the classical expression for the electrical conductivity of a plasma. However, the conductivity of the plasma in strong fields can actually differ very sharply from this classical expression. To understand this, we refer to Fig. 3, which shows the frictional force f experienced by an electron in Coulomb collisions with ions as a function of the electron velocity v. The occurrence of the maximum on the curve is qualitatively explained as follows.

The frictional force represents the loss of momentum of the particle per unit time:

$$f \sim \frac{mv}{\tau} \sim \nu_{ei} mv. \tag{1.5}$$

The form of (1.5) implies that in every collision the particle loses momentum on the same order of magnitude as its initial momentum. In actuality collisions occur with a small transfer of momentum. Essentially (1.5) deals with collisions for which the change of momentum is appreciable, i.e., where the potential $e^2/r$ is of the same order as the kinetic energy:

$$\frac{mv^2}{2} \sim \frac{e^2}{r};$$
$$r \sim \frac{2e^2}{mv^2}; \; \sigma \sim \pi r^2 \sim \frac{4\pi e^4}{m^2 v^4}; \; \nu_{ei} \sim nv\sigma \sim \frac{4\pi n e^4}{m^2 v^3}. \tag{1.6}$$

When the drift velocity of the electron is much smaller than its thermal velocity the effective collision rate is determined by the thermal velocity $\nu_{ei} \sim \frac{4\pi n e^4}{m_e^2 v_{Te}^3}$ and, consequently, the frictional force is proportional to the velocity. But if the drift velocity of the electrons is greater than $v_{Te}$, the effective number of collisions is determined by the directed velocity v, and the frictional force begins to decrease with increasing velocity. It is a simple matter to estimate the maximum frictional force by inserting the value of $v_{Te}$ in place of v in (1.5):

$$f_{\max} \sim \frac{4\pi n e^4}{m_e v_{Te}^2} \simeq \frac{\omega_{0e}^2}{v_{Te}^2} e^2 = \frac{e^2}{\lambda_{De}^2} = eE_D. \tag{1.7}$$

Here $\omega_{0e}^2 = 4\pi n e^2/m_e$ is the square of the so-called plasma frequency, $\lambda_{De} = v_{Te}/\omega_{0e}$ is the Debye radius, and $E_D$ is the Dreicer electric field [4]

$$E_D \simeq \frac{e}{\lambda_{De}^2}. \tag{1.8}$$

The field $E_D$ is the magnitude of the field generated by a point electron at a distance equal to the Debye radius [the more precise expression for $E_D$ differs from the rough estimate (1.8) by a logarithmic factor]:

The occurrence of the maximum on the curve of Fig. 3 has important physical consequences. Specifically, if a field $E < E_D$ is applied to the plasma, equilibrium can occur between acceleration and deceleration, and we get the mean velocity (1.4), whereas for $E > E_D$ the frictional forces cannot balance the force imposed by the external field. Under these conditions, as might be apparent at first sight, the electrons could be freely accelerated in the plasma, continuously gaining energy. This conclusion, however, is false. It would appear that the resistance of the plasma should decline sharply for $E > E_D$, whereas in fact it increases considerably. This is usually referred to as the anomalous resistance of the plasma. The situation in this case is to a certain extent analogous to the anomalous diffusion described earlier. Just as for diffusion, here collisions between particles play a relatively small part; collective effects are predominant.

When the electron velocity exceeds $v_{Te}$, instability begins to set in, causing the self-consistent fields in the plasma to increase [15]. Subsequently an electron suffers, roughly speaking, collisions, not with individual ions, but with these growing self-consistent fields, by which it is scattered and thereby suffers an additional loss of momentum. It is impossible to describe such effects as anomalous resistance without analyzing the nonlinear effects. We note that the values of $E_D$ are relatively small, especially in a high-temperature or low-density plasma. Let us take an extreme case, when the temperature of the electrons is on the order of 100 keV, and the plasma density $n \sim 10^{11}$ $cm^{-3}$. Then $E_D$ is only $10^{-5}$ or $10^{-6}$ V/cm. This example shows that in a high-temperature plasma even very small fields can excite instabilities, and make the resistance of the plasma anomalously high. In a nonisothermal plasma, $T_e \gg T_i$, instability can occur even if the drift velocity of the electrons relative to the ions is much smaller than the mean thermal velocity of the electrons (but much larger than the mean thermal velocity of the ions). This instability produces anomalous resistance in the plasma for much smaller currents than in an isothermal plasma.

A similar situation involving the onset of anomalous resistance takes place for other types of influence acting on the plasma, for example, the action of high-frequency fields. In this case, however, it must be remembered that the maximum velocity that the charge can acquire in the absence of collisions depends on the frequency of the field, because the alternating field causes the charge to execute periodic motion. It is obvious that

$$v_{\max} \sim \frac{eE}{m\omega}. \tag{1.9}$$

As $\omega \to 0$ (constant field), $v_{max} \to \infty$. The condition $v_{max} > v_{Te}$ is particularly stringent at high frequencies. Consequently, if we want to have $v_{max} > v_{Te}$ at very high frequencies, say laser frequencies, E must be very large. For example, at a temperature on the order of 1 eV the field E must be on the order of $10^7$ V/cm. For $v_{max} > v_{Te}$ a special kind of instability arises in a time-varying field; this instability is to a certain extent analogous to that in a constant field [16].

In a time-varying field there is also a specific instability arising from the fact that when the plasma is acted upon by a large periodic wave field the nonlinear interaction of various wave modes with each other becomes significant (see Chap. V). Suppose that a fairly intense wave is incident on the plasma. Through thermal fluctuations or other causes other modes exist in the plasma, but their amplitudes are extremely small. The nonlinear effects of interaction of the primary wave with these small-amplitude waves can lead to the pumping of energy into the latter, thus altering the energy of the primary wave. Because the other wave modes have increased, nonlinear effects also become important for them. As a result of this redistribution of energy the primary wave can become "trapped" in the plasma, an effect that may be clearly interpreted as an effective nonlinear absorption of high-frequency power. These problems are important not only for high-frequency heating, but also to the confinement and acceleration of plasma by high-frequency fields, etc.

Thus we have come to understand the fundamental role played by nonlinear effects in plasma physics, and we are in a position to ascertain its basic macroscopic parameters, such as the diffusion coefficient and resistance.

We have been concerned above largely with the problems of the physics of laboratory plasma. But no less important is the role of nonlinear effects in a plasma in the milieu of outer space. It is important to bear in mind that a large part of the matter known to us (neglecting photons, neutrinos, etc.) exists in the ionized state, i.e., it comprises a plasma. Astrophysicists have gradually come to realize that typical plasma phenomena occur in interstellar and even in intergalactic space, not to mention the stars and galactic nuclei. Although the interstellar plasma is extremely rarefied ($\sim 1$ particle/cm$^3$), it is by no means "imperceptible," considering its colossal dimensions. In space we find explosions of an astounding magnitude, which are accompanied by the ejection of plasma and act as sources of nonstationary plasma (explosions of supernovae or explosions on stars, i.e., chromospheric flares, and even the explosions of galaxies or parts of them). The instabilities of the plasma in this case appear as a tremendous variety of cosmic phenomena. Although the development of astrophysics leads to extremely interesting results with each passing year, a great deal is left unresolved in the interpretation of these phenomena. There is little doubt, therefore, concerning the aid that can come form the investigation of the above-mentioned problems of plasma physics. It is important to realize, however, that without knowledge of the amplitudes of the fields developed as the result of instabilities it is often difficult to make even crude estimates.

Let us confine our attention for the moment to a single example, which bears on a very important astrophysical problem, namely the origin of cosmic radiation. Today the idea is becoming more firmly implanted that only by the detailed investigation of plasma will it be possible to shed light on the fundamental problems of the origin of cosmic rays [17, 18]. The main observational fact is that a plasma very effectively generates fast particles, whose energy can become exceedingly large while they are relatively few in number. In general this occurs under conditions in which instabilities can develop in the plasma. Fast particles occur in an unstable plasma, not only under cosmic conditions, but also in laboratory equipment [19]. In space, however, the phenomena are encountered on a larger scale. Thus, the energy density of cosmic rays turns out to be on the same order as other forms of energy, e.g., the plasma energy, magnetic field, etc. Cosmic rays can even affect the evolution of the galaxies, as they represent a factor of cosmological order. It is now becoming more and more apparent that the origin of the cosmic rays is closely linked with the ability of an unstable plasma to generate fast particles.

Now let us consider the physical mechanism of this generation. Let us examine, for example, an individual fast particle entering an unstable plasma. As mentioned before, instabilities cause the generation in the plasma of fields whose magnitudes are regulated by nonlinear effects. Suppose that a certain distribution of self-consistent fields has been established by the development of instability, corresponding, say, to randomly distributed domains, inside which the field is greater than outside. A fast particle impinging on the plasma suffers collisions with these domains, inside which are the self-consistent fields mentioned. These collisions do not necessarily produce an appreciable change in the momentum of the particle in each event. All that matters is that, in traversing these domains, on average the particle gradually increases its energy.

In fact, the latter phenomenon is readily understood if we modify somewhat Fermi's celebrated arguments concerning the mechanism of statistical acceleration [20, 21]. We pose the following problem: can the individual particle, on colliding with the domains in which the self-consistent fields exist, alter them significantly? The self-consistent fields are created by a large collective group of particles and, in general, are much larger than the fields that accompany the individual particle. Consequently, the self-consistent fields must not be altered appreciably in such a collision. Roughly speaking, this indicates that the self-consistent fields effectively have a large mass, or, in other words, the collision of the individual particle with the self-consistent fields is more or less equivalent to the collision of a light particle with a very heavy one. The general tendency toward a statistical equipartition of energy will necessarily produce an effective increase in energy or acceleration of the lighter particle [22, 21].

It is clear from the above that the investigation of nonlinear effects will help to elucidate the origin of cosmic rays. It is also clear that without knowing the amplitudes of the self-consistent fields it would be impossible to estimate even the efficiency and rate of acceleration of cosmic rays, let alone to explain the more subtle, but no less important, problems such as their energy spectrum and isotropy.

Next we examine a problem that should have applications both in astrophysics and in the laboratory. Specifically we are concerned with the transmission of electromagnetic waves through a plasma. This inveterate problem, which has been the object of many investigations, has still not been worked out in sufficient detail from the point of view of nonlinear effects. We have already given some consideration to the transmission of strong electromagnetic waves through a plasma. We now examine the transmission of very weak waves of small amplitude, which cannot act on the plasma to alter its state significantly. One might expect this to be precisely a situation in which the linear approximation would be valid. However, this is not always the case. We pose the following question: does the propagation of a weak electromagnetic signal differ in a stable and an unstable plasma? An unstable plasma differs from one that is stable in the development of the self-consistent fields. How do the latter affect the transmission of an electromagnetic wave? Roughly speaking, we can assert that the occurrence of fields in the plasma associated with instability is also accompanied by an increase in the density fluctuations and inhomogeneities of the plasma. A weak wave can be scattered by these inhomogeneities, so that the energy of the wave decreases (more precisely, the intensity of the wave in its initial direction of motion), i.e., effective absorption takes place. Moreover, it follows from the general dispersive properties of any medium that absorption must be accompanied by a change in the index of refraction for the wave. In a number of instances these collective absorption and scattering effects can greatly exceed those due to collisions. It is also clear that their magnitude is determined on the whole by the amplitude and spectrum of the self-consistent fields produced in the unstable plasma (Chaps. V and IX).

We note at this point that the problem of the transmission of weak electromagnetic waves through a plasma is to a certain extent analogous to the problem considered above of the acceleration of cosmic rays. Here we are concerned with the "trajectory" of an electromagnetic wave in the self-consistent plasma fields, whereas before we were concerned with the "trajectory" of charged particles. Lately there has been an extensive growth of experimental studies utilizing combination scattering of waves in a plasma for plasma diagnostics (Chap. V).

We conclude with one final comment. Naturally the development of plasma physics began with linear effects. At first it seemed that the objective of, for example, the theory of plasma confinement was to look for those ranges of physical parameters for which the plasma would be stable. Actually, in the range of stability the fields in the plasma do not grow, and the application of the linear approximation is justified. However, research eventually disclosed, at first on paper and subsequently by experiment, a continual chain of new plasma instabilities. As a result researchers began to realize that a plasma was highly prone to become unstable and that the presence of instabilities was its most characteristic attribute as a state of matter. Consequently, the objective became the physical investigation of the real states of a plasma in which it was most likely to be encountered. It was soon clear that nonlinear effects were the most important factor in comprehending the physical processes in a plasma.

With this we conclude our discussion of the role of nonlinear effects in the basic problems of plasma physics, although the above examples by no means exhaust all the important applications, among which we could include the theory of plasma waveguides, the acceleration of particles in plasma, and a host of other topics. The above examples were chosen not only for the importance of the related problems, but also for our ability at the present time to sketch a rough outline of the physical phenomena involved.

Many of the problems touched on are still far from solution, and in many cases we have only special solutions. It is of the utmost importance, therefore, to be able to estimate some of the various elementary nonlinear interactions and to formulate physical ideas for these interactions. The clear interpretation of elementary nonlinear interactions should facilitate their application to complex real-life conditions. Below we first of all consider some of the simplest nonlinear interactions in plasma and the consequences that they produce. Secondly we give examples that have an important bearing on certain applications.

## 2. CLASSIFICATION OF THE NONLINEARITIES OF A PLASMA

The nonlinear effects of a plasma are extremely diverse. Consequently, before embarking on a discussion of the problems of nonlinear interactions, we need to give at least a crude classification of those interactions. Nonlinear processes can be classified according to the magnitude of the electromagnetic fields, the characteristic times in which the interactions become pronounced, and, finally, the characteristic lengths.

The distinction between weak and strong nonlinearities is extremely important. Weak nonlinearities are defined as those which can be described by the first terms of the amplitude expansion of the electromagnetic field associated with a given type of process. As a rule the fields in a plasma are time-varying. Therefore, the field can have a certain characteristic frequency $\omega$ assigned to it. A plasma particle moving in this field will execute periodic oscillations. The velocity amplitude of these oscillations are easily derived from the equation

$$\frac{dv}{dt} \sim \frac{eE_0}{m} e^{-i\omega t}.$$

We have

$$v = v_{\sim} e^{-i\omega t}; \quad v_{\sim} = \frac{eE_0}{m\omega}.$$

An important quantity is the ratio of $v_{\sim}$ to the characteristic velocity $v_*$ of the investigated process in the plasma. If we consider a certain wave process, it can be characterized by the phase velocity $v_* = v_\varphi$. In the presence of thermal motion of the particles (in the event the given process depends on the thermal motion) the characteristic velocity might be interpreted as the mean thermal velocity of the particles $v_* = v_T = \sqrt{T/m}$. The weak nonlinearity condition is usually expressed in terms of smallness of the parameter $v_\sim/v_*$:

$$v_\sim/v_* \ll 1.$$

If $v_\sim$ happens to be comparable with $v_*$, the nonlinearity is said to be strong. It is clear that $v_\sim$ can be very large, in excess of $v_T$. Some of the first investigations of nonlinear processes [23, 24] in fact dealt with the case of strong nonlinearities:

$$v_\sim \gg v_T, \quad v_\sim \sim v_\varphi.$$

In these studies travelling waves of the form $E = E\left(\frac{x}{v_\varphi} - t\right)$ having a constant phase velocity were investigated. The simpler case when the field E depends only on time, i.e., $v_\varphi = \infty$, has been treated in more detail by Silin [16]. Clearly, strongly nonlinear waves can be unstable (this is strikingly exhibited in the case $v_\varphi = \infty$ [16]). In addition, for finite $v_\varphi$, a particularly strong instability occurs for $v_\varphi > v_\sim$. It is important to note that the phase velocity of the waves, as shown by Fainberg, is a function of their amplitude when nonlinearities are taken into account.

Weak nonlinearities have been studied in far more detail to date. This is dictated by the fact that simple methods are available in this case for the nonlinear processes by expansion in the field amplitudes. We will be primarily concerned below with weak nonlinearities.

Another classification involves the characteristic time of the nonlinear process. We denote this time by $\tau$. It may be larger or smaller than the characteristic times $\tau_*$ of a process in the plasma. An important role is played by the time between two collisions $\tau = 1/\nu$, where $\nu$ is the effective collision frequency. If nonlinear interactions take place during a time much smaller than the collision time, they are most effective. Nonlinear interactions of this kind are called nondissipative. On the other hand, in the case when the nonlinear interactions are weak and $\tau \gg 1/\nu$, the nonlinearities are dissipative. The most important factor in the latter case is the dependence of the effective collision frequency on the amplitude of the fields [25]. Dissipative nonlinearities were studied first in connection with radio wave propagation problems and have been analyzed in detail by Ginzburg and Gurevich [25]. Our primary concern below will be collisionless nonlinearities with $\tau \ll 1/\nu$.

The next important point is the comparison of the characteristic lengths $l$, over which the nonlinear interaction (for example, conversion of one wave mode into another) takes place, with the characteristic length of the system $l_*$. A possible candidate for $l_*$ would be, for example, a dimension of the plasma if we are dealing with a confined plasma. If $l \ll l_*$, the presence of the boundaries are of little consequence, and the nonlinear interactions proceed as though the boundaries were absent, i.e., as in an unconfined plasma. We should point out that $l$ decreases with increasing field amplitude. Our main interest, of course, lies in values of $l_*$ such that $l \ll l_*$. We will be primarily concerned below with nonlinear effects in an unconfined plasma.

It must also be emphasized that nonlinear interaction processes are significantly modified when the characteristic lengths of certain interacting waves are greater than the typical dimension of the system (as may happen in the case of transverse modes).

## Literature Cited

1. V. L. Ginzburg, Propagation of Electromagnetic Waves in a Plasma, Gordon and Breach, New York (1962).
2. V. D. Shafranov, "Electromagnetic waves in a plasma," Reviews of Plasma Physics, Vol. 3., Consultants Bureau, New York (1967).
3. V. P. Silin and A. A. Rukhadze, Electromagnetic Properties of a Plasma and Plasma-like Media. Atomizdat (1961).
4. T. H. Stix, Theory of Plasma Waves. McGraw-Hill, New York (1962).
5. Ya. B. Zel'dovich and N. D. Novikov, "Relativistic astrophysics," Uspekhi Fiz. Nauk, 84:377 (1964).
6. V. L. Ginzburg and L. N. Ozernoi, "Role of coherent plasma radio emission for quasars and remnants of supernova stars," Izv. V. Uch. Zav., Radiofizika, 9:221 (1966).
7. S. A. Kaplan, "A possible interpretation of the radio emission of cosmic sources of small dimensions," Astron. Zh. 44:102 (1967).
8. V. V. Zheleznyakov, Radio Emission of the Sun and Planets. Izd. "Nauka" (1964).
9. Ya. V. Fainberg, "Interaction of beams of charged particles with a plasma," Atomnaya Énergiya, 11:313 (1961); Ya. B. Fainberg, et al., "Interaction of intense electron beams with a plasma," Atomnaya Énergiya, 11:493 (1961); 14:249 (1963); 18:315 (1965).

10. E. K. Zavoiskii, "Collective interactions and the problem of producing a high-temperature plasma," Atomnaya Énergiya, 14:57 (1963); M. V. Babykin, P. P. Gavrin, E. K. Zavoiskii, L. I. Rudakov, and V. A. Skoryupin, "Turbulent heating of a plasma," Zh. Éksp. Teor. Fiz., 43:411 (1962); "Trapping and confinement of a turbulently heated plasma in a magnetic trap," Zh. Éksp. Teor. Fiz., 43:1976 (1962); "Observations of two-stream ionic instability in the turbulent heating of a plasma," Zh. Éksp. Teor. Fiz., 43:1547 (1962); M. V. Babykin, E. K. Zavoiskii, L. I. Rudakov, et al., "New results on the turbulent heating of a plasma," Zh. Éksp. Teor. Fiz., 46:511 (1964).
11. D. Bohm, et al., The Characteristics of Electrical Discharges in Magnetic Fields. New York (1949).
12. A. B. Mikhailovskii, "Oscillations of an inhomogeneous plasma," Problems in Plasma Theory, Vol. 3. Atomizdat (1963), p. 141.
13. B. B. Kadomtsev, Plasma Turbulence, Academic Press, New York (1965).
14. H. Dreicer, "Electron and ion runaway in a fully ionized gas," Phys. Rev., 117:329 (1960).
15. O. Buneman, "Dissipation of currents in ionized media," Phys. Rev., 115:503 (1959).
16. V. P. Silin, "Parametric resonance in a plasma," Zh. Éksp. Teor. Fiz., 40:1679 (1965).
17. V. N. Tsytovich, "Mechanisms of the onset of plasma turbulence and the acceleration of cosmic rays," Astron. Zh., 42:33 (1965).
18. V. L. Ginzburg, "Cosmic rays and plasma phenomena in the galaxy and metagalaxy," Astron. Zh., 42:1449 (1965).
19. L. A. Artsimovich, Controlled Thermonuclear Reactions, Gordon and Breach, New York (1964).
20. V. L. Ginzburg and S. I. Syrovatskii, Origin of Cosmic Radiation. Pergamon, New York.
21. E. Fermi, "On the origin of the cosmic radiation," Phys. Rev., 75:1169 (1949).
22. V. N. Tsytovich, "Acceleration of particles by radiation in the presence of a medium," Dokl. Akad. Nauk SSSR, 142:319 (1962).
23. A. I. Akhiezer and G. Ya. Lyubarskii, "Nonlinear theory of the oscillations of an electronic plasma," Dokl. Akad. Nauk SSSR, 80:193 (1951); A. I. Akhiezer, G. Ya. Lyubarskii, and Ya. B. Fainberg, "Nonlinear theory of oscillations in a plasma," Uch. Zap. Khar'kovsk. Gos. Univ., 6:73 (1955).
24. R. Z. Sagdeev, "Collective processes and shock waves in a rarefied plasma," Problems in Plasma Theory, Vol. 4. Atomizdat (1964), p. 20.
25. V. L. Ginzburg and A. V. Gurevich, "Nonlinear effects in a plasma in a variable magnetic field," Uspekhi Fiz. Nauk, 70:201, 1393 (1960).

*Chapter II*

# Weak Nonlinearities in Cold Plasma

## 1. NONLINEAR EQUATION FOR THE FIELD IN A PLASMA

We begin our discussion with the simplest case of a cold plasma in order to illustrate certain important characteristics of wave interaction in plasma as an example of elementary nonlinear interaction. First of all we must define what we mean by the term "cold plasma." A plasma can be called cold when the thermal motion of the particles is negligibly small. This means that we must specify a reference quantity against which the mean thermal velocity may be neglected. Let the given nonlinear effects be weak, so that in the first approximation we can speak of the interaction of linear waves. Linear waves may be characterized by their phase velocities

$$v_{\varphi} = \omega/k, \tag{2.1}$$

where $\omega$ is the wave frequency, and k is the wave number. Plasma particles executing thermal motion have certain mean thermal velocities $v_T$. A plasma is said to be cold if the phase velocities of the given waves greatly exceed the mean thermal velocities of the plasma particles:

$$v_{\varphi} \gg v_{\mathrm{T}}. \tag{2.2}$$

We assume then that (2.2) is satisfied for all waves in whose interaction we are interested and for all plasma particles (electrons and ions).

Let there be a certain number of waves present in the plasma. We denote their fields by $\mathbf{E}_1$, $\mathbf{E}_2$, . . . . These fields produce currents $\mathbf{j}$ in the plasma by setting its particles in motion. To a first approximation the motion is linear with respect to the field

$$\mathbf{j} = \mathbf{j}^{(1)} = \sigma \mathbf{E}, \tag{2.3}$$

where $\mathbf{E} = \mathbf{E}_1 + \mathbf{E}_2 + \ldots$. The superscript (1) indicates that Eq. (2.3) corresponds to a first-order effect with respect to the field. We note that in the linear approximation the superposition principle applies, i.e., the fields $\mathbf{E}_1$, $\mathbf{E}_2$, . . . . induce currents independently in the plasma, and the resultant current is equal to the sum of these individual currents. However, (2.3) corresponds only to the first approximation in the field.

We write the expansion for the current generated in the plasma in powers of the field in the form

$$\mathbf{j} = \mathbf{j}^{(1)} + \mathbf{j}^{(2)} + \mathbf{j}^{(3)} + \ldots \tag{2.4}$$

The superscript (2) indicates that this part of the current contains quadratic terms with respect to the total field, (3) indicates cubic terms, etc. Even for quadratic field effects the superposition principle is violated. The term $\mathbf{j}^{(2)}$ indicates in particular to what extent the current induced by, say, the field $\mathbf{E}_1$ in the presence of $\mathbf{E}_2$ differs from that induced by $\mathbf{E}_1$ in the absence of $\mathbf{E}_2$. The solution of specific electrodynamic problems can be obtained by substituting the expression (2.4) for the nonlinear current into Maxwell's equations. Eliminating the magnetic field $\mathbf{H}$ from Maxwell's equations

$$\operatorname{rot} \mathbf{E} = -\frac{1}{c}\frac{\partial \mathbf{H}}{\partial t}; \quad \operatorname{rot} \mathbf{H} = \frac{4\pi}{c}\mathbf{j} + \frac{1}{c}\frac{\partial \mathbf{E}}{\partial t} \tag{2.5}$$

allows them to be conveniently written in the form

$$\operatorname{rot}\operatorname{rot} \mathbf{E} + \frac{1}{c^2}\frac{\partial^2 \mathbf{E}}{\partial t^2} = -\frac{4\pi}{c^2}\frac{\partial \mathbf{j}}{\partial t}. \tag{2.6}$$

Substituting Eq. (2.4) into (2.6), we obtain nonlinear equations in $\mathbf{E}$ describing the electrodynamic effects of wave interaction. Under conditions when it is permissible to expand the nonlinear current in powers of the electric field $\mathbf{E}$ we call the nonlinearities weak.

## 2. THE LINEAR APPROXIMATION

We now recall the results of the linear ∂ approximation for **j**. Since **j** = en**v**, it follows that $\partial \mathbf{j}^{(1)}/\partial t = en\partial \mathbf{v}/\partial t$, and $\partial v/\partial t = eE/m_e$ i.e.,

$$\frac{\partial \mathbf{j}^{(1)}}{\partial t} = \frac{e^2 n}{m_e}\mathbf{E} = \frac{\omega_{0e}^2}{4\pi}\mathbf{E}, \tag{2.7}$$

where

$$\omega_{0e}^2 = \frac{4\pi n e^2}{m_e} \tag{2.8}$$

is the plasma frequency of the electrons. By inserting (2.7) into Maxwell's equations (2.6), we can derive a single equation from the three equations (2.6) for longitudinal waves (rot **E** = 0):

$$\frac{\partial^2 \mathbf{E}}{\partial t^2} + \omega_{0e}^2 \mathbf{E} = 0 \tag{2.9}$$

and two equations for transverse waves (div **E** = 0):

$$\Delta \mathbf{E} - \frac{1}{c^2}\frac{\partial^2 \mathbf{E}}{\partial t^2} - \frac{1}{c^2}\omega_{0e}^2 \mathbf{E} = 0. \tag{2.10}$$

These correspond to the three possible orientations of the polarization field vector **e** of the wave relative to the wave vector. In the cold-plasma approximation the longitudinal waves become simply oscillations with a frequency $\omega_{0e}$.

For the propagation of plane waves the dispersion equations (dependence of frequency on wave number) are easily derived by substituting $\mathbf{E} = \mathbf{E}_0 e^{-i(\omega t - \mathbf{kr})}$ into Eqs. (2.9) and (2.10):

$$\omega^l = \omega_{0e};\ \omega^t = \sqrt{\omega_{0e}^2 + k^2 c^2}. \tag{2.11}$$

Here the superscripts $l$ and t refer to longitudinal and transverse waves. The general expression for an arbitrary field satisfying the linear equations (2.9) and (2.10) comprises a set of superimposed plane waves

$$\begin{aligned} \mathbf{E}^l &= \int \mathbf{E}_\mathbf{k}^l \exp\left[-i\omega^l(\mathbf{k})\, t + i\mathbf{kr}\right] d\mathbf{k}; \\ \mathbf{E}^t &= \int \mathbf{E}_\mathbf{k}^t \exp\left[-i\omega^t(\mathbf{k})\, t + i\mathbf{kr}\right] d\mathbf{k}. \end{aligned} \tag{2.12}$$

Here the transverse and longitudinal waves do not interact and cannot be transformed into one another. If, for example, the field is initially transverse, then in the linear approximation it remains the same for all subsequent times.

## 3. NONLINEAR OSCILLATIONS OF COLD PLASMA CHARGES

We now turn to the consideration of nonlinear effects. We confine the discussion here to those effects which are described by a current which is quadratic with respect to the field **E** in the case when (2.2) is satisfied. Since the linear relation between **j** and **E** in the linear approximation arises solely from the displacement of the plasma electrons, it is reasonable to assume also that only the plasma electrons contribute to nonlinear effects. If the inequality (2.2) is satisfied, this assumption is justified for the nonlinear current $\mathbf{j}^{(2)}$. In a cold plasma the motion of the electrons under the influence of external fields **E** and **H** may be described by the fluid equations

$$\frac{\partial \mathbf{v}}{\partial t} + (\mathbf{v}\nabla)\,\mathbf{v} = \frac{e}{m_e}\left(\mathbf{E} + \frac{1}{c}\,[\mathbf{vH}]\right); \tag{2.13}$$

$$\frac{\partial n}{\partial t} + \operatorname{div} n\mathbf{v} = 0 \tag{2.14}$$

The plasma current **j** can be found for a given n and **v**:

$$\mathbf{j} = en\mathbf{v}. \tag{2.15}$$

The set of equations (2.13) and (2.14) is nonlinear and can be used to find the nonlinear current to any order in the field **E**. This system is conveniently described in another equivalent form by expanding all variables in Fourier series. For example,

$$\begin{aligned} \mathbf{v} &= \int \mathbf{v}_k e^{-i(\omega t - \mathbf{kr})}\, dk; \\ \mathbf{E} &= \int \mathbf{E}_k e^{-i(\omega t - \mathbf{kr})}\, dk; \quad dk = d\mathbf{k}\, d\omega. \end{aligned} \tag{2.16}$$

It is important to remember at this point that the use of expansions of the type (2.16) does not imply any assumption about superposition. The difference between (2.16) and (2.12) lies in the fact that each term of the sum (integral) (2.12), but not of (2.16), is a solution of the linear equation. This is expressed in the fact that the plane-wave frequency is a completely defined function of the wave vector **k**. In (2.16) the frequency is arbitrary, and the integration is carried out with respect to it. The descriptions of the behavior of the electrons by means of either $\mathbf{v}(\mathbf{r}, t)$ or $\mathbf{v}_k$ are entirely equivalent, because $\mathbf{v}(\mathbf{r}, t)$ can always be found in terms of $\mathbf{v}_k$ and vice versa:

$$\mathbf{v}_k = \frac{1}{(2\pi)^4}\int \mathbf{v}(\mathbf{r}, t)\, e^{i(\omega t - \mathbf{kr})}dx; \quad dx = d\mathbf{r}dt. \tag{2.17}$$

In order to transform Eqs. (2.13)-(2.15) into a form that contains only $\mathbf{v}_k$, $\mathbf{E}_k$, . . . , it is helpful to use the following property of the transformation of a product of two functions A and B:*

$$(AB)_k = \int A_{k_1}B_{k-k_1}dk_1. \tag{2.18}$$

The relation (2.18) is easily written in symmetric form, using the delta function:†

$$B_{k-k_1} = \int B_{k_2}dk_2\delta(k - k_1 - k_2). \tag{2.20}$$

The delta function has the property that it has a nonzero value only if its argument is equal to zero, and the integral of the delta function is equal to unity.

Substituting (2.19) into (2.18), we obtain

$$(AB)_k = \int A_{k_1}B_{k_2}d\lambda, \tag{2.21}$$

where

$$d\lambda = dk_1dk_2\delta(k - k_1 - k_2). \tag{2.22}$$

---

* Equation (2.18) can be proved by the following simple method. From (2.17)

$$(AB)_k = \frac{1}{(2\pi)^4}\int ABe^{i(\omega t - \mathbf{kr})}\,dx =$$

$$= \frac{1}{(2\pi)^4}\int A_{k_1}e^{-i(\omega_1 t - \mathbf{k}_1\mathbf{r})}\,dk_1\cdot B_{k_2}e^{-i(\omega_2 t - \mathbf{k}_2\mathbf{r})}\,dk_2 e^{i(\omega t - \mathbf{kr})}\,dx = \int A_{k_1}B_{k_2}d\lambda.$$

Here we have replaced A and B with expansions of the type (2.16) and integrated over x, using

$$\frac{1}{(2\pi)^4}\int dx e^{i(\omega - \omega_1 - \omega_2)t - i(\mathbf{k} - \mathbf{k}_1 - \mathbf{k}_2)\mathbf{r}} = \delta(k - k_1 - k_2). \tag{2.19}$$

The last equation is readily verified by substituting $v_k = \delta(k)$ into a relation of the type (2.16); we obtain as a result $\mathbf{v}(\mathbf{r}, t) = 1$, which on substitution into (2.17) yields (2.19).

† Equation (2.20) is best verified as follows: Since the integrands of (2.20) have nonzero values only for $k_2 = k - k_1$, $B_{k_2}$ may be replaced by $B_{k-k_1}$ and the latter taken outside the integral; the remaining integral is equal to unity.

For the transformation of Eqs. (2.13) and (2.15) it is sufficient to use (2.21) and the fact that $\partial/\partial t$ is equivalent to $-i\omega$, and for a plane wave $\nabla$ to $i\mathbf{k}$. We obtain

$$-i\omega \mathbf{v}_k + i\int (\mathbf{v}_{k_1}\mathbf{k}_2)\,\mathbf{v}_{k_2} d\lambda = \frac{e}{m_e}\mathbf{E}_k + \frac{e}{m_e c}\int [\mathbf{v}_{k_1}\mathbf{H}_{k_2}]\,d\lambda; \qquad (2.23)$$

$$\omega n_k = \mathbf{k}\int n_{k_1}\mathbf{v}_{k_2} d\lambda. \qquad (2.24)$$

Now we use Maxwell's equation

$$\frac{1}{c}\frac{\partial \mathbf{H}}{\partial t} = -\operatorname{rot}\mathbf{E} \qquad (2.25)$$

to express the magnetic field $\mathbf{H}_k$ in terms of the electric field $\mathbf{E}_k$:

$$\frac{1}{c}\mathbf{H}_k = \left[\frac{\mathbf{k}}{\omega}\mathbf{E}_k\right]. \qquad (2.26)$$

The Lorentz force now has the form

$$\left[\frac{\mathbf{v}_{k_1}}{c}\mathbf{H}_{k_2}\right] = \frac{1}{\omega_2}[\mathbf{v}_{k_1}[\mathbf{k}_2\mathbf{E}_{k_2}]] = \frac{\mathbf{k}_2}{\omega_2}(\mathbf{v}_{k_1}\mathbf{E}_{k_2}) - \frac{\mathbf{E}_{k_2}}{\omega_2}(\mathbf{k}_2\mathbf{v}_{k_1}). \qquad (2.27)$$

Substituting the last expression into Eq. (2.23), we obtain

$$\mathbf{v}_k = \frac{ie}{\omega m_e}\mathbf{E}_k + \frac{ie}{\omega m_e}\int \frac{\mathbf{k}_2}{\omega_2}(\mathbf{v}_{k_1}\mathbf{E}_{k_2})\,d\lambda + \frac{1}{\omega}\int (\mathbf{k}_2\mathbf{v}_{k_1})\left(\mathbf{v}_{k_2} - \frac{ie}{m_e\omega_2}\mathbf{E}_{k_2}\right)d\lambda. \qquad (2.28)$$

Note that all terms on the right-hand side except the first describe nonlinear effects. The first, on the other hand, describes the familiar small oscillations of a free charge in a wave field $\mathbf{E}_k$. Assuming the nonlinear effects are small and neglecting them in the first approximation, we have

$$\mathbf{v}_k = \mathbf{v}_k^{(0)} + \mathbf{v}_k^{(1)}; \qquad (2.29)$$

$$\mathbf{v}_k^{(0)} = 0; \quad \mathbf{v}_k^{(1)} = \frac{ie}{\omega m_e}\mathbf{E}_k. \qquad (2.30)$$

To find the second-order effect with respect to the external field, for which we are looking, we substitute the first approximation (2.30) into the nonlinear term (2.28). We see at once that the last term on the right-hand side of (2.28) becomes zero, and obtain

$$\mathbf{v}_k^{(2)} = -\frac{e^2}{m_e^2\omega}\int \frac{\mathbf{k}_2}{\omega_1\omega_2}(\mathbf{E}_{k_1}\mathbf{E}_{k_2})\,d\lambda = -\frac{e^2}{2m_e^2\omega}\int \frac{\mathbf{k}}{\omega_1\omega_2}(\mathbf{E}_{k_1}\mathbf{E}_{k_2})\,d\lambda. \qquad (2.31)$$

Here we have symmetrized the result for $\mathbf{k}_2 \to (\mathbf{k}_1 + \mathbf{k}_2)/2 = \mathbf{k}/2$; this is always possible, because the remaining expressions are symmetric with respect to the indices 1 and 2.

How straightforward is the interpretation of this electron velocity, which is nonlinear with respect to the field? We recall that an electron moving in the field of a plane monochromatic wave in general executes a rather complex figure-of-eight type of motion. This is due to the action of the Lorentz force of the wave. At first the electron begins to move in the direction of the electric field of the wave. The acquisition of velocity in the direction of the electric field brings the Lorentz force into play, so that a force and displacement are induced in the direction perpendicular to **H** and **E**. The action of several waves on the particle produces a similar result. Above we investigated an arbitrary field, which can always be represented as a sum of an infinite number of waves.

For simplicity let us examine two monochromatic waves. The nonlinear effect of their interactions results from one of the waves forcing the electrons to move with a velocity $\mathbf{v}$ along $\mathbf{E}_{k_1}$, the field of the other $\mathbf{E}_{k_2}$ acting on this velocity through the Lorentz force. We saw earlier that it is the Lorentz force which contributes to the nonlinear velocity (2.31). In this approximation, however, the Lorentz force contributes only partly, because now another nonlinear effect occurs. This arises because a plasma element, caused to move with velocity $\mathbf{v}$ by the field $\mathbf{E}_{k_1}$, experiences the field $\mathbf{E}_{k_2}$ at an altered frequency owing to the Doppler effect. This effect partially offsets the action of the Lorentz force. It is important to bear in mind that this compensation causes the vector $\mathbf{v}_k^{(2)}$, apart from its dependence on the character of the fields $\mathbf{E}_{k_1}$ and $\mathbf{E}_{k_2}$, to be in the direction of $\mathbf{k}$, i.e., $\mathbf{v}^{(2)}$ represents a superimposition of only longitudinal plane waves.

Thus, in the nonlinear approximation, longitudinal electron velocity waves occur in the plasma and can be excited by any fields, transverse fields in particular, i.e., transverse and longitudinal waves are no longer independent in the nonlinear approximation.

## 4. NONLINEAR OSCILLATIONS OF PLASMA DENSITY

The electron velocities and densities are related to one another by the continuity equation. Consequently, as well as velocity waves there are also density waves. They can be found with the help of Eq. (2.24), by expanding the density $n_k$ in terms of the field:

$$n_k = n_k^{(0)} + n_k^{(1)} + n_k^{(2)} + \dots \tag{2.32}$$

The zeroth approximation corresponds to a homogeneous plasma $n_0(\mathbf{r}, t) = n_0 = \text{const}$. Hence

$$n_k^{(0)} = \frac{1}{(2\pi)^4}\int n_0(\mathbf{r}, t)\, e^{i(\omega t - \mathbf{kr})} dx = n_0 \delta(k), \tag{2.33}$$

which follows immediately from the definition of the delta function (2.19). Moreover, from Eq. (2.24) we obtain in the first approximation

$$\omega n_k^{(1)} = \mathbf{k}\int \left(n_{k_1}^{(0)}\mathbf{v}_{k_2}^{(1)} + n_{k_1}^{(1)}\mathbf{v}_{k_2}^{(0)}\right) d\lambda, \tag{2.34}$$

and, since $\mathbf{v}_k^{(0)} = 0$,

$$\omega n_k^{(1)} = \mathbf{k}\int dk_1 dk_2 n_0 \delta(k_1)\, \mathbf{v}_{k_2}^{(1)} \delta(k - k_2) = n_0 \mathbf{k}\mathbf{v}_k^{(1)},$$

or, substituting the value found for $\mathbf{v}_k^{(1)}$ from (2.30), we obtain

$$n_k^{(1)} = \frac{n_0 ie}{m_e \omega^2}(\mathbf{kE}_k). \tag{2.35}$$

It is apparent from (2.35) that in the first approximation only longitudinal waves generate density waves. Transverse waves, for which $(\mathbf{kE}_k) = 0$, do not stimulate density waves.

In the next-higher approximation

$$\omega n_k^{(2)} = \mathbf{k}\int \left(n_{k_1}^{(1)}\mathbf{v}_{k_2}^{(1)} + n_{k_1}^{(0)}\mathbf{v}_{k_2}^{(2)}\right) d\lambda.$$

For transverse waves, for which $n_{k_1}^{(1)} = 0$,

$$n_k^{(2)} = n_0 \frac{\mathbf{kv}_k^{(2)}}{\omega} = -\frac{e^2 n_0 k^2}{2m_e^2 \omega^2}\int \frac{(\mathbf{E}_{k_1}\mathbf{E}_{k_2})}{\omega_1 \omega_2}\, d\lambda. \tag{2.36}$$

Therefore, transverse waves generate plasma density waves in the nonlinear approximation.

## 5. NONLINEAR PLASMA CURRENT

The plasma current may be written in the form

$$\mathbf{j}_k = e(n\mathbf{v})_k = e\int n_{k_1}\mathbf{v}_{k_2} d\lambda. \tag{2.37}$$

We expand with respect to the field:

$$\mathbf{j}_k = \mathbf{j}_k^{(1)} + \mathbf{j}_k^{(2)} + \ldots$$

In the first approximation we have

$$\mathbf{j}_k^{(1)} = e \int n_{k_1}^{(0)} \mathbf{v}_{k_2}^{(1)} d\lambda = en_0 \int \delta(k_1)\, \delta(k - k_1 - k_2)\, \mathbf{v}_{k_2}^{(1)} dk_1 dk_2 = en_0 \mathbf{v}_k^{(1)}, \quad (2.38)$$

or, substituting the value found for $\mathbf{v}_k^{(1)}$,

$$\mathbf{j}_k^{(1)} = \frac{ie^2 n_0}{\omega m_e} \mathbf{E}_k = i \frac{\omega_{0e}^2}{4\pi\omega} \mathbf{E}_k, \quad (2.39)$$

which naturally agrees with the earlier expression (2.7).

The unknown current $\mathbf{j}_k^{(2)}$ is

$$\mathbf{j}_k^{(2)} = e \int \left( n_{k_1}^{(1)} \mathbf{v}_{k_2}^{(1)} + n_{k_1}^{(0)} \mathbf{v}_{k_2}^{(2)} \right) d\lambda =$$

$$= e \int \left[ \frac{n_0 ie}{m_e \omega_1^2} (\mathbf{k}_1 \mathbf{E}_{k_1}) \frac{ie}{\omega_2 m_e} \mathbf{E}_{k_2} - \frac{n_0 e^2 \mathbf{k}}{2m_e^2\, \omega\omega_1\omega_2} (\mathbf{E}_{k_1} \mathbf{E}_{k_2}) \right] d\lambda. \quad (2.40)$$

After symmetrization of the first term with respect to the indices 1 and 2 we finally obtain

$$\mathbf{j}_k^{(2)} = -\frac{\omega_{0e}^2 e}{8\pi m_e} \int \frac{d\lambda}{\omega_1\omega_2} \left[ \frac{\mathbf{k}}{\omega} (\mathbf{E}_{k_1}\mathbf{E}_{k_2}) + \mathbf{E}_{k_1} \frac{(\mathbf{k}_2 \mathbf{E}_{k_2})}{\omega_2} + \mathbf{E}_{k_2} \frac{(\mathbf{k}_1 \mathbf{E}_{k_1})}{\omega_1} \right]. \quad (2.41)$$

The last two terms of (2.41) are associated with the plasma density waves $n_k^{(1)}$. They are not concerned with transverse waves, which do not generate density waves in the first approximation. According to (2.41) transverse waves excite waves involving longitudinal plasma currents.

## 6. NONLINEAR EQUATIONS FOR TRANSVERSE AND LONGITUDINAL ELECTROMAGNETIC FIELDS

The nonlinear equations for the components $\mathbf{E}_k$ of the expansion of the field in plane waves are also useful. Recognizing that $\partial/\partial t \to -i\omega$, and $\nabla \to i\mathbf{k}$, we obtain

$$\operatorname{rot}\operatorname{rot}\mathbf{E} = [\nabla[\nabla\mathbf{E}]] \to -[\mathbf{k}[\mathbf{k}\mathbf{E}_k]];$$

$$\frac{\partial^2 \mathbf{E}}{\partial t^2} \to -\omega^2 \mathbf{E}_k.$$

Consequently, the equation

$$\operatorname{rot}\operatorname{rot}\mathbf{E}+\frac{1}{c^2}\frac{\partial^2\mathbf{E}}{\partial t^2}=-\frac{4\pi}{c^2}\frac{\partial}{\partial t}(\mathbf{j}^{(1)}+\mathbf{j}^{(2)}) \tag{2.42}$$

assumes the simple form (remembering that $\frac{\partial \mathbf{j}^{(1)}}{\partial t}=\frac{\omega_{0e}^2}{4\pi}\mathbf{E}$ )

$$c^2[\mathbf{k}\,[\mathbf{k}\mathbf{E}_k]]+(\omega^2-\omega_{0e}^2)\,\mathbf{E}_k=\frac{i\omega_{0e}^2 e}{2m_e}\int\frac{\omega}{\omega_1\omega_2}\left\{\frac{\mathbf{k}}{\omega}(\mathbf{E}_{k_1}\mathbf{E}_{k_2})+\right.$$

$$\left.+\mathbf{E}_{k_1}\frac{(\mathbf{k}_2\mathbf{E}_{k_2})}{\omega_2}+\mathbf{E}_{k_2}\frac{(\mathbf{k}_1\mathbf{E}_{k_1})}{\omega_1}\right\}dk_1dk_2\delta\,(k-k_1-k_2). \tag{2.43}$$

Thus the final result is an integral equation.

One other remark is in order concerning Eq. (2.43). As mentioned, the nonlinearity in (2.43) was assumed to be small, as only the nonlinear current $\mathbf{j}^{(2)}$ was considered. Furthermore, it was assumed that the temperature was zero. For $T \neq 0$ corrections are needed for both the linear and the nonlinear parts of Eq. (2.43). However, since the nonlinear effect is small, the corrections associated with $T \neq 0$ may be neglected, while the corrections for thermal motion in the linear part of the equation, although small, can be of the same order as the nonlinear effect, and in many instances must be included. Thermal motion can be accounted for in the linear part (left-hand side) of Eq. (2.43) by a simple extension of the equation. Notice that the second term on the left-hand side, $(\omega^2-\omega_{0e}^2)\mathbf{E}_k$, has the form $\omega^2(1-\omega_{0e}^2/\omega^2)\mathbf{E}_k = \omega^2\varepsilon\mathbf{E}_k$, where $\varepsilon$ is the dielectric constant of the plasma at $T = 0$. Allowance for the thermal motion implies substitution of $\varepsilon$ into (2.43) with regard for the thermal motion. Now, of course, $\varepsilon$ is a tensor, i.e., in Eq. (2.43) the expression $(\omega^2-\omega_{0e}^2)(\mathbf{E}_k)_i$ (the subscript i denoting the projection of $\mathbf{E}$ on the appropriate coordinate axis 1, 2, or 3, corresponding to x, y, or z) must be replaced by $\omega^2\varepsilon_{ij}(\mathbf{E}_k)_j$.

The tensor nature of the dielectric constant simply reflects the fact that when the thermal motion is included, the direction of motion of the electrons under the influence of an external field does not coincide with the direction of the field. As we know, in an isotropic plasma the tensor $\varepsilon_{ij}$ has only two distinct components, expressed in terms of $\varepsilon^t$ and $\varepsilon^l$. In the linear approximation these components describe the propagation of transverse and longitudinal waves:

$$\varepsilon_{ij} = \varepsilon^l \frac{k_i k_j}{k^2} + \varepsilon^t \left(\delta_{ij} - \frac{k_i k_j}{k^2}\right). \tag{2.44}$$

Expanding the double vector product in the first term on the left-hand side of Eq. (2.43), we obtain for the left-hand side of (2.43)

$$\left\{- c^2 k^2 \left(\delta_{ij} - \frac{k_i k_j}{k^2}\right) + \omega^2 \varepsilon^l \frac{k_i k_j}{k^2} + \omega^2 \varepsilon^t \left(\delta_{ij} - \frac{k_i k_j}{k^2}\right)\right\} E_{kj}. \tag{2.45}$$

We can now readily obtain equations which, in the linear approximation, reduce to the transverse and longitudinal wave equations. For this we multiply the left-hand side of (2.43) by $k_i$ and $k_j \delta_{is} - k_i \cdot k_s / dk^2$, respectively. Introducing the transverse field $E^t$ and the longitudinal field $E^l$

$$\mathbf{E}^l = \frac{\mathbf{k}(\mathbf{kE})}{k^2}; \quad \mathbf{E}^t = \mathbf{E} - \frac{\mathbf{k}(\mathbf{kE})}{k^2}, \tag{2.46}$$

we obtain

$$\omega^2 \varepsilon^l \mathbf{E}_k^l = \frac{i\omega\omega_{0e}^2 e}{2m_e} \int \frac{d\lambda}{\omega_1 \omega_2} \left\{\frac{\mathbf{k}}{\omega} (\mathbf{E}_{k_1} \mathbf{E}_{k_2}) + \frac{\mathbf{k}(\mathbf{kE}_{k_1})}{k^2} \left(\frac{\mathbf{k}_2}{\omega_2} \mathbf{E}_{k_2}^l\right) + \frac{\mathbf{k}(\mathbf{kE}_{k_2})}{k^2} \left(\frac{\mathbf{k}_1}{\omega_1} \mathbf{E}_{k_1}^l\right)\right\} \equiv \mathbf{j}_{\mathbf{k}\omega}^l; \tag{2.47}$$

$$(k^2 c^2 - \omega^2 \varepsilon^t) \mathbf{E}_k^t = - \frac{i\omega\omega_{0e}^2 e}{2m_e} \int \frac{d\lambda}{\omega_1 \omega_2} \left\{\left(\frac{\mathbf{k}_2}{\omega_2} \mathbf{E}_{k_2}^l\right)\left(\mathbf{E}_{k_1} - \frac{\mathbf{k}(\mathbf{kE}_{k_1})}{k^2}\right) + \left(\frac{\mathbf{k}_1}{\omega_1} \mathbf{E}_{k_1}^l\right)\left(\mathbf{E}_{k_2} - \frac{\mathbf{k}(\mathbf{kE}_{k_2})}{k^2}\right)\right\} \equiv \mathbf{j}_{\mathbf{k}\omega}^t. \tag{2.48}$$

These are the nonlinear equations required. If the nonlinear effects are neglected, the left-hand sides of (2.47) and (2.48) give $\varepsilon^l = 0$ and $k^2 c^2 = \omega^2 \varepsilon$, which lead to the dispersion equations (2.11) for cold plasma. Equations (2.47) and (2.48) describe, for example, the generation of longitudinal by transverse waves. This has the physical interpretation that one of the transverse waves modulates the electromagnetic parameters of the plasma, in particular the density and velocity of its electrons, while the other is scattered by the inhomogeneities created by the first wave and radiates new waves.

## 7. EQUATIONS FOR THE WAVE AMPLITUDES

Next we consider the solution of the system (2.47) and (2.48). The problem appears as though it could be very difficult, considering the complexity of the equations. Notice, however, that we do not require the exact solution of these equations, as from the outset we have been concerned only with weak nonlinearity. We saw earlier that the longitudinal wave amplitude can change as the result of interaction with transverse waves. By the presumed smallness of the nonlinearity this change in amplitude is necessarily very slow, or "quasi-adiabatic". In other words, there is physical justification for the notion that the fields are very nearly linear, except that their amplitudes vary slowly in space and time.

As we saw above, linear fields can be described by Eqs. (2.12). The assumption is that the fields described by the nonlinear equations can be approximately represented in the form

$$\mathbf{E}(\mathbf{r}, t) = \int \mathbf{E}_{\mathbf{k}}^{l}(\mathbf{r}, t)\, e^{-i(\omega^{l}(\mathbf{k})t-\mathbf{kr})} d\mathbf{k} + \int \mathbf{E}_{\mathbf{k}}^{t}(\mathbf{r}, t)\, e^{-i(\omega^{t}(\mathbf{k})t-\mathbf{kr})} d\mathbf{k}. \quad (2.49)$$

The amplitudes $\mathbf{E}_{\mathbf{k}}^{l}(\mathbf{r}, t)$ and $\mathbf{E}_{\mathbf{k}}^{t}(\mathbf{r}, t)$ are slowly varying functions of **r** and t, for which only the first derivatives with respect to **r** and t need be included. This method of solving the nonlinear equations has been used by Van der Pol [1] and in nonlinear optics by Bloembergen [2], and it has been mathematically justified by Bogolyubov [3]. It is reasonable that the derivatives of the amplitudes need to be included only in the left-hand, linear, side of Eqs. (2.49), because the right-hand side represents a small nonlinear effect, and with the insertion of (2.47) and (2.48) in that side the amplitudes may be regarded as constant.

By some straightforward calculations given in Appendix 1 we obtain the following equations for the amplitudes $\mathbf{E}_{\mathbf{k}}^{l}(\mathbf{r}, t)$ and $\mathbf{E}_{\mathbf{k}}^{t}(\mathbf{r}, t)$:

$$\left(\frac{\partial}{\partial t} + \mathbf{v}_{\mathrm{rp}}^{l} \frac{\partial}{\partial \mathbf{r}}\right) \mathbf{E}_{\mathbf{k}}^{l}(\mathbf{r}, t) = -i \frac{1}{\left(\frac{\partial}{\partial\omega}\omega^{2}\varepsilon^{l}\right)\Big|_{\omega=\omega^{l}(\mathbf{k})}} \int e^{-i\omega t + i\omega^{l}(\mathbf{k})t}\, \mathbf{j}_{\mathbf{k}\omega}^{l}\, d\omega; \quad (2.50)$$

$$\left(\frac{\partial}{\partial t} + \mathbf{v}_{\mathrm{rp}}^{t} \frac{\partial}{\partial \mathbf{r}}\right) \mathbf{E}_{\mathbf{k}}^{t}(\mathbf{r}, t) = i \frac{1}{\left(\frac{\partial}{\partial\omega}\omega^{2}\varepsilon^{t}\right)\Big|_{\omega=\omega^{t}(\mathbf{k})}} \int e^{-i\omega t + i\omega^{t}(\mathbf{k})t}\, \mathbf{j}_{\mathbf{k}\omega}^{t}\, d\omega. \quad (2.51)$$

Here $\mathbf{v}_{\mathrm{gr}}^{l,t} = d\omega^{l,t}(\mathbf{k})/d\mathbf{k}$ represents the group velocities of the longitudinal and transverse waves.

## 8. NONLINEAR INTERACTION OF WEAKLY DAMPED AND WEAKLY GROWING WAVES

Of special interest are the interactions of waves which are weakly damped or growing in the linear approximation. For such waves the growth rate (or damping rate) of the oscillations is much smaller than their frequency. The solution of the dispersion equation for $\omega(\mathbf{k})$ is in general complex:

$$\omega(\mathbf{k}) = \Omega(\mathbf{k}) + i\gamma_{\mathbf{k}}, \tag{2.51}$$

where $\Omega(\mathbf{k})$ is the real part of the frequency, and $\gamma_{\mathbf{k}} = \operatorname{Im} \omega(\mathbf{k})$ is the oscillation growth rate, characterizing the variation in amplitude of the oscillations:

$$e^{-i\omega(\mathbf{k})t} = e^{\gamma_{\mathbf{k}}t} e^{-i\Omega(\mathbf{k})t}. \tag{2.52}$$

Thus, for the waves in question

$$|\gamma_{\mathbf{k}}| \ll \Omega(\mathbf{k}). \tag{2.53}$$

The only significant growth or damping of the oscillations takes place over many wave periods. We note that a slow change of amplitude can occur for two reasons, through instability (or damping) or by nonlinear effects. In cases when these amplitude changes are in opposite sense, it becomes possible to have nonlinear quasistationary states with excited self-consistent fields. We first consider the limit $\gamma_{\mathbf{k}} \to 0$, when the waves are neither damped nor growing, and then show the modifications to the equations for $\gamma_{\mathbf{k}} \neq 0$. The field $\mathbf{E}_k$ on the right-hand side of the equations is easily expressed in terms of the amplitude $\mathbf{E}_{\mathbf{k}}(\mathbf{r}, t)$. Since the equations describe a weak nonlinear effect, it may be assumed in the first approximation that, for purposes of substitution into the right-hand side of the equations, the amplitudes $\mathbf{E}_{\mathbf{k}}(\mathbf{r}, t)$ are constant. Then approximately

$$\begin{aligned}\mathbf{E}_k^l &= \frac{1}{(2\pi)^4} \int \mathbf{E}(\mathbf{r}, t)\, e^{-i\mathbf{k}\mathbf{r}+i\omega t}\, d\mathbf{r}\, dt = \\ &= \frac{1}{(2\pi)^4} \int \mathbf{E}_{\mathbf{k}'}^l e^{-i\Omega^l(\mathbf{k}')t+i\mathbf{k}'\mathbf{r}-i\mathbf{k}\mathbf{r}+i\omega t}\, d\mathbf{k}'\, d\mathbf{r}\, dt = \\ &= \int \delta(\mathbf{k}-\mathbf{k}')\,\delta(\omega - \Omega^l(\mathbf{k}'))\,\mathbf{E}_{\mathbf{k}'}^l\, d\mathbf{k}' = \mathbf{E}_{\mathbf{k}}^l \delta(\omega - \Omega^l(\mathbf{k})).\end{aligned} \tag{2.54}$$

Substituting (2.54) into the first term on the right-hand side of Eq. (2.47) gives the following result:

$$\left(\frac{\partial}{\partial t} + \mathbf{v}^l_{\mathrm{gr}}\frac{\partial}{\partial \mathbf{r}}\right)\mathbf{E}^l_{\mathbf{k}} = \frac{e\omega_{0e}^2}{2m_e}\int \frac{d\mathbf{k}_1\, d\mathbf{k}_2 \delta(\mathbf{k}-\mathbf{k}_1-\mathbf{k}_2)}{\left(\frac{\partial}{\partial\omega}\omega^2\varepsilon^l\right)\Big|_{\omega=\Omega^l(\mathbf{k})}} e^{i\Omega^l(\mathbf{k})t} \times$$
$$\times\left(\mathbf{k}\frac{(\mathbf{E}^t_{\mathbf{k}_1}\mathbf{E}^t_{\mathbf{k}_2})\exp\left(-i\left[\Omega^t(\mathbf{k}_1)+\Omega^t(\mathbf{k}_2)\right]t\right)}{\Omega^t(\mathbf{k}_1)\,\Omega^t(\mathbf{k}_2)} + \right.$$
$$+ 2\mathbf{k}\frac{(\mathbf{E}^t_{\mathbf{k}_1}\mathbf{E}^l_{\mathbf{k}_2})\exp\left(-i\left[\Omega^t(\mathbf{k}_1)+\Omega^l(\mathbf{k}_2)\right]t\right)}{\Omega^t(\mathbf{k}_1)\,\Omega^l(\mathbf{k}_2)} +$$
$$\left. + \mathbf{k}\frac{(\mathbf{E}^l_{\mathbf{k}_1}\mathbf{E}^l_{\mathbf{k}_2})\exp\left(-i\left[\Omega^l(\mathbf{k}_1)+\Omega^l(\mathbf{k}_2)\right]t\right)}{\Omega^l(\mathbf{k}_1)\,\Omega^l(\mathbf{k}_2)}\right). \tag{2.55}$$

The three terms on the right-hand side of (2.55) correspond to three different processes. The first term describes the interaction of two transverse and one longitudinal wave, the second describes the interaction of two longitudinal and one transverse wave, and the third describes the interaction of three longitudinal waves. Each of these processes may in general be treated independently. In order not to complicate the picture, let us limit the discussion to one process, namely the interaction of two transverse waves and one longitudinal wave. As we shall find out below, there are conditions under which the processes are strictly separable. For example, the transverse waves can have frequencies falling in an interval such that processes involving more than one longitudinal wave are virtually prohibited. Hence, in what follows we use the nonlinear equations for longitudinal waves

$$\left(\frac{\partial}{\partial t} + \mathbf{v}^l_{\mathrm{gr}}\frac{\partial}{\partial \mathbf{r}}\right)\mathbf{E}^l_{\mathbf{k}} = \frac{e\omega_{0e}^2\mathbf{k}}{2m_e\frac{\partial}{\partial\omega}\omega^2\varepsilon^l\Big|_{\omega=\Omega^l(\mathbf{k})}}\int\frac{d\mathbf{k}_1 d\mathbf{k}_2\,(\mathbf{E}^t_{\mathbf{k}_1}\mathbf{E}^t_{\mathbf{k}_2})}{\Omega^t(\mathbf{k}_1)\,\Omega^t(\mathbf{k}_2)}\times$$
$$\times e^{i\left(\Omega^l(\mathbf{k})-\Omega^t(\mathbf{k}_1)-\Omega^t(\mathbf{k}_2)\right)t}\,\delta(\mathbf{k}-\mathbf{k}_1-\mathbf{k}_2) \tag{2.56}$$

and the similarly derived equation for transverse waves

$$\left(\frac{\partial}{\partial t} + \mathbf{v}^t_{\mathrm{gr}}\frac{\partial}{\partial \mathbf{r}}\right)\mathbf{E}^t_{\mathbf{k}} = \frac{e\omega_{0e}^2}{m_e\frac{\partial}{\partial\omega}\omega^2\varepsilon^t\Big|_{\omega=\Omega^t(\mathbf{k})}}\int\frac{\left[\Omega^t(\mathbf{k}_1)+\Omega^l(\mathbf{k}_2)\right]}{\Omega^t(\mathbf{k}_1)\left[\Omega^l(\mathbf{k}_2)\right]^2}\times$$
$$\times(\mathbf{k}_2\mathbf{E}^l_{\mathbf{k}_2})\left(\mathbf{E}^t_{\mathbf{k}_1}-\frac{\mathbf{k}\,(\mathbf{k}\mathbf{E}_{\mathbf{k}_1})}{k^2}\right)e^{i\left(\Omega^t(\mathbf{k})-\Omega^t(\mathbf{k}_1)-\Omega^l(\mathbf{k}_2)\right)t}\,\delta(\mathbf{k}-\mathbf{k}_1-\mathbf{k}_2)\,d\mathbf{k}_1 d\mathbf{k}_2. \tag{2.57}$$

We now find what changes are needed in these equations in order to allow for $\gamma_{\mathbf{k}} \neq 0$. Here it is useful to introduce the "true" amplitude $\mathbf{A}_{\mathbf{k}}$ $(\mathbf{r}, t)$:

$$\mathbf{E}^l(\mathbf{r}, t)\,e^{\gamma^l t} = \mathbf{A}^l(\mathbf{r}, t);\ \mathbf{E}^t(\mathbf{r}, t)\,e^{\gamma^t t} = \mathbf{A}^t(\mathbf{r}, t). \tag{2.58}$$

The left-hand sides of Eqs. (2.56) and (2.57) now take the form

$$\left(\frac{\partial}{\partial t} - \gamma_{\mathbf{k}} + \mathbf{v}_{gr} \frac{\partial}{\partial \mathbf{r}}\right) \mathbf{A}_{\mathbf{k}}(\mathbf{r}, t).$$

In right-hand side $\mathbf{E}_{\mathbf{k}}$ must be replaced by $\mathbf{A}_{\mathbf{k}}$. The validity of this statement is easily checked if at the very beginning solutions of the following form are substituted into Eqs. (2.47) and (2.48):

$$\mathbf{E} = \int \mathbf{A}_{\mathbf{k}}(\mathbf{r}, t)\, e^{-i\Omega(\mathbf{k})t + i\mathbf{k}\mathbf{r}} d\mathbf{k}. \tag{2.59}$$

Consequently, in the presence of growth or damping it is required formally to add the term $-\gamma_{\mathbf{k}} \mathbf{E}_{\mathbf{k}}$ to the left-hand sides of Eqs. (2.57) and (2.56).

To facilitate our further analysis of the equations let us introduce some suitable notation. We introduce unit vectors $\mathbf{e}_{\mathbf{k}}$ characterizing the wave polarization:

$$\mathbf{A}^{t}_{\mathbf{k}} = A^{t}_{\mathbf{k}} \mathbf{e}_{\mathbf{k}}; \quad \mathbf{A}^{l}_{\mathbf{k}} = A^{l}_{\mathbf{k}} \frac{\mathbf{k}}{k}. \tag{2.60}$$

We make use of the fact that $\Omega(\mathbf{k})$, representing the solution of the dispersion equation, has the property*

$$\Omega(-\mathbf{k}) = -\Omega(\mathbf{k}). \tag{2.61}$$

Since $\mathbf{E}$, as given by (2.59), is real, we obtain the following, replacing $\mathbf{k}$ in (2.59) by $-\mathbf{k}$ and making use of (2.61):

$$A^{t}_{\mathbf{k}}(\mathbf{r}, t) = A^{t*}_{-\mathbf{k}}(\mathbf{r}, t); \quad \mathbf{e}_{-\mathbf{k}} = \mathbf{e}^{*}_{\mathbf{k}}. \tag{2.62}$$

We note that, since the unit vector for longitudinal waves $\mathbf{k}/k$ changes sign in going from $\mathbf{k}$ to $-\mathbf{k}$,

$$A^{l}_{\mathbf{k}}(\mathbf{r}, t) = -A^{l*}_{-\mathbf{k}}(\mathbf{r}, t). \tag{2.63}$$

Replacing $\mathbf{k}_2$ by $-\mathbf{k}_2$ in (2.56) and using (2.60)-(2.62), we find that Eqs. (2.56) may be written in the form

$$\left(\frac{\partial}{\partial t} - \gamma^{l}_{\mathbf{k}} + \mathbf{v}^{l}_{gr} \frac{\partial}{\partial \mathbf{r}}\right) A^{l}_{\mathbf{k}} = -\int \alpha^{l}_{\mathbf{k}, \mathbf{k}_1, \mathbf{k}_2} A^{t}_{\mathbf{k}_1} A^{t*}_{\mathbf{k}_2} e^{-i\Delta\Omega t} d\mathbf{k}_1 d\mathbf{k}_2, \tag{I}$$

* The validity of (2.61) is best illustrated directly by means of the dispersion equation. Thus, for longitudinal waves $\varepsilon^{l}(\Omega(\mathbf{k}), \mathbf{k}) = 0$, but $\varepsilon^{l}(\omega, \mathbf{k}) = \varepsilon^{l*}(-\omega, -\mathbf{k})$. If $\varepsilon^{l}$ is real, then, by inserting $\omega = \Omega(\mathbf{k})$ into the last equation, we have $\varepsilon^{l}(-\Omega(\mathbf{k}), -\mathbf{k}) = 0$. On the other hand, the solution of $\varepsilon^{l}(\omega, -\mathbf{k})$ is $\omega = \Omega(-\mathbf{k})$.

with the following notation:

$$\Delta\Omega = \Delta\Omega_{\mathbf{k},\,\mathbf{k}_1,\,\mathbf{k}_2} = \Omega^t(\mathbf{k}_1) - \Omega^t(\mathbf{k}_2) - \Omega^l(\mathbf{k}). \tag{2.64}$$

The quantity $\Delta\Omega$ will play a very important part later on. Roughly speaking, it characterizes the frequency separation of the interacting waves. Moreover,

$$\alpha^l_{\mathbf{k},\,\mathbf{k}_1,\,\mathbf{k}_2} = \frac{e\omega_{0e}^2 k\delta(\mathbf{k} - \mathbf{k}_1 + \mathbf{k}_2)(\mathbf{e}_{\mathbf{k}_1}\mathbf{e}^*_{\mathbf{k}_2})}{2m_e \frac{\partial}{\partial\omega}(\omega^2\varepsilon^l(\omega, \mathbf{k}))\Big|_{\omega=\Omega^l(\mathbf{k})} \Omega^t(\mathbf{k}_1)\,\Omega^t(\mathbf{k}_2)}. \tag{2.65}$$

The equation for $A^t_{\mathbf{k}}$ is obtained similarly. It is conveniently written in two forms:

$$\left(\frac{\partial}{\partial t} - \gamma^t_{\mathbf{k}_1} + \mathbf{v}^t_{\mathrm{gr}_1}\frac{\partial}{\partial \mathbf{r}}\right) A^t_{\mathbf{k}_1} = \int \alpha^t_{1,\,\mathbf{k},\,\mathbf{k}_1,\,\mathbf{k}_2} A^l_{\mathbf{k}} A^t_{\mathbf{k}_2} e^{i\Delta\Omega t} d\mathbf{k}\, d\mathbf{k}_2 \tag{II}$$

and

$$\left(\frac{\partial}{\partial t} - \gamma^t_{\mathbf{k}_2} + \mathbf{v}^t_{\mathrm{gr}_2}\frac{\partial}{\partial \mathbf{r}}\right) A^t_{\mathbf{k}_2} = -\int \alpha^t_{2,\,\mathbf{k},\,\mathbf{k}_1,\,\mathbf{k}_2} A^{l*}_{\mathbf{k}} A^t_{\mathbf{k}_1} e^{-i\Delta\Omega t} d\mathbf{k}\, d\mathbf{k}_1, \tag{III}$$

where

$$\alpha^t_{1,\,\mathbf{k},\,\mathbf{k}_1,\,\mathbf{k}_2} = \frac{e\omega_{0e}^2 k(\Omega^l(\mathbf{k}) + \Omega^t(\mathbf{k}_2))\,\delta(\mathbf{k} - \mathbf{k}_1 + \mathbf{k}_2)(\mathbf{e}^*_{\mathbf{k}_1}\mathbf{e}_{\mathbf{k}_2})}{m_e \frac{\partial}{\partial\omega}\omega^2\varepsilon^t(\omega, \mathbf{k}_1)\Big|_{\omega=\Omega^t(\mathbf{k}_1)} \Omega^t(\mathbf{k}_2)(\Omega^l(\mathbf{k}))^2}, \tag{2.66}$$

$$\alpha^t_{2,\,\mathbf{k},\,\mathbf{k}_1,\,\mathbf{k}_2} = \frac{e\omega_{0e}^2 k(\Omega^t(\mathbf{k}_1) - \Omega^l(\mathbf{k}))\,\delta(\mathbf{k} - \mathbf{k}_1 + \mathbf{k}_2)(\mathbf{e}^*_{\mathbf{k}_2}\mathbf{e}_{\mathbf{k}_1})}{m_e \frac{\partial}{\partial\omega}\omega^2\varepsilon^t(\omega, \mathbf{k}_2)\Big|_{\omega=\Omega^t(\mathbf{k}_2)} \Omega^t(\mathbf{k}_1)(\Omega^l(\mathbf{k}))^2}. \tag{2.67}$$

Although (II) and (III) are equivalent, it is convenient to use the complete system (I)-(III), particularly when the transverse wave spectrum contains two components with nonzero amplitudes $A^t_{\mathbf{k}_1}$ and $A^t_{\mathbf{k}_2}$.

## 9. SOME CONSEQUENCES OF THE DERIVED EQUATIONS

Let us conclude by examining some characteristics of the nonlinear interactions described by (I)-(III).

First, the right- and left-hand sides of the equations contain dissimilar wave types. In other words, the transformation of one type into others is significant. It is natural therefore to expect that nonlinear effects can limit the maximum wave amplitude. In fact, the right-hand sides of the equations become significant only for sufficiently large wave amplitudes. However, when this happens there is a transformation of one wave type into others. Can this transformation limit the amplitude? This can happen if, for example, waves of one type grow, while waves of another type decay. Even if the latter do not decay, it is clear that the conversion of the excited waves into others may reduce their amplitude (for example, those into which the excited waves are converted may move outside the range of generation, thereby minimizing the reverse conversion process, etc.). We see that wave conversion can in general be used to stabilize a plasma instability.

Second, interaction between waves from different parts of the spectrum may be important. For example, it is obvious from Eq. (II) that the transverse wave $\mathbf{k}_2$ may, as a result of its interaction with a longitudinal wave, change into, or excite, a second transverse wave which has, in general, a different value of $\mathbf{k} = \mathbf{k}_1$ and thus a different frequency.

Normally instability appears only in certain parts of the spectrum, while in other parts there is damping. Consequently, nonlinear effects can result in the transformation of waves from a region of instability into one of damping. We see that some conclusions concerning the physical mechanisms underlying the nonlinear stabilization of plasma instabilities can be drawn merely from the structure of the equations.

Finally, it is important to mention the analogy between the equations derived and the familiar kinetic equation for the particles of a plasma:

$$\left(\frac{\partial}{\partial t} + \mathbf{v}\,\frac{\partial}{\partial \mathbf{r}}\right) f_p = \int \Lambda\,(f_p f_{p_1})\,\sigma\,(p,\, p_1)\, dp_1. \tag{2.81}$$

Here $f_p$ is the particle distribution function, and v their velocity; the right-hand side of (2.81) has a quadratic dependence on the distribution function, as represented symbolically by the notation $\Lambda\,(f_p f_{p_1})$ , and describes the collisions of particles with each other. In Eqs. (I)-(III) the role of the velocity is played by the group velocity of the waves, and the role of the collisions by the nonlinear interaction of the waves with its quadratic dependence on the field amplitudes. We shall see later that this analogy is not merely fortuitous, and show that under certain conditions the waves can be characterized by wave distribution functions, and their nonlinear interaction described as "collisions" of waves.

However, the kinetic description of wave interaction is not always possible. In the next chapter we turn our attention to explore when in fact such a description is possible.

## Literature Cited


1. B. Van der Pol, Nonlinear Theory of Electrical Oscillations [Russian translation]. Svyaz'izdat (1935).
2. J. A. Armstrong, N. Bloembergen, J. Ducuing, and P. S. Pershan, "Interactions between light waves in a nonlinear dielectric," Phys. Rev., 127:1918 (1962).
3. N. N. Bogolyubov and O. A. Mitropol'skii, Asymptotic Methods in the Theory of Nonlinear Oscillations. Gordon and Breach, New York (1962).

*Chapter III*

# Nonlinear Interaction of Three Wave Packets

## 1. WAVE PACKETS

A wave packet is a set of superimposed plane monochromatic waves having close values of the wave numbers $\mathbf{k}$ and, hence, close frequencies $\Omega(\mathbf{k})$.

Suppose that a wave packet has a certain mean wave number $\mathbf{k}_0$, with a spread $\delta\mathbf{k}$ about this mean, and that the corresponding mean frequency is $\Omega(\mathbf{k}_0) = \Omega_0$, with a spread $\delta\Omega$. Thus,

$$\delta\mathbf{k} \ll \mathbf{k}_0; \quad \delta\Omega \ll \Omega_0. \tag{3.1}$$

The field of this packet may be written in the form

$$\mathbf{E} = \int \mathbf{A}_{\mathbf{k}} e^{-i\Omega(\mathbf{k})t + i\mathbf{k}\mathbf{r}}\, d\mathbf{k}, \tag{3.2}$$

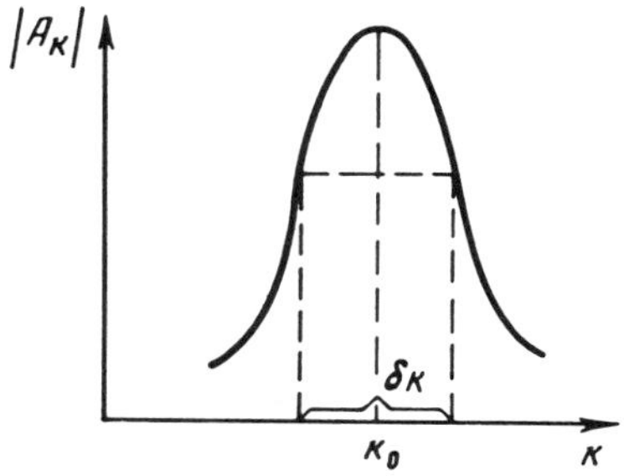


Fig. 4. A narrow wave packet.

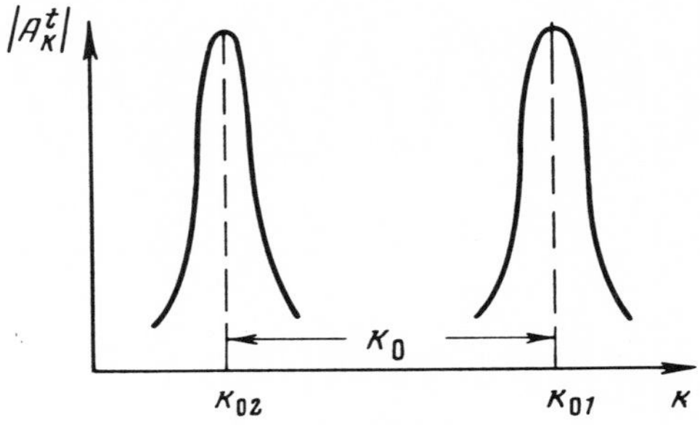


Fig. 5. Two narrow wave packets.

where $\mathbf{A}_k$ has a maximum about $\mathbf{k} = \mathbf{k}_0$ (Fig. 4).

The analysis of the interaction of wave packets is important, first because packets of, say, transverse waves are easily produced experimentally and, secondly because the growth of plasma instabilities is frequently greatest near certain values of $\mathbf{k} = \mathbf{k}_0$, i.e., the waves build up predominantly around $\mathbf{k}_0$.

Nonlinear wave interaction can alter both the shape of the wave packet, i.e., the distribution of the waves with respect to the wave number and frequency, and the mean amplitude and mean energy of the wave packet. The nonlinear equations found above describe the interactions of two transverse waves and one longitudinal wave. Accordingly we assume that there are three wave packets, two of transverse waves and one of longitudinal waves. We also assume that the transverse wave packets do not overlap, i.e., they have no wavelengths in common (Fig. 5). For effective nonlinear interaction the mean values of the wave numbers of the packets cannot be arbitrary. According to Eqs. (I)-(III) they must satisfy

$$\mathbf{k}_{10} - \mathbf{k}_{20} \approx \mathbf{k}_0. \qquad (3.3)$$

This equation need not be satisfied exactly, because effective interaction requires only that each of the three packets contain waves satisfying the condition

$$\mathbf{k}_1 - \mathbf{k}_2 = \mathbf{k}. \qquad (3.4)$$

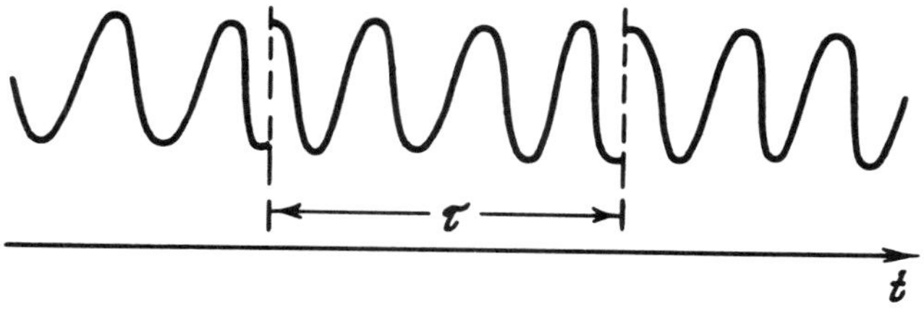


Fig. 6. Shifts of the wave phase.

## 2. PHASE RELATIONS FOR WAVE PACKETS

Before taking up the analysis of wave interaction, we must consider the phases of the interacting waves. Here we are concerned with whether the phases of the waves in a given packet are random or nonrandom. Interactions of random waves turn out to be significantly different from those of nonrandom waves. Let us elaborate on what we mean when we say that the phase is random. The amplitude $\mathbf{A_k}$ of the wave packet is in general complex:

$$\mathbf{A_k} = |\mathbf{A_k}|\, e^{i\varphi_\mathbf{k}}, \tag{3.5}$$

where $|\mathbf{A_k}|$ is the wave amplitude, and $\varphi_\mathbf{k}$ is its phase. Let the field in question comprise a wave with frequency $\Omega_0$ for a finite time interval, then let the phase slowly vary, and afterwards let oscillations again occur with the same frequency $\Omega_0$. These phase shifts occur consecutively after mean time intervals $\tau$. In other words, the field represents a sequence of wave trains (Fig. 6).

Let us assume that the phases vary randomly. Suppose that we are making measurements of the field strength of the wave at the same point in space after equal time intervals $t_0$ which are multiples of $2\pi/\Omega$. If the phase did not change, the measurements would give the same value for $\mathbf{E}$. If $t_0 \gg \tau$, with random phases we obtain completely different values for the field. If the mean value of the field thus measured is zero for a time $t \gg \tau$, this is an example of a random field.

It is clear from these arguments that the concept of randomness of a field involves the "measurement technique." Thus, if in the example given we take $t_0 \ll \tau$, the measured fields could exhibit regular, rather than random, values. In physical problems the "observation time" $t_0$ is normally determined by the characteristic time of the physical process. Consequently, the same field might be random for some processes and nonrandom for others. For nonlinear processes the characteristic time might be, for example, the period of nonlinear transfer of energy from one oscillatory mode to another, or the e-folding time of linear theory. These are the times which must be compared with $\tau$, the characteristic time of the phase shift of the oscillations, to determine whether the oscillations are random or not. Moreover, the answer to this question can change with the evolution of the system. For example, the characteristic time for the nonlinear transfer of energy from one mode to another depends on the oscillation amplitudes, which can change with time. Similarly, the linear growth rates will depend on a number of plasma parameters, which, as mentioned above, vary through nonlinear effects.

A field comprising a set of wave trains will not be a monochromatic wave. However, it can be represented as a superposition of fields of the type (3.2). If $\tau \gg 2\pi/\Omega_0$, the spectrum $|\mathbf{A_k}|$ has a maximum around $\mathbf{k} = \mathbf{k}_0$. If we assume for simplicity that all the waves have the same direction, the width $\delta\Omega$ of the packet is related to $\delta\mathbf{k}$ by

$$\delta\Omega \approx \frac{d\Omega(\mathbf{k})}{d\mathbf{k}}\bigg|_{\mathbf{k}=\mathbf{k}_0} \delta\mathbf{k} \simeq \mathbf{v}_{gr}(\mathbf{k}_0)\,\delta\mathbf{k}.$$

It is well known that the width of the spectrum of a wave packet representing a set of wave trains of duration $\tau$ is of order $\delta\Omega \sim 1/\tau$. Hence,

$$\delta\mathbf{k} \sim \frac{1}{\mathbf{v}_{gr}(\mathbf{k}_0)\,\tau}.$$

The condition $t_0 \gg \tau$ for the field to be random can be written

$$\delta\Omega \gg \frac{1}{t_0}. \tag{3.6}$$

Here $t_0$ is the characteristic time for the given process. The inequality (3.6) shows that a narrow spectrum cannot correspond to a random field.

If $1/t_0 \ll \Omega_0$, i.e., if the characteristic times of the process associated with nonlinear interaction are much longer than the oscillation period, as we assumed in deriving the equations describing weak nonlinearities, then according to (3.6) it is possible to have

$$\frac{1}{t_0} \ll \delta\Omega \ll \Omega_0, \tag{3.7}$$

i.e., both the nonlinear interactions and the phase shifts can occur over many periods of the field oscillations. If inequality (3.7) is satisfied, the phase of the waves can change repeatedly before the nonlinear effects will change the spectrum or mean wave amplitudes. But, if the inequality

$$\delta\Omega \ll \frac{1}{t_0} \ll \Omega_0 \tag{3.8}$$

is fulfilled, the nonlinear interactions will be felt before any change in the phase relations of the waves. It is clear, therefore, that the nonlinear interaction of wave packets can be entirely different for conditions (3.7) and (3.8).

The nonlinear interaction equations describe the dynamics of the variation of both the amplitudes and the phases of the waves, i.e., they can answer the following question: if one of the inequalities (3.7) or (3.8) holds initially for a certain wave packet, will it be violated later on and change to the opposite inequality? It must be stressed that all we have the right to assume in advance are certain initial properties of the wave packets. Their subsequent variation is completely described by the nonlinear equations. As a result of the development of nonlinear interaction the inequality (3.8) can be violated and revert to (3.7). This can happen, for example, if the nonlinear effects only slightly alter the mean wave intensities, but cause rapid broadening of their spectra. If the characteristic interaction time is determined by the wave intensities and depends only slightly on their spectral widths, the broadening of the wave spectrum only slightly affects $t_0$, while an increase in $\delta\Omega$ violates (3.8). The reverse situation is also possible, when as a result of nonlinear action the intensity of one of the waves is abruptly altered, for example, as the result of energy being pumped into it from other waves, while the width of the wave spectrum changes only slightly. In this case $t_0$ can decrease, and a transition occurs from the inequality (3.7) to (3.8).

To be sure, these examples are crude, but they have been chosen to illustrate how the dynamics of the variation of both the phases and amplitudes of the waves are completely described by the nonlinear equations. However, the investigation of such problems is often far from simple. It is simpler to treat the problem on the assumption that either (3.7) or (3.8) is satisfied throughout the entire duration of wave interaction. It is also possible to en-

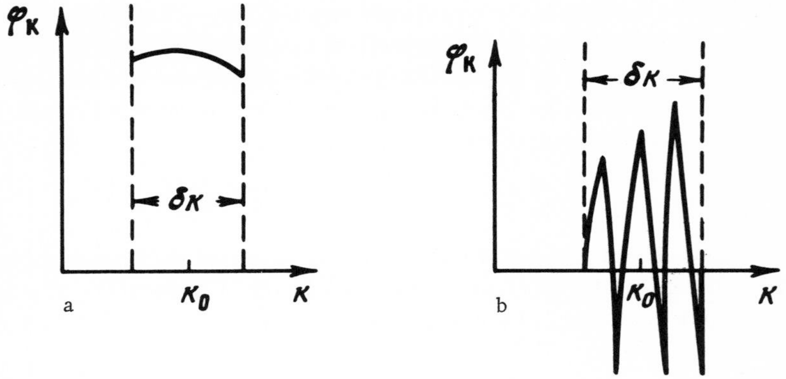


Fig. 7. Wave phase versus wave number for waves of (a) stationary and (b) random phase.

counter the situation in which (3.7) is satisfied for some interacting wave packets while (3.8) is satisfied for others.

Finally, it must be pointed out that not all the waves of a given packet need participate in nonlinear interaction, because the condition (3.4) might not be met for all the waves. In this case the width of the packet must be interpreted as that $\delta\mathbf{k}$ which covers those waves capable of interacting. In connection with the change of the wave spectra in nonlinear interaction the interval $\delta\mathbf{k}$ of the waves capable of interacting can also change. Above we illustrated the moduli of the amplitudes for wave packets (Fig. 5). A smooth variation in amplitude is possible either for packets having completely defined phases (which we call "fixed phases") or for packets whose phases are random. The variation of $\varphi_k$ in (3.5) within the limits of the packet is different in these two cases. In the former the phase of the oscillations varies smoothly (Fig. 7a), while in the latter there are repeated sudden variations with changes of sign (Fig. 7b).* It is clear that if the ever-increasing intervals of $\mathbf{k}$ begin to play a role as the result of interaction, for small effective $\delta\mathbf{k} \ll \delta\mathbf{k}_0$ the phases may also be regarded as fixed, but not when $\delta\mathbf{k} \sim \delta\mathbf{k}_0$.

* Strictly speaking, Fig. 7b shows the phase for one possible realization of the statistical ensemble.

Next we consider some simple problems of nonlinear interaction, assuming that the phases of all the waves are either fixed or random (random variables are discussed in more detail below).

## 3. INTERACTION OF FIXED-PHASE WAVE PACKETS

For waves such that $|\mathbf{A_k}|$ and $\varphi_\mathbf{k}$ vary only slightly within the limits of the packet it is convenient to introduce the mean amplitude and mean phase by means of the relation

$$\int |\mathbf{A_k}| e^{i\varphi_\mathbf{k}} d\mathbf{k} = |\overline{\mathbf{A}}_{\mathbf{k}_0}| e^{i\varphi_{\mathbf{k}_0}} = \overline{\mathbf{A}}_{\mathbf{k}_0}; \overline{A}_{k_0} = A_{k_0}\,\delta k \tag{3.9}$$

We assume that the mean values of $\mathbf{k}_0$ of the three wave packets satisfy the relation (3.3), and for simplicity that all quantities vary only with time.

Integrating each of the fundamental equations (I), (II), and (III) derived in the preceding chapter over $\mathbf{k}$, $\mathbf{k}_1$, and $\mathbf{k}_2$, respectively, we obtain

$$\left(\frac{\partial}{\partial t} - \gamma^l\right) A^l = -\beta^l A_1^t A_2^{t*} e^{-i\Delta\Omega t}; \tag{3.10}$$

$$\left(\frac{\partial}{\partial t} - \gamma_1^t\right) A_1^t = \beta_1^t A^l A_2^t e^{i\Delta\Omega t}; \tag{3.11}$$

$$\left(\frac{\partial}{\partial t} - \gamma_2^t\right) A_2^t = -\beta_2^t A^{l*} A_1^t e^{-i\Delta\Omega t}. \tag{3.12}$$

Here $A^l = \overline{A}^l_{\mathbf{k}_0}$; $\overline{A}^t_{1,2} = \overline{A}^t_{\mathbf{k}_{10,20}}$; $\Delta\Omega = -\Omega^l(\mathbf{k}_{10} - \mathbf{k}_{20}) + \Omega^t(\mathbf{k}_{10}) - \Omega^t(\mathbf{k}_{20})$, and the quantities $\beta$ are related to the coefficients $\alpha$ of Eqs. (I), (II), and (III) of Chap. 2 by the expressions

$$\alpha_{\mathbf{k}_0^l \mathbf{k}_{10}^t \mathbf{k}_{20}^t} = \beta\delta(\mathbf{k}_0 - \mathbf{k}_{10} + \mathbf{k}_{20}). \tag{3.13}$$

Thus, the problem has been reduced to a set of equations for the mean amplitudes and phases of the interacting waves. Strong interaction is possible only if the right-hand sides of Eqs. (3.10), (3.11), and (3.12) do not repeatedly change sign during the characteristic time $t_0$ for nonlinear interation, i.e.,

$$\Delta\Omega t_0 \ll 1. \tag{3.14}$$

This condition is often used for the analysis of problems in nonlinear optics and is called the coherence condition. Roughly speaking, it implies that during the time of interaction the phase difference between the waves cannot change so much as to allow the sign of nonlinear interaction to change. We observe that $\Delta\Omega$ is larger than or of the same order as the sums of the widths of the interacting packets, for each of which the following condition holds in the case of fixed phases:

$$\delta\Omega t_0 \ll 1. \tag{3.15}$$

This is the reason that (3.14) can be satisfied.

If the coherence condition is satisfied, the equations become

$$\left(\frac{\partial}{\partial t} - \gamma^l\right) A^l = -\beta^l A_1^t A_2^{t*}; \tag{3.16}$$

$$\left(\frac{\partial}{\partial t} - \gamma_1^t\right) A_1^t = \beta_1^t A^l A_2^t; \tag{3.17}$$

$$\left(\frac{\partial}{\partial t} - \gamma_2^t\right) A_2^t = -\beta_2^t A^{l*} A_1^t. \tag{3.18}$$

These equations are very similar in structure to those investigated in nonlinear optics [1-3]. It is seen at once that if two of the three waves are initially intense, the amplitude of the third can grow. The same type of nonlinear generation is familiar in nonlinear optics. For example, if two intense laser beams are transmitted through a crystal, it is possible to generate a third wave.

## 4. INTERACTION OF UNDAMPED WAVE PACKETS IN AN UNSTABLE PLASMA

Let $\gamma^l = \gamma_1^t = \gamma_2^t = 0$, i.e., let there be neither damping nor growth of the waves. In this case energy is transferred from one wave to another and back again. This transfer is reversible, the period of the oscillations being determined by the initial wave amplitudes [4]. To solve the set of equations

$$\frac{\partial}{\partial t} A^l = -\beta^l A_1^t A_2^{t*}; \tag{3.19}$$

$$\frac{\partial}{\partial t} A_1^t = \beta_1^t A^l A_2^t; \tag{3.20}$$

$$\frac{\partial}{\partial t} A_2^t = -\beta_2^t A^{l*} A_1^t \tag{3.21}$$

it is convenient to use the conservation laws ($\mathrm{Im}\,\beta = 0$)

$$x_1 + x_2 = \text{const}; \quad x_1 + x^l = \text{const};$$
$$\Omega^l x^l + \Omega_1^t x_1 + \Omega_2^t x_2 = \text{const}, \tag{3.22}$$

where

$$x_1 = \frac{|A_1^t|^2}{\beta_1^t}; \quad x_2 = \frac{|A_2^t|^2}{\beta_2^t}; \quad x^l = \frac{|A^l|^2}{\beta^l}. \tag{3.23}$$

The last integral of (3.23) depends on the first two of (3.22), as it can be derived from them using the condition

$$\Omega_1^t = \Omega^l + \Omega_2^t.$$

The conservation laws (3.22) are easily verified by multiplying (3.19)-(3.21) by $A^*$, $A_1^*$, $A_2^*$, respectively, and finding the real part of the expressions obtained by the addition of (3.19) and (3.20).

Some simple results may be inferred from the conservation laws. For example, suppose that the intensity of the wave $A_1^t$ is large and the others $A_2^t$, $A^l$ are small. It follows from the first conservation law (3.22) that the intensity of the wave $A_2^t$, due to nonlinear energy transfer, cannot exceed a certain maximum value determined by the initial intensity of the wave $A_1^t$. A similar result can be deduced for longitudinal waves from the second conservation law (3.22). From two of the integrals of (3.22) we can obtain a solution to the system (3.19) and (3.21) in the special case when the phase difference between the interacting waves is zero:

$$\varphi = \varphi^l - \varphi_1^t + \varphi_2^t = 0.$$

Multiplying (3.20) by $A_1^*$ and taking advantage of the fact that the right-hand side of the resulting equation is real, we obtain

$$\frac{\partial}{\partial t} x_1 = 2AA_2A_1^*. \tag{3.24}$$

Squaring (3.24), we have

$$\left(\frac{\partial}{\partial t} x_1\right)^2 = 4\beta_1^t\beta_2^t\beta^l x^l x_1 x_2 = 4\beta_1^t\beta_2^t\beta^l x_1 (x_0^l + x_{10} - x_1)(x_{10} + x_{20} - x_1). \tag{3.25}$$

Here we have used the conservation laws to express $x_2$ and $x^l$ in terms of $x_1$. The subscript 0 corresponds to the initial values of the corresponding variables. Equation (3.25) is trivially solved by separating the variables:

$$2\sqrt{\beta^l\beta_1^t\beta_2^t}\,t = \int\limits_{x_{10}}^{x_1} \frac{dx_1}{\sqrt{x_1(x_0^l + x_{10} - x_1)(x_{10} + x_{20} - x_1)}}. \tag{3.26}$$

It follows from (3.25) alone that $x_1$ cannot exceed the values $x_0^l + x_{10}$ and $x_{10} + x_{20}$.

After reaching $x_{max} = \min\{x_0^l + x_{10}, x_{10} + x_{20}\}$, the variable $x_1$ begins to vary in the opposite direction until it reaches $x_1 = 0$ ($x_1 < 0$ is impossible for the same reason). In order to find the time interval $\tau$ in which $x_1$ changes from 0 to $\min\{x_0^l + x_{10}, x_{10} + x_{20}\} = x_{max}$, it is sufficient to insert $x_1 = 0$ and $x_1 = x_{max}$ into (3.26) and to subtract one from the other. If $x_0^l > x_{20}$,

$$\xi = \frac{x_1}{x_{10} + x_{20}}; \quad \varkappa = \frac{x_{10} + x_{20}}{x_0^l + x_{10}};$$
$$\sqrt{x_0^l + x_{10}}\; 2\sqrt{\beta_1^t\beta_2^t\beta^l}\;\tau = \int\limits_0^1 \frac{d\xi}{\sqrt{\xi(1-\xi)(1-\varkappa\xi)}}. \tag{3.27}$$

Let the initial intensity of the packet $A_1^t$ be much larger than all the others, $x_{10} \gg x_0^l$, $x_{10} \gg x_{20}$. The integration of (3.27) is then simple.

The time taken for the energy of the packet $A_1^t$ to be transferred to the packets $A_2^t$ and $A^l$ is given by

$$\tau = \frac{1}{2\sqrt{|A_0^t|^2 + \frac{\beta_1^t}{\beta^l}|A_0^l|^2}}\,\frac{1}{\sqrt{\beta_2^t\beta^l}}\ln\frac{16x_{10}}{x_0^l - x_{20}}. \tag{3.28}$$

Substituting the values of $\beta_1^t$, $\beta_2^t$, and $\beta^l$ into this, we obtain an estimate of $\tau$ for $\Omega_{1,2} \gg \omega_{0e}$:

$$\tau \simeq \frac{1}{\omega_{0e}}\,\frac{\ln\dfrac{16x_{10}}{x_0^l - x_{20}}}{2\sqrt{\dfrac{|E_0^t|^2}{4\pi} + \dfrac{\Omega_1^t}{\Omega^l}\dfrac{|E_0^l|^2}{4\pi}}}\sqrt{8m_e n v_\varphi^2\left(\frac{\Omega_1^t}{\Omega^l}\right)^3}. \tag{3.29}$$

If $|E_0^l|^2 \ll \frac{\omega_{0e}}{\Omega_2^t}|E_0^t|^2$, then the characteristic time is determined from the relation[2]

$$\tau \simeq \frac{1}{\omega_{0e}}\sqrt{\frac{2m_e n v_\varphi^2}{W^t}\left(\frac{\Omega_1^t}{\omega_{0e}}\right)^{3/2}}\ln\frac{8\omega_{0e}W^t}{\Omega_1^t W_0^l}; \; v_\varphi = \frac{\omega_{0e}}{k^l}, \tag{3.30}$$

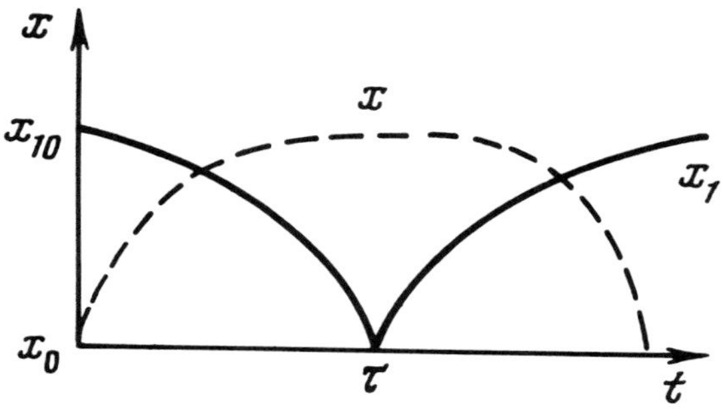


Fig. 8. Intensity of fixed-phase longitudinal and transverse waves as a function of time; $\tau$ is the characteristic time of nonlinear wave interaction.

where $W^t$ and $W_0^l$ are the initial energy densities of the packets $\mathbf{k}_1$ and $\mathbf{k}^l$, respectively. Notice the time $\tau$ is inversely proportional to the first power of the wave amplitude or $(W^t)^{1/2}$.

The time $\tau$ is similarly found for the case $\varphi \neq 0$, when the nonlinear energy transfer of the waves is accompanied by a change in the phase difference. The time dependence of the intensity of the packets is shown in Fig. 8 for $\varphi = 0$ and $x_{10} \gg x_{20}, x_0^l$ [4].

It is not too difficult to show that for the nonlinear interaction of the wave packets with $\varphi \neq 0$

$$x_1 x_2 x^l \sin^2 \varphi = \text{const.} \tag{3.31}$$

From this we can find the characteristic time for nonlinear conversion of waves whose initial phase difference is not zero [1, 2, 4].

## 5. INSTABILITY OF INTENSE WAVES IN A PLASMA

Consider the transmission of an intense transverse wave $A_1$ through a plasma. We assume that, in addition to this wave, there are weak waves $A_2$ and $A$ initially present in the plasma. According to the above result the energy of the original wave is transferred to the waves $A_2$ and $A$. If the intensities of the latter are small (compared with $A_1$), this transfer will appear as an instability. For $A_1^t = \text{const} = A_1^t(0)$ Eqs. (3.10)-(3.12) become linear:

$$\frac{\partial A^l}{\partial t} = -\beta^l A_1^t(0)\, A_2^{t*} e^{-i\Delta\Omega t};$$
$$\frac{\partial A_2^{t*}}{\partial t} = -\beta_2^t A_1^{t*}(0)\, A^l e^{i\Delta\Omega t}$$

or

$$\frac{\partial^2 A^l}{\partial t^2} + i\Delta\Omega \frac{\partial A^l}{\partial t} = \beta^l \beta_2^t \,|A_1^t(0)|^2 A^l; \tag{3.32}$$

$$A^l = A_0^l e^{-i\omega t}; \quad \omega = \frac{\Delta\Omega}{2} \pm i\sqrt{\gamma_{\text{nlr}}^2 - \frac{(\Delta\Omega)^2}{4}}; \tag{3.33}$$

$$\gamma_{\text{nlr}}^2 = \frac{e^2\omega_{0e}^4 k^2\, |\overset{*}{e}_{\mathbf{k}_1} e_{\mathbf{k}_2}|^2\, |A_1^t(0)|^2}{2m_e^2 \Omega^t(\mathbf{k}_1)\, \Omega^t(\mathbf{k}_2)\, (\Omega^l(\mathbf{k}))^2 \left.\frac{\partial \varepsilon^l \omega^2}{\partial \omega}\right|_{\omega=\Omega^l(\mathbf{k})} \left.\frac{\partial}{\partial \omega}\, \omega^2 \varepsilon^t\right|_{\omega=\Omega^t(\mathbf{k}_2)}}. \tag{3.34}$$

For $\Delta\Omega \ll \gamma_{\text{nlr}}$ we have

$$\omega = \pm i\gamma_{\text{nlr}}; \quad \gamma_{\text{nlr}} = \frac{1}{\tau},$$

where $\tau$ is of the same order of magnitude as the $\tau$ in (3.30). We see, therefore, that the characteristic nonlinear interaction times can be found by analyzing the instability of intense waves. Of course, the entire preceding discussion holds for any other waves satisfying the conditions

$$\Omega_1 = \Omega_2 + \Omega,$$
$$\mathbf{k}_1 = \mathbf{k}_2 + \mathbf{k}.$$

The method described here was first used by Sagdeev and Oraevskii [3]. The term "decay instability" often used for such processes inaccurately portrays the nature of the phenomenon, because what is actually involved is a transfer of energy from one wave to another. Notice that this transfer is reversible (provided there are no collisions or other dissipative events). The interest in decay instabilities has arisen partly through the attempt to find irreversible collective processes involving the dissipation of large-amplitude waves in plasma. Processes of this type can result in the conversion of large-amplitude waves propagating in plasma into shock waves.* In the example given the waves $A_2$ and A could have grown from the thermal fluctuation level, i.e., they could have

* Shock waves cannot occur in the absence of dissipative processes, because the entropy changes at the wave front. According to Sagdeev's hypothesis [3] the dissipation mechanism in shock waves propagating in a plasma results from collective processes, rather than binary particle collisions.

been random in the sense that the phase jump time $\tau_\varphi$ is short in comparison with certain processes having a long duration. However, they must be nonrandom compared with the nonlinear interaction in question, i.e.,

$$\gamma_{\text{nlr}}\tau_\varphi \gg 1. \tag{3.35}$$

We shall see presently that when (3.35) fails the interaction can change and become irreversible.

## 6. RANDOM-PHASE WAVE PACKETS

If the phase, and hence the sign, of the field changes repeatedly during the nonlinear interaction time, then, roughly speaking, only the mean values of the physical variables are meaningful where nonlinear effects are concerned. Only the mean squared value of the fields will be nonzero. We now clarify in broad terms the meaning of phase-averaging [5, 6].

Let us find the mean value of the complex amplitude of the electric field:

$$A_{\mathbf{k}} = |A_{\mathbf{k}}|\, e^{i\varphi_{\mathbf{k}}}. \tag{3.36}$$

The time average is the same as the average over the statistical ensemble. In the present case we can investigate an ensemble of systems differing from one another only in the phases $\varphi_{\mathbf{k}}$. The distribution of the phases $\varphi_{\mathbf{k}}$ is random:

$$\langle A_{\mathbf{k}}\rangle = |A_{\mathbf{k}}|\,(\langle\cos\varphi_{\mathbf{k}}\rangle + i\,\langle\sin\varphi_{\mathbf{k}}\rangle). \tag{3.37}$$

The mean value (3.36) is zero, since $\langle \cos\varphi_{\mathbf{k}} \rangle = \langle \sin \varphi_{\mathbf{k}} \rangle = 0$.

It is easily verified that only the mean value of the product $\langle A_{\mathbf{k}} A^*_{\mathbf{k}'}\rangle$ for $\mathbf{k}' = \mathbf{k}$, which is independent of $\varphi_{\mathbf{k}}$, will have a nonzero value. If $\mathbf{k} \neq \mathbf{k}'$, averaging yields zero. Therefore, it is often useful to write the result of averaging in the form

$$\langle A_{\mathbf{k}} A^*_{\mathbf{k}'}\rangle = |A_{\mathbf{k}}|^2\,\delta(\mathbf{k} - \mathbf{k}'). \tag{3.38}$$

Equation (3.38) is essentially a definition of $|A_{\mathbf{k}}|^2$, as it can be proved that this proportionality between $\langle A_{\mathbf{k}} A^*_{\mathbf{k}'}\rangle$ and the delta function $\delta(\mathbf{k} - \mathbf{k}')$ is unique. Moreover, $|A_{\mathbf{k}}|^2$ has a simple physical interpretation in the definition (3.38) in that it measures the spectral density of the mean energy of the electric field in the wave packet.

Thus, to find* $\langle E^2\rangle/4\pi$:

$$\langle E^2\rangle = \int \langle A^*_{\mathbf{k}} A_{\mathbf{k}'}\rangle \exp\left[i\left(\Omega(\mathbf{k}) - \Omega(\mathbf{k}')\right)t - i(\mathbf{k} - \mathbf{k}')\mathbf{r}\right] d\mathbf{k}d\mathbf{k}' =$$

$$= \int |A_{\mathbf{k}}|^2 \delta(\mathbf{k} - \mathbf{k}') \exp\left[i\left(\Omega(\mathbf{k}) - \Omega(\mathbf{k}')\right)t - i(\mathbf{k} - \mathbf{k}')\mathbf{r}\right] d\mathbf{k}d\mathbf{k}'. \quad (3.39)$$

Owing to the presence of the delta function we can set $\mathbf{k} = \mathbf{k}'$ in the integrand, so that

$$\langle E^2\rangle = \int |A_{\mathbf{k}}|^2\, d\mathbf{k}. \quad (3.40)$$

This shows that the energy density of the field $\langle E^2\rangle$ is the sum (integral) of the $|A_{\mathbf{k}}|^2$, i.e., $|A_{\mathbf{k}}|^2$ may be interpreted as the spectral energy density of waves having $\mathbf{k}$ in an interval $d\mathbf{k}$. We note that according to (3.40) the intensities of all the other waves add, as must be the case when interference effects are unimportant (i.e., the phases are random).

Let the phases of all three wave packets be random initially. It is reasonable to believe that if the characteristic time for build-up of the wave amplitude caused by nonlinear effects is much longer than the phase-shift time, the nonlinear effects will alter only slightly the phase relations of the wave, resulting merely in slight phase correlation. In this case the phases may be approximately regarded as random at every stage of nonlinear interaction. For each of the random waves the following relation is fulfilled:

$$\delta\Omega t_0 \gg 1. \quad (3.41)$$

This means, however that the coherence condition cannot be satisfied for these waves, because $\Delta\Omega = \Omega_1 - \Omega_2 - \Omega$ will have a frequency spread no narrower than each of the frequencies $\Omega_1$, $\Omega_2$, and $\Omega$, i.e.,

$$\Delta\Omega t_0 \gg 1. \quad (3.42)$$

Consequently, the random-phase approximation corresponds to physical conditions contrary to those which prevail for fixed-phase waves. It is reasonable to expect, therefore, that the interaction of the waves in this case will also be quite different, as we shall have frequent occasion to realize later. If (3.42) is satisfied, the quantity $e^{i\Delta\Omega t}$ will change sign repeatedly during the time of interaction. In the present case this is not too important because

* The expression for the energy density of an electromagnetic field in a plasma differs from $\langle E^2\rangle/4\pi$ when regard is taken of $\varepsilon$ (see below).

the amplitude $A_{\mathbf{k}}^{l}$ of the random wave will also change sign frequently within the limits $\delta\mathbf{k}$ of one packet, this change being more important than the variation of $e^{i\Delta\Omega t}$. It is necessary, therefore, to find an equation for the mean-square combinations of fields $\langle A_{\mathbf{k}}A_{\mathbf{k}'}^{*}\rangle$ or the related quantities $|A_{\mathbf{k}}|^2$. The factor $e^{i\Delta\Omega t}$ drops out of the equations describing the variation of $|A_{\mathbf{k}}|^2$, so that after averaging there are no rapidly oscillating terms. In order to obtain the equations needed to describe the interaction of random-phase waves we must multiply Eqs. (I), (II), and (III) by $A_{\mathbf{k}}^{l*}$, $A_{\mathbf{k}_1}^{t*}$, and $A_{\mathbf{k}_2}^{t*}$, respectively, and average over the phases.

Here we briefly outline the basic reasoning, referring the reader to Appendix 2 for the details. In order to average over the phases we need to form the second-order field combinations. For this the appropriate equations must be multiplied by $A_{\mathbf{k}'}$ and then averaged according to (3.38). If the left-hand side includes expressions involving the amplitude to the second power, the right-hand side will contain third-order terms of the type $\langle A_{\mathbf{k}}A_{\mathbf{k}_1}A_{\mathbf{k}_2}\rangle$. If the phases are independent, then $\langle A_{\mathbf{k}}A_{\mathbf{k}_1}A_{\mathbf{k}_2}\rangle = \langle A_{\mathbf{k}}\rangle\langle A_{\mathbf{k}_1}\rangle\langle A_{\mathbf{k}_2}\rangle = 0$. This means that the slight phase correlation due to nonlinear interaction must be taken into account. Each of the amplitudes may be written in the form

$$A = A_0 + \delta A, \tag{3.43}$$

where $A_0$ represents the completely random fields, and $\delta A$ is a small correction associated with phase correlation. Now

$$\langle AA_1A_2\rangle \approx \langle A_0A_{10}A_{20}\rangle + \langle \delta A A_{10}A_{20}\rangle + \langle A_0\delta A_{10}A_{20}\rangle + \\ + \langle A_0A_{10}\delta A_{20}\rangle = \langle AA_{10}A_{20}\rangle + \langle A_0A_1A_{20}\rangle + \langle A_0A_{10}A_2\rangle. \tag{3.44}$$

Use has been made here of the fact that

$$\langle A_0A_{10}A_{20}\rangle = 0. \tag{3.45}$$

To find the small corrections due to phase correlation we use Eqs. (I), (II), and (III) to express A in terms of $A_{10}A_{20}$. The corresponding approximate expressions are given in Appendix 2.

The mean value of the product of four amplitudes can roughly be expanded into the product of the mean products of two amplitudes defined by (3.38). As a result the equation we want takes the following form after averaging over the transverse-wave polarizations (see Appendix 2, Sec. 1):

$$\left(\frac{\partial}{\partial t} - \gamma_{\mathbf{k}}^{l} + \mathbf{v}_{\mathrm{gr}}\frac{\partial}{\partial \mathbf{r}}\right) | A_{\mathbf{k}}^{l} |^2 = \frac{e^2\omega_{0e}^4 k^2}{2m_e^2}\int d\mathbf{k}_1 d\mathbf{k}_2 \left(1 + \frac{(\mathbf{k}_1\mathbf{k}_2)^2}{k_1^2 k_2^2}\right) \times$$

$$\left\{\frac{| A_{\mathbf{k}_1}^{t} |^2 | A_{\mathbf{k}_2}^{t} |^2}{4\omega_1^2\omega_2^2\left(\frac{\partial}{\partial\omega}\,\omega^2\varepsilon^{l}\right)^2_{\omega=\Omega^{l}(\mathbf{k})}} - \frac{| A_{\mathbf{k}_2}^{t} |^2 | A_{\mathbf{k}}^{l} |^2}{2\omega_2^2\omega^2\left(\frac{\partial}{\partial\omega}\,\omega^2\varepsilon_1^{t}\right)_{\omega=\Omega^{t}(\mathbf{k}_1)}\left(\frac{\partial}{\partial\omega}\,\omega^2\varepsilon^{l}\right)_{\omega=\Omega^{l}(\mathbf{k})}} + \right.$$

$$\left.\frac{| A_{\mathbf{k}_1}^{t} |^2 | A_{\mathbf{k}}^{l} |^2}{2\omega_1^2\omega^2\left(\frac{\partial}{\partial\omega}\,\omega^2\varepsilon_2^{t}\right)_{\omega=\Omega^{t}(\mathbf{k}_2)}\left(\frac{\partial}{\partial\omega}\,\omega^2\varepsilon^{l}\right)_{\omega=\Omega^{l}(\mathbf{k})}}\right\} \pi\delta(\Delta\Omega)\,\delta(\mathbf{k} - \mathbf{k}_1 + \mathbf{k}_2);$$

$$\omega_{1,2} = \Omega^{t}(\mathbf{k}_{1,2}). \tag{3.46}$$

The form of the equation is greatly simplified by the introduction of the following notation:

$$N_{\mathbf{k}}^{l} = \frac{\pi^2}{\omega^2}\left(\frac{\partial}{\partial\omega}\,\omega^2\varepsilon^{l}\right)_{\omega=\Omega^{l}(\mathbf{k})} | A_{\mathbf{k}}^{l} |^2; \tag{3.47}$$

$$N_{\mathbf{k}_1}^{t} = \frac{\pi^2}{2\omega_1^2}\left(\frac{\partial}{\partial\omega_1}\,\omega^2\varepsilon_1^{t}\right)_{\omega_1=\Omega^{t}(\mathbf{k}_1)} | A_{\mathbf{k}_1}^{t} |^2;\; N_{\mathbf{k}_2}^{t} = \frac{\pi^2}{2\omega_2^2}\left(\frac{\partial}{\partial\omega_2}\,\omega^2\varepsilon_2^{t}\right)_{\omega_2=\Omega^{t}(\mathbf{k}_2)} | A_{\mathbf{k}_2}^{t} |^2; \tag{3.48}$$

$$w_{\mathbf{k}}(\mathbf{k}_1, \mathbf{k}_2) = \frac{e^2\omega_{0e}^2 k^2\left(1 + \frac{(\mathbf{k}_1\mathbf{k}_2)^2}{k_1^2k_2^2}\right)\delta(\mathbf{k} - \mathbf{k}_1 + \mathbf{k}_2)(2\pi)^6}{2\pi^2 m_e^2\left(\frac{\partial}{\partial\omega}\,\omega^2\varepsilon^{l}(\omega, \mathbf{k})\right)_{\omega=\Omega^{l}(\mathbf{k})}\left(\frac{\partial}{\partial\omega}\,\omega^2\varepsilon_1^{t}(\omega, \mathbf{k}_1)\right)_{\omega=\Omega^{t}(\mathbf{k}_1)}\left(\frac{\partial}{\partial\omega}\,\omega^2\varepsilon_2^{t}(\omega, \mathbf{k}_2)\right)_{\omega=\Omega^{t}(\mathbf{k}_2)}}. \tag{3.49}$$

The meaning of these representations will become clear later: $N_{\mathbf{k}}^{l}$, $N_{\mathbf{k}_1}^{t}$, and $N_{\mathbf{k}_2}^{t}$ are the number densities of longitudinal and transverse-wave quanta, multiplied by $\hbar$, and $w_{\mathbf{k}}(\mathbf{k}_1, \mathbf{k}_2)$ is the probability, divided by $\hbar$, that a quantum $N_{\mathbf{k}_1}^{t}$ will emit a quantum $N_{\mathbf{k}}^{l}$.

With the notation (3.47)-(3.49) the equations describing the interaction of random-phase waves assume the simple form [7]

$$\left(\frac{\partial}{\partial t} - \gamma_{\mathbf{k}}^{l} + \mathbf{v}_{\mathrm{gr}}^{l}\frac{\partial}{\partial \mathbf{r}}\right) N_{\mathbf{k}}^{l} = \int w_{\mathbf{k}}(\mathbf{k}_1, \mathbf{k}_2)\frac{d\mathbf{k}_1 d\mathbf{k}_2}{(2\pi)^6} \times \{N_{\mathbf{k}_1}^{t}N_{\mathbf{k}_2}^{t} - N_{\mathbf{k}_2}^{t}N_{\mathbf{k}}^{l} + N_{\mathbf{k}_1}^{t}N_{\mathbf{k}}^{l}\}\,\pi\delta(\Delta\Omega). \tag{3.50}$$

Two more equations are obtained analogously:

$$\left(\frac{\partial}{\partial t} - \gamma_{\mathbf{k}_1}^{t} + \mathbf{v}_{\mathrm{gr}}^{t}\frac{\partial}{\partial \mathbf{r}}\right) N_{\mathbf{k}_1}^{t} = -\int \widetilde{w}\,\frac{d\mathbf{k}\,d\mathbf{k}_2}{(2\pi)^6}(N_{\mathbf{k}_1}^{t}N_{\mathbf{k}_2}^{t} - N_{\mathbf{k}_2}^{t}N_{\mathbf{k}}^{l} + N_{\mathbf{k}_1}^{t}N_{\mathbf{k}}^{l}); \tag{3.51}$$

$$\left(\frac{\partial}{\partial t} - \gamma_{\mathbf{k}_2}^{t} + \mathbf{v}_{\mathrm{gr}}^{t}\frac{\partial}{\partial \mathbf{r}}\right) N_{\mathbf{k}_2}^{t} = \int \widetilde{w}\,\frac{d\mathbf{k}\,d\mathbf{k}_1}{(2\pi)^6}(N_{\mathbf{k}_1}^{t}N_{\mathbf{k}_2}^{t} - N_{\mathbf{k}_2}^{t}N_{\mathbf{k}}^{l} + N_{\mathbf{k}_1}^{t}N_{\mathbf{k}}^{l});$$

$$\widetilde{w} = w_{\mathbf{k}}(\mathbf{k}_1, \mathbf{k}_2)\,\pi\delta(\Omega^{l} - \Omega_1^{t} + \Omega_2^{t}) = w_{\mathbf{k},\mathbf{k}_1,\mathbf{k}_2}. \tag{3.52}$$

These equations differ from the fixed-phase case in that they contain only the squares of the wave amplitudes.

Let us now examine the characteristic time for nonlinear energy transfer between the waves for $\gamma_{\mathbf{k}} = 0$ and $\partial N/\partial \mathbf{r} = 0$. For example, let the wave intensity of one of the packets $N^t_{\mathbf{k}_1}$ be much greater than that of the others, $N^t_{\mathbf{k}_2}$ and $N^l_{\mathbf{k}}$, where $N^t_{\mathbf{k}_2} \ll N^l_{\mathbf{k}}$. Then for the increase in $N^l_{\mathbf{k}}$ we have

$$\frac{\partial N^l_{\mathbf{k}}}{\partial t} = N^l_{\mathbf{k}} \int \tilde{w} N^t_{\mathbf{k}_1} \frac{d\mathbf{k}_1 d\mathbf{k}_2}{(2\pi)^6}. \tag{3.53}$$

In the initial stage of nonlinear interaction the wave intensity of the packet $N^t_{\mathbf{k}_1}$ may be regarded as approximately constant. Then (3.53) gives

$$\frac{\partial N^l_{\mathbf{k}}}{\partial t} = \gamma^l_{\text{nlr}} N^l_{\mathbf{k}}, \tag{3.54}$$

where

$$\gamma^l_{\text{nlr}} = \int \tilde{w} N^t_{\mathbf{k}_1} \frac{d\mathbf{k}_1 d\mathbf{k}_2}{(2\pi)^6} \tag{3.55}$$

is logically referred to as the nonlinear growth rate characterizing the buildup of longitudinal waves by transverse waves. The increment $N^t_{\mathbf{k}_1}$, is proportional to $\gamma_{\mathbf{k}}$, i.e., to the square of the amplitude of the transverse waves or their energy density. The characteristic transfer time is therefore inversely proportional to the square of the wave amplitude, rather than to the first power as in the case of fixed-phase waves. This marked distinction stems from the fact that the random-phase approximation corresponds to physical conditions

$$\Delta\Omega t_0 \gg 1, \tag{3.56}$$

which are quite unlike those in the coherent situation associated with the interaction of fixed-phase waves. We also notice that the interactions of random-phase wave packets differ from those of fixed-phase waves in other respects.

## 7. QUASISTATIONARY EQUILIBRIUM STATES AND STABILITY OF RANDOM-PHASE WAVES

We saw above that there is a continuous reversible transfer of energy between wave modes for fixed-phase wave packets in the absence of the increments $\gamma_{\mathbf{k}}^{l}$, $\gamma_{\mathbf{k}_1}^{t}$, and $\gamma_{\mathbf{k}_2}^{t}$. For random phase another result is found: even in the absence of linear wave growth and damping it is possible to arrive at an equilibrium or, more precisely, "quasi-equilibrium" state.

This is best illustrated in a one-dimensional model. Let the directions of all the waves, both within the limits of each packet in the different packets, coincide. Let us assume also that the widths of the transverse-wave packets $\delta\Omega^t$ are smaller than $\omega_{0e}$, i.e., that the two transverse-wave packets have no **k** in common, and the difference between their mean frequencies is close to $\omega_{0e}$. The equations for

$$N_k = \int N_{\mathbf{k}} d\mathbf{k}_{\perp} \tag{3.57}$$

become (see Appendix 2, Sec. 2)

$$\left(\frac{\partial}{\partial t} - \gamma^l\right) N^l = \widetilde{\beta}^l \left(N_1^t N_2^t - N_2^t N^l + N_1^t N^l\right); \tag{3.58}$$

$$\left(\frac{\partial}{\partial t} - \gamma_1\right) N_1^t = -\widetilde{\beta}^t \left(N_1^t N_2^t - N_2^t N^l + N_1^t N^l\right); \tag{3.59}$$

$$\left(\frac{\partial}{\partial t} - \gamma_2\right) N_2^t = \widetilde{\beta}^t \left(N_1^t N_2^t - N_2^t N^l + N_1^t N^l\right). \tag{3.60}$$

The mean wave numbers of the waves described by $N_2^t$ and $N_1^t$ are uniquely determined from the conservation laws for a given longitudinal wave number **k** (see Appendix 2):

$$\widetilde{\beta}^t = \frac{e^2\omega_{0e}^3}{8\pi m_e^2 (\Omega^t)^2 c^2}; \quad \widetilde{\beta}^t = \widetilde{\beta}^l \left(\frac{\omega_{0e}}{\Omega^t}\right)^3; \quad \Omega^t \gg \omega_{0e}. \tag{3.61}$$

These equations show that even for $\gamma = 0$ equilibrium is possible for nonzero intensities of the packets. This requires that

$$N_1^t N_2^t - N_2^t N^l + N_1^t N^l = 0. \tag{3.62}$$

For fixed phases the right-hand side of the equations, which describes the nonlinear interaction, was proportional to the product of the amplitudes of the interacting waves and vanished only when the intensity of one of the waves became zero. In other words, for random waves equilibrium states with strong self-consistent fields are possible.

It is also possible to have equilibrium states with damping or growth. In this case the intensities of the wave packets are determined from the set of equations

$$-\gamma^l N^l = \tilde{\beta}^l (N_1^t N_2^t - N^l N_2^t + N_1^t N^l); \quad (3.63)$$

$$-\gamma_1 N_1^t = -\tilde{\beta}^t (N_1^t N_2^t - N^l N_2^t + N_1^t N^l); \quad (3.64)$$

$$-\gamma_2 N_2^t = \tilde{\beta}^t (N_1^t N_2^t - N^l N_2^t + N^l N_1^t). \quad (3.65)$$

According to (3.63)-(3.65)

$$\frac{\gamma_1 N_1^t}{\tilde{\beta}^t} = -\frac{\gamma^l N^l}{\tilde{\beta}^l} = -\frac{\gamma_2 N_2^t}{\tilde{\beta}^t},$$

and since $N_1^t$, $N_2^t$, and $N^l$ are positive, equilibrium is possible only if: 1) $\gamma_1 < 0$, $\gamma^l$, $\gamma_2 > 0$; 2) $\gamma_1 > 0$, $\gamma^l$, $\gamma_2 < 0$, i.e., some waves must decay, while others must grow.

The existence of the equilibrium state (3.62) for random waves raises the following questions: will the system tend to this state if the initial state is far from equilibrium? Will the equilibrium state be stable?

The first question is answered for $\gamma_k = 0$ by solving the set of equations (3.58)-(3.60) [8]. It is readily seen that these equations have two integrals, i.e., there are two conservation laws:

$$N_1^t + N_2^t = \text{const} = N_{10}^t + N_{20}^t;$$

$$\frac{N_1^t}{\tilde{\beta}^t} + \frac{N^l}{\tilde{\beta}^l} = \text{const} = \frac{N_{10}^t}{\tilde{\beta}^t} + \frac{N_0^l}{\tilde{\beta}^l}. \quad (3.66)$$

Here the subscript 0 indicates the initial intensity of the corresponding packets. The conservation laws enable us to express $N_2^t$ and $N^l$ in terms of $N_1^t$ and to obtain a single equation for $N_1^t$:

$$\frac{1}{\tilde{\beta}^t}\frac{\partial N_1^t}{\partial t} = -N_1^t (N_{20}^t + N_{10}^t - N_1^t) + (N_{20}^t + N_{10}^t - 2N_1^t)\left(N_0^l + \frac{N_{10}^t \tilde{\beta}^l}{\tilde{\beta}^t} - \frac{\tilde{\beta}^l}{\tilde{\beta}^t} N_1^t\right). \quad (3.67)$$

It is an elementary matter to find the solution of Eq. (3.67):

$$-\tilde{\beta}^t\, dt = \int \frac{dN_1^t}{N_1^t (N_{20}^t + N_{10}^t - N_1^t) - (N_{20}^t + N_{10}^t - 2N_1^t)\left(N_0^l + \frac{N_{10}^t \tilde{\beta}^l}{\tilde{\beta}^t} - \frac{\tilde{\beta}^l N_1^t}{\tilde{\beta}^t}\right)}. \quad (3.68)$$

A very simple result is obtained for the case when the transverse-wave frequency $\Omega^t$ is much greater than the longitudinal-wave frequency:

$$N_1^t = \frac{\widetilde{N}_{20} - \widetilde{N}_{10}\,\dfrac{N_{10}^t - \widetilde{N}_{20}}{N_{10}^t - \widetilde{N}_{10}}\,e^{\tau}}{1 - \dfrac{N_{10}^t - \widetilde{N}_{20}}{N_{10}^t - \widetilde{N}_{10}}\,e^{\tau}}, \tag{3.69}$$

where

$$\tau = 2\widetilde{\beta}^l\,(\widetilde{N}_{20} - \widetilde{N}_{10})t;\ \widetilde{N}_{20} = \max\left\{\frac{N_{10}^t + N_{20}^t}{2},\quad N_{10}^t + \frac{\widetilde{\beta}^t}{\beta^l}N_0^l\right\};$$
$$\widetilde{N}_{10} = \min\left\{\frac{N_{10}^t + N_{20}^t}{2},\quad N_{10}^t + \frac{\widetilde{\beta}^t}{\widetilde{\beta}^l}N_0^l\right\}. \tag{3.70}$$

Note that the solution (3.69) can be found by another route, namely by first neglecting the terms $N_1^t N_2^t$ in the initial equations. This is equivalent to saying that the spectral density of the transverse waves $N_1^t$ and $N_2^t$ is much smaller than for the longitudinal waves $N^l$. This is already apparent from the second conservation law (3.66), which shows that if $N_{10}^t$ is large, $N^l$ must be of order $N_1^t\frac{\widetilde{\beta}^l}{\widetilde{\beta}^t} \simeq N_1^t\left(\frac{\Omega^t}{\omega_{0e}}\right)^3$. The latter does not imply that the energy density (not the spectral density, but the actual density) of the transverse waves is smaller than the longitudinal energy density. On the contrary, the transverse-wave energy density can be much larger than that of the longitudinal waves.

In fact, the spectral width $\Delta k^t$ of the transverse waves is much larger than that $\Delta k^l$ of the longitudinal waves. Recognizing that

$$\Omega^l \simeq \omega_{0e},\quad \Omega^t \simeq \sqrt{k^2c^2 + \omega_{0e}^2},$$

we have

$$\Omega^l - \Omega^t(k_1) + \Omega^t(k_1 - k) = \omega_{0e} - \sqrt{k_1^2c^2 + \omega_{0e}^2} + \sqrt{(k_1 - k)^2c^2 + \omega_{0e}^2} = 0. \tag{3.71}$$

Differentiating this equation, we obtain

$$dk_1\left(\frac{k_1}{\sqrt{k_1^2c^2 + \omega_{0e}^2}} - \frac{k_1 - k}{\sqrt{(k_1 - k)^2c^2 + \omega_{0e}^2}}\right) = dk\,\frac{k_1 - k}{\sqrt{(k_1 - k)^2c^2 + \omega_{0e}^2}}.$$

Since $k_1 \gg k$, $k_1 c \gg \omega_{0e}$ we obtain

$$dk_1\,\frac{k}{c^2}\,\frac{\omega_{0e}^2}{k_1^3} \approx dk.$$

From (3.71) we have approximately $ck \simeq \omega_{0e}$. Hence,

$$\Delta k = \left(\frac{\omega_{0e}}{\Omega^t}\right)^3 \Delta k_1.$$

Consequently,

$$N^l\Delta k \sim N^t\left(\frac{\Omega^t}{\omega_{0e}}\right)^3 \Delta k \simeq N^t\Delta k_1.$$

The definition of $N^t$ and $N^l$ implies that they are of the same order as the spectral energy density divided by the frequency. Therefore, the energy density is of order

$$W \sim N\omega\Delta k(2\pi)^{-3}.$$

i.e.,

$$W^l \sim \frac{\omega_{0e}}{\Omega^t} W^t,$$

and, since the transverse-wave frequency is much larger than the longitudinal-wave frequency, $W^l \ll W^t$.

Thus we find that, although $N^t \ll N^l$, the transverse-wave energy density can be much larger than the longitudinal-wave energy density.

We now introduce the critical longitudinal-wave density

$$W^l_{\text{cr}} = \frac{\omega_{0e}}{\Omega^t} W^t. \tag{3.72}$$

There is a significant distinction between the nonlinear wave interaction in the case when the following relation holds for the initial wave intensities:

$$W^l \ll W^l_{\text{cr}}, \tag{3.73}$$

and in the opposite case.

Suppose that (3.73) holds for t = 0 and the intensity of the wave packet $N^t_{k_1}$ is much larger than that of the wave packet $N^t_{k_2}$. Then

$$\widetilde{N}_{20} = N^t_{10} + \frac{\widetilde{\beta}^t}{\widetilde{\beta}^l} N^l_0\,;\ \widetilde{N}_{10} = \frac{N^t_{10}}{2}.$$

It follows from the solution (3.69) that $N_1(t) \to \widetilde{N}_{10}$ as $\tau \to \infty$, i.e., the intensity of the packet $N^t_{k_1}$ is halved. By the conservation law (3.66) $N^t_1 + N^t_2 \simeq N^t_{10}$, and the intensity of the transverse waves $N^t_{k_2}$ also proves to be approximately half the initial intensity of the packet $N^t_{k_1}$. It follows from the second conservation law that $N^l \to \frac{\widetilde{\beta}^l}{\widetilde{\beta}^t} N^t_{10}$ as $t \to \infty$, i.e., the longitudinal-wave energy reaches the

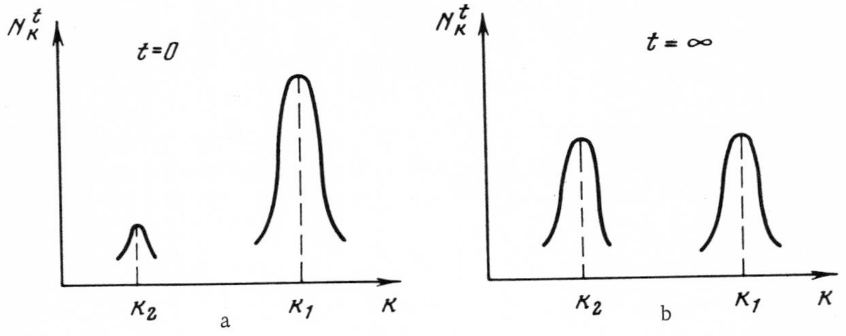


Fig. 9. Interaction of two random-phase transverse-wave packets with plasma. The higher-frequency wave is the stronger one: a) Initial state; b) final state.

critical value $W^l \sim \frac{\omega_{0e}}{\Omega^t} W_1^t(0)$. Since this value is small compared with $W_1^t$, we can say the energy is approximately redistributed between the two transverse-wave packets. This redistribution tends to equalize the intensities of the transverse-wave packets (Fig. 9). In this situation, therefore, the main process is spectral energy transfer between the waves. Naturally, if the transverse-wave frequency is of the same order as the longitudinal-wave frequency, the energy of the excited longitudinal waves will be of the same order as the transverse-wave energy.

Suppose now that the lower-frequency wave is initially the stronger one, i.e., that $N_{20}^t \gg N_{10}^t$ and $W^l \ll W_{cr}^l$. Then $\tilde{N}_{20} \simeq N_{20}^t/2$,

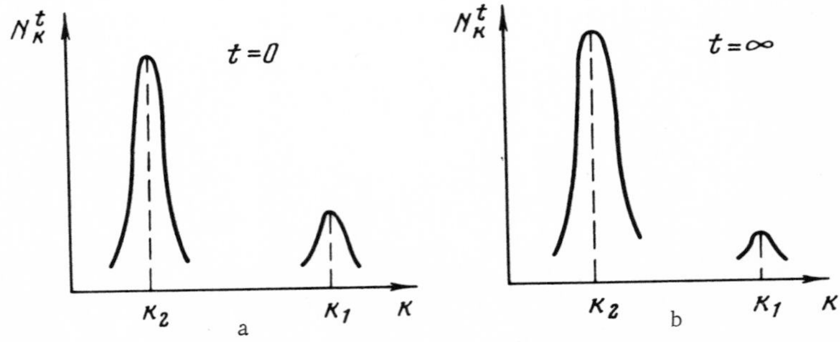


Fig. 10. Interaction of two random-phase transverse-wave packets in plasma. The lower-frequency wave is the stronger one; a) Initial state; b) final state.

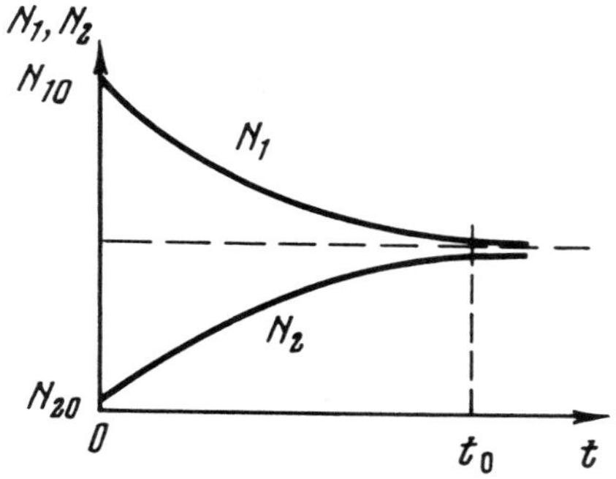


Fig. 11. Number of quanta of two packets of interacting transverse waves as a function of time.

and $\widetilde{N}^t_{10} \simeq N^t_{10} \simeq 0$, i.e., $N_1(t) \to \widetilde{N}_{10} \simeq N^t_{10}$ as $t \to \infty$, $N_2 \simeq N^t_{20}$ (from the conservation law), and the wave intensities hardly change (Fig. 10). This example demonstrates the asymmetric and irreversible behavior of the energy transfer between random-phase waves and the marked difference from the interaction of fixed-phase waves. The relaxation process of intensity equalization for $N^t_{k_1} \gg N^t_{k_2}$ is shown in Fig. 11. The characteristic relaxation time $t_0$ is easily expressed in terms of the initial intensity of the transverse-wave packet $N^t_{k_1}$, if we assume that the term containing the exponential in the denominator of (3.69) is of order unity, i.e.,

$$\tau_1 \simeq \ln \left| \frac{N^t_{10} - \widetilde{N}_1}{N^t_{10} - \widetilde{N}_2} \right| \simeq \ln \frac{N^t_{10}}{2 \frac{\widetilde{\beta}^t}{\widetilde{\beta}^l} N^l_0} = 2\widetilde{\beta}^l \frac{N^t_{10}}{2} t_0 .$$

Expressing $N_{10}$ in terms of the initial energy density of the packet, $W^t_{10}$, we obtain

$$t_0 = \frac{1}{\gamma} \ln \frac{W^t_{10} \frac{\omega_{0e}}{\Omega^t}}{2W^l_0}, \tag{3.74}$$

where

$$\gamma = \frac{\pi}{2} \omega_{0e} \frac{W^t_{10}}{n m_e c^2} \frac{\omega_{0e}}{\delta \Omega^t} . \tag{3.75}$$

Note that $\gamma$ corresponds to the nonlinear growth rate (3.55) introduced above [9], and that Eq. (3.74) could have been derived by using (3.54) to find the time taken for the longitudinal-wave intensity to increase from $W^l_0$ to the final value $W^l_{cr} = \frac{\omega_{0e}}{\Omega^t} W^t_{10}$. Of course,

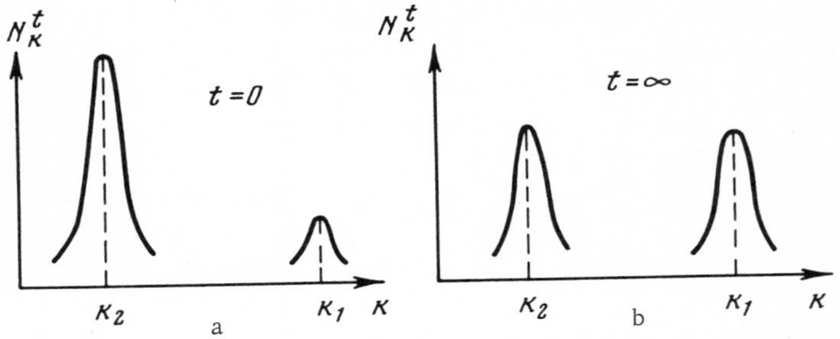


Fig. 12. Interaction of two transverse-wave packets with a plasma in which strong longitudinal waves have been excited: a) Initial state; b) final state.

the approximate formula (3.74) does not describe the saturation effects implicit in (3.69).

Let us now examine the case in which the initial longitudinal-wave intensity is fairly high:

$$W_0^l \gg W_{cr}^l. \tag{3.76}$$

First suppose that the initial wave intensity of the packet $N_{k_2}^t$ is much larger than $N_{k_1}^t$. Then $\widetilde{N}_{20} \simeq \dfrac{\widetilde{\beta}^t}{\widetilde{\beta}^l} N_0^l$, $\widetilde{N}_{10} \simeq N_{20}^t/2$. For $\tau \to \infty$ we have $N_1(t) \to \widetilde{N}_{10} = N_{20}^t/2$ so that $N_2^t \to N_{20}^t/2$, i.e., the intensities of the transverse-wave packets are equalized (Fig. 12). But if initially $N_{k_1}^t \gg N_{k_2}^t$, then $\widetilde{N}_{20} = \dfrac{\widetilde{\beta}^t}{\widetilde{\beta}^l} N_0^l$; $\widetilde{N}_{10} = N_{10}^t/2$, i.e., the intensities are equalized (see Fig. 9). The characteristic time for the intensity equalization process for the condition (3.76) is estimated in a way similar to (3.74):

$$\tau_0' \simeq 2\widetilde{\beta}^t t_0' N_0^l \approx 1,$$

or

$$\frac{1}{t_0'} = \gamma' \simeq \frac{\pi}{2} \frac{W_0^l}{n m_e c^2} \frac{\Omega^t}{\delta\Omega^t} \omega_{0e}.$$

The characteristic time $t_0' \ll t_0$, i.e., in the presence of intense longitudinal waves the energy transfer is much more rapid than in the case of intense transverse waves only.

If there is no growth or damping of the oscillations, the system approaches the equilibrium state asymptotically (see Fig. 11). It is reasonable to expect this state to be stable.

Let us assume that the N's differ only slightly from the quasi-equilibrium values. We then obtain linearized equations. Substituting into them solutions of the form const. exp. [− (iΩt − iϰx)], we obtain a dispersion relation for Ω [10]:

$$(\Omega - v^l_{\mathrm{cr}}\varkappa)\left\{\left[-i(\Omega-\varkappa) + N^t_{10}\frac{\widetilde{\beta}^l}{2}\right]\left[-i(\Omega-\varkappa) + N^t_{10}\frac{\widetilde{\beta}^l}{2} - i\omega_{0e}\left(\frac{\omega_{0e}}{\Omega^t}\right)^3\right] - \left(N^t_{10}\frac{\widetilde{\beta}^l}{2}\right)^2\right\} = 0. \tag{3.77}$$

It has solutions

$$\Omega = \varkappa v^l_{\mathrm{cr}};\quad \Omega = \varkappa - i\widetilde{\beta}^l N^t_{10};\quad \Omega = \varkappa - i\frac{\omega_{0e}}{2}\left(\frac{\omega_{0e}}{\Omega^t}\right)^3.$$

Notice that all three solutions correspond to damped oscillations.

We now investigate the equilibrium states in the case when growth and damping of the oscillations occur in the linear approximation. Assuming that $N^t_1$ is unstable and $N^t_2$, $N^l$ are stable, we obtain for equilibrium

$$|\gamma|\, N^l = \widetilde{\beta}^l N^l (N^t_1 - N^t_2);\quad |\gamma_1|\, N^t_1 = \widetilde{\beta}^t N^l (N^t_1 - N^t_2);$$
$$|\gamma_2|\, N^t_2 = \widetilde{\beta}^t N^l (N^t_1 - N^t_2). \tag{3.77'}$$

Here we have assumed that $N^l \gg N^t_1, N^t_2$, which is legitimate for $\Omega^t \gg \omega_{0e}$. This is also obvious from the solution of (3.77'):

$$N^l = \frac{\gamma_1\gamma_2}{\widetilde{\beta}^t(\gamma_2-\gamma_1)};\quad N^t_1 = \frac{\gamma_2\gamma}{\widetilde{\beta}^l(\gamma_2-\gamma_1)};\quad N^t_2 = \frac{\gamma_1\gamma}{\widetilde{\beta}^l(\gamma_2-\gamma_1)}. \tag{3.78}$$

The ratio of $N^l$ to $N^t_1$ contains the large factor $\left(\frac{\widetilde{\beta}^l}{\widetilde{\beta}^t}\right) = \left(\frac{\Omega^t}{\omega_{0e}}\right)^3$, i.e., it is supposed that $\gamma$ is not appreciably greater than $\gamma_1$ or $\gamma_2$, or, more precisely, $\frac{\gamma}{\gamma_1} \ll \left(\frac{\Omega^t}{\omega_{0e}}\right)^3$; $\frac{\gamma}{\gamma_2} \ll \left(\frac{\Omega^t}{\omega_{0e}}\right)^3$.

For deviations from the equilibrium values of (3.78) we obtain the dispersion equation

$$-i\Omega = \frac{(\widetilde{\beta}^l)^2\gamma_1\gamma_2(-2i\Omega+\gamma_2-\gamma_1)(-i\Omega+\gamma)}{(\widetilde{\beta}^t)^2(\gamma_2-\gamma_1)(i\Omega+\gamma_1)(-i\Omega+\gamma_2)}. \tag{3.79}$$

The frequency-dependent factor on the right-hand side varies from $\frac{(\gamma_2-\gamma_1)\gamma}{\gamma_1\gamma_2}$ to 2 as Ω varies from 0 to ∞. Since the right-hand side of (3.79) contains the large factor $\left(\frac{\widetilde{\beta}^l}{\beta^t}\right)^2 \sim \left(\frac{\Omega^t}{\omega_{0e}}\right)^6$, Ω must also be large. Assuming, therefore, on the right-hand side that $\Omega \gg \gamma, \gamma_1, \gamma_2$, we obtain

$$i\Omega = \frac{2(\tilde{\beta}^l)^2}{(\tilde{\beta}^t)^2}\frac{\gamma_1\gamma_2}{\gamma_2 - \gamma_1}, \tag{3.80}$$

which corresponds to strong damping. The large factor on the right-hand side of (3.79) can also be compensated if the numerator on the right-hand side is assumed to be near zero, so that

$$i\Omega = \gamma; \quad i\Omega = \frac{\gamma_2 - \gamma_1}{2}, \tag{3.81}$$

i.e., for $\gamma_2 > \gamma_1$, when the equilibrium state (3.78) exists, it is stable. The equilibrium relaxation time is determined in this case by the smaller of $\gamma$ and $(\gamma_2 - \gamma_1)/2$.

On a final note we again stress the sharp qualitative distinction between the interaction of random waves and waves with fixed phases.

## 8. EXCITATION OF LONGITUDINAL WAVES BY STRONG RANDOM TRANSVERSE WAVES

Suppose now that the random-phase transverse waves are strong, and that their intensity variation can be neglected in the initial stages of the nonlinear process. We make no assumptions regarding the phases of the longitudinal modes. With the problem thus stated we can find how the characteristic times for longitudinal-wave generation depends upon the transverse-wave intensity.

We also assume that the transverse-wave frequencies are much larger than those of the longitudinal-waves:

$$\Omega^t \gg \omega_{0e},$$

which allows us to assume $N^t \ll N^l$. We have used above two different approaches to describe random and nonrandom phases, namely for fixed phases we have been concerned with the behavior of the waves themselves, i.e., we have used a dynamical description, whereas for random phases we have invoked certain mean values, i.e., we have used a statistical description. Here we must adopt a combined approach, using the statistical description only for the strong transverse waves, i.e., averaging all variables only over the phases of these particular waves. The required averaging is described in Appendix 2.

Assuming that all variables are independent of the coordinates, we expand the amplitudes as a Fourier series:

$$A_{\mathbf{k}}(t) = \int A_{\mathbf{k}\omega} e^{-i\omega t} d\omega. \tag{3.82}$$

Since the transverse-wave amplitudes are constant, $A^t_{\mathbf{k}\omega} = A^t_{\mathbf{k}}\,\delta(\omega)$. The equation obtained for $A^l_{\mathbf{k}\omega}$ with these assumptions turns out to be linear, i.e., $A^l_{\mathbf{k}\omega}$ drops out, and a dispersion equation is obtained linking the frequencies and wave numbers of the longitudinal waves ($\delta \to +0$):

$$-i\omega - \gamma^l_{\mathbf{k}} = i \int \frac{w_{\mathbf{k}\mathbf{k}_2}}{2} \frac{N^t_{\mathbf{k}_2} - N^t_{\mathbf{k}_2-\mathbf{k}}}{\omega - \Delta\Omega_{\mathbf{k},\,\mathbf{k}-\mathbf{k}_2,\,-\mathbf{k}_2} + i\delta} d\mathbf{k}_2, \tag{3.83}$$

where

$$w_{\mathbf{k},\,\mathbf{k}_2} = w_{\mathbf{k},\,\mathbf{k}_2-\mathbf{k},\,\mathbf{k}_2} = \frac{e^2\omega_{0e}\mathbf{k}^2}{8\pi^2 m_e^2 \left|\Omega^t(\mathbf{k}_2)\,\Omega^t(\mathbf{k}_2-\mathbf{k})\right|}\left(1 + \frac{(\mathbf{k}_2,\,\mathbf{k}_2-\mathbf{k})^2}{\mathbf{k}_2^2(\mathbf{k}_2-\mathbf{k})^2}\right); \tag{3.84}$$

$N^t_{\mathbf{k}_2}$ is interpreted as the number of quanta and is related to the square of the field amplitude by Eq. (3.48); $\Delta\Omega_{\mathbf{k},\mathbf{k}_1,\mathbf{k}_2} = \Omega^t(\mathbf{k}_1) - \Omega^t(\mathbf{k}_2) - \Omega^l(\mathbf{k})$; $\Delta\Omega = -\Delta\,\Omega_{\mathbf{k},\mathbf{k}-\mathbf{k}_2,-\mathbf{k}_2}$.

For $|\omega| \ll |\Delta\Omega|$, i.e., for a small transverse-wave intensity, (3.83) yields the expression for the damping rate for the random waves. For $k \ll k_2$ we have $\Delta\Omega = \Omega^l - \mathbf{k}\mathbf{v}^t_{gr}$, where $\mathbf{v}^t_{gr}$ is the transverse-wave group velocity. Then in (3.83)

$$\mathrm{Im}\frac{1}{\omega + \Delta\Omega + i\delta} \simeq \mathrm{Im}\frac{1}{\Delta\Omega + i\delta} = -i\pi\delta(\Delta\Omega),$$

and for $k \ll k_2$ this gives $-i\pi\delta(\Omega^l - \mathbf{k}\mathbf{v}_{gr})$. Consequently,

$$\mathrm{Im}\,\omega = -\gamma^l_{\mathbf{k}} + \frac{\pi}{2}\int w_{\mathbf{k}\mathbf{k}_2}\,\delta(\Delta\Omega)\,(N^t_{\mathbf{k}_2} - N^t_{\mathbf{k}_2-\mathbf{k}})\,d\mathbf{k}_2. \tag{3.85}$$

The first term is the linear growth rate, the second is a nonlinear term corresponding to the estimate (3.55) [the term $N^t_{\mathbf{k}_2-\mathbf{k}}$ is dropped from (3.55), as we have assumed that the intensity of the second transverse-wave packet is small].

By analogy with the well-known case of plasma particles and their kinetic instabilities, which are proportional to $\delta(\Omega^l - \mathbf{k}\mathbf{v})$, we call the nonlinear instabilities (3.85) "kinetic" instabilities. The result (3.85) shows that for $|\omega| \ll \Delta\Omega$ (in the given case $\gamma^{nlr}_{\mathbf{k}} \leqslant \Delta\,\Omega$) the longitudinal wave grows with the nonlinear growth rate given by (3.55).

Suppose that for a given $\mathbf{k}_2 = \mathbf{k}_{20}$ we have found from the transverse-wave packet a value $\mathbf{k} = \mathbf{k}_0$ for which $\mathbf{k}_0\mathbf{v}^t_{gr} = \Omega^l(\mathbf{k}_0)$. Let us investigate another wave packet $\mathbf{k}_2 = \mathbf{k}_{20} + \delta\mathbf{k}_2$. The difference ($\Delta\Omega$) for this wave and the wave $\mathbf{k}_{20}$ is $\mathbf{k}_0\delta\mathbf{v}_{gr}^t$:

$$\delta\mathbf{v}_{gr}^t = \frac{\delta\mathbf{k}_2}{\sqrt{k_{20}^2 + \frac{\omega_{0e}^2}{c^2}}} - \frac{\mathbf{k}_{20}(\mathbf{k}_{20}\delta\mathbf{k}_2)}{\left(k_{20}^2 + \frac{\omega_{0e}^2}{c^2}\right)^{3/2}}.$$

If the packet is one-dimensional and $\Omega^t \gg \omega_{0e}$, then

$$\delta\mathbf{v}_{gr} \sim \frac{\omega_{0e}^2}{(\Omega^t)^2}\frac{\delta\Omega^t}{\Omega^t},$$

and, since $v_{gr} \simeq c$,

$$k \sim \frac{\omega_{0e}}{c} \quad ; \quad k_0\delta v_{gr} \sim \left(\frac{\omega_{0e}}{\Omega^t}\right)^3 \delta\Omega^t.$$

Consequently, for the one-dimensional case the condition for (3.85) to apply becomes

$$\gamma^{nlr} \simeq \omega \ll \left(\frac{\omega_{0e}}{\Omega^t}\right)^3 \delta\Omega^t. \tag{3.86}$$

We observe also that for a one-dimensional transverse-wave spectrum with $\delta\Omega^t < \omega_{0e}$ the conservation laws cannot be fulfilled simultaneously for the terms of (3.85) containing $N_{\mathbf{k}_2}^t$ and $N_{\mathbf{k}_2-\mathbf{k}}^t$, hence the second term of (3.85) describes another wave packet. In the three-dimensional case this is violated for $\Delta\Theta \gg (\omega_{0e}/\Omega^t)^{3/2}$, where $\Delta\Theta$ is the angular spread in the direction of the transverse waves in the packet, and both $N_{\mathbf{k}_2}^t$, and $N_{\mathbf{k}_2-\mathbf{k}}^t$ can describe the same packet. The increment is now smaller by a factor of approximately $1/(\Delta\Theta)^2 \cdot (\omega_{0e}/\Omega^t)^{3/2}$ compared with the one-dimensional increment [11]. For $\Delta\Theta \gg (\omega_{0e}/\Omega^t)^{3/2}$ the corresponding criterion has the form ($\mathbf{k}_{20}$ and $\mathbf{k}_0$ are in the same direction, and $\Delta\Theta \ll 1$)

$$\gamma_{1\text{-}d}^{nlr} \ll (\Delta\theta)^2\,\delta\Omega^t. \tag{3.87}$$

We now examine the limit of large growth, when $|\omega| \gg \Delta\Omega$. The imaginary part of the denominator of (3.83) is negligible. Replacing $\mathbf{k}_2 - \mathbf{k}$ by $\mathbf{k}_2'$ in the second term and denoting $\mathbf{k}_2'$ by $\mathbf{k}_2$, we obtain the dispersion relation

$$-(\omega - i\gamma_{\mathbf{k}}^l) = \frac{1}{2}\int\left\{\frac{w_{\mathbf{k}\mathbf{k}_2}}{\omega - \Delta\Omega_{\mathbf{k},\,\mathbf{k}-\mathbf{k}_2,\,-\mathbf{k}_2}} - \frac{w_{\mathbf{k},\,\mathbf{k}_2+\mathbf{k}}}{\omega - \Delta\Omega_{\mathbf{k},\,-\mathbf{k}_2,\,-\mathbf{k}-\mathbf{k}_2}}\right\} N_{\mathbf{k}_2}^t d\mathbf{k}_2. \tag{3.88}$$

Equation (3.88) can also result in instabilities, which are appropriately called hydrodynamic instabilities by analogy with particle beams. Consider the case $k \ll k_2$, $k_2 \gg \omega_{0e}/c$. Expanding in the two small parameters $\Delta\Omega/\omega$ and $k/k_2$, we obtain

$$-(\omega - i\gamma_{\mathbf{k}}^{l}) = \frac{1}{2}\int\left\{w_{\mathbf{k}\mathbf{k}_2}\frac{\Delta\Omega_{\mathbf{k},\,\mathbf{k}-\mathbf{k}_2,-\mathbf{k}_2} - \Delta\Omega_{\mathbf{k},\,-\mathbf{k}_2,\,-\mathbf{k}-\mathbf{k}_2}}{\omega^2} - \frac{1}{\omega}\mathbf{k}\,\frac{\partial w_{\mathbf{k}\mathbf{k}_2}}{\partial \mathbf{k}_2}\right\}N_{\mathbf{k}_2}^{t}d\mathbf{k}_2. \tag{3.89}$$

We have

$$\Delta\Omega_{\mathbf{k},\,\mathbf{k}-\mathbf{k}_2,\,-\mathbf{k}_2} - \Delta\Omega_{\mathbf{k},-\mathbf{k}_2,-\mathbf{k}-\mathbf{k}_2} = 2\Omega^t(\mathbf{k}_2) - \Omega^t(\mathbf{k}_2 - \mathbf{k}) -$$
$$- \Omega^t(\mathbf{k}_2 + \mathbf{k}) \simeq - k_i k_j\,\frac{\partial^2\Omega^t}{\partial k_{2i}\partial k_{2j}} = \frac{c^4}{\Omega^t}\left(\frac{k^2}{c^2} - \frac{(\mathbf{k}_2\mathbf{k})^2}{(\Omega^t)^2}\right),$$

which for zero angle between the vectors $\mathbf{k}_2$ and $\mathbf{k}$ is of order $k^2c^2\omega_{0e}^2/(\Omega^t)^3$, for angles $\Theta \gg (\omega_{0e}/\Omega^t)^2$ is of order $c^2k^2\Theta^2/\Omega^t$, and finally reaches a maximum value $\sim c^2k^2/\Omega^t$ for $\Theta \sim 1$.

This estimate enables us to compare the first and second terms of the expansion (3.89). The first is of order $k^2/\omega^2\Omega^t$, the second of order $k/\omega\Omega^t$, because $\partial w_{\mathbf{k}\mathbf{k}_2}/d\mathbf{k}_2$ is of order $c\, w_{\mathbf{k}_2}/\Omega^t$. Consequently, for $\Theta \sim 1$, i.e., for a packet of transverse waves having a wide angular spread, with $\omega \gg kc$ and for $\Theta \ll (\omega_{0e}/\Omega^t)^2$ with $\omega \gg (\omega_{0e}/\Omega^t)^2 kc$ we obtain

$$\omega - i\gamma_{\mathbf{k}}^{l} = \frac{1}{2\omega}\int \mathbf{k}\,\frac{\partial w_{\mathbf{k}\mathbf{k}_2}}{\partial \mathbf{k}_2}\,d\mathbf{k}_2 N_{\mathbf{k}_2}^{t}. \tag{3.90}$$

For $\Theta \sim 1$, $\omega \ll kc$ and for $\Theta \ll (\omega_{0e}/\Omega^t)^2$, $\omega \ll (\omega_{0e}/\Omega^t)^2 kc$ we have

$$-(\omega - i\gamma_{\mathbf{k}}^{l}) = \frac{1}{2\omega^2}\int w_{\mathbf{k}\mathbf{k}_2}\,\frac{\left(k^2 - \dfrac{(\mathbf{k}_2\mathbf{k})^2}{(\Omega^t)^2}\right)}{\Omega^t}\,N_{\mathbf{k}_2}^{t}\,d\mathbf{k}_2. \tag{3.91}$$

Substituting the expression (3.84) into (3.90), we obtain

$$(\omega - i\gamma_{\mathbf{k}}^{l})\,\omega = -\frac{e^2\omega_{0e}k^2}{4\pi^2 m_e^2}\int \frac{(\mathbf{k}\mathbf{k}_2)}{(\Omega^t)^4}\,N_{\mathbf{k}_2}^{t}\,d\mathbf{k}_2 c^2. \tag{3.92}$$

If $\gamma_{\mathbf{k}}^{l} = 0$, i.e., if the waves neither decay nor grow in the linear approximation, energy transfer takes place for waves propagating in the direction of the transverse waves. For waves propagating against the transverse waves of the packet nonlinear damping occurs. An estimate of the nonlinear increment can be obtained by allowing for the fact that the energy density $W^t$ is

$$W^t = \int \frac{|A_{\mathbf{k}_2}^{t}|^2}{8\pi}\,d\mathbf{k}_2 \simeq \int \frac{2c k_2 N_{\mathbf{k}_2}^{t}\,d\mathbf{k}_2}{(2\pi)^3}. \tag{3.93}$$

According to (3.92) we have [14]

$$\gamma_{\mathbf{k}}^{\mathrm{nlr}} \simeq \omega_{0e}\sqrt{\frac{\omega_{0e}^4}{4(\Omega^t)^4}\,\frac{W^t c}{n m_e v_\varphi^3}}. \tag{3.94}$$

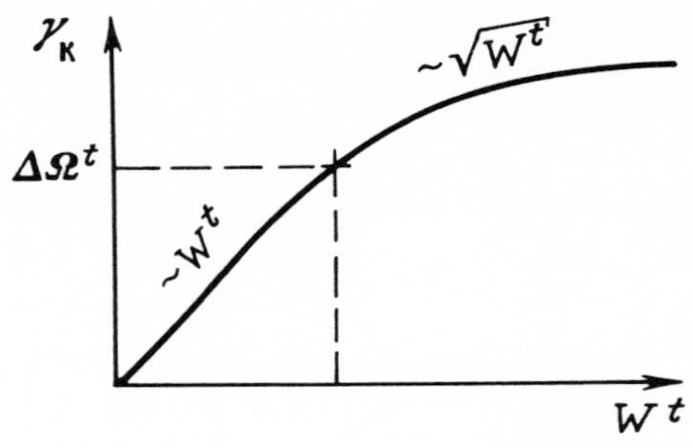


Fig. 13. Nonlinear growth rate for longitudinal-wave generation as a function of the transverse-wave intensity.

Here $\Omega^t = k_2c$ is the characteristic frequency of the transverse waves, and $v_\varphi = \omega_{0e}/k$ is the longitudinal-wave phase velocity. The growth rate of the instability (3.91) for $\Theta \sim 1$ can be estimated in precisely the same manner:*

$$\gamma_k \simeq \omega_{0e}\left(\left(\frac{\omega_{0e}}{\Omega^t}\right)^4 \frac{W^t c^2}{8nm_e v_\varphi^4}\right)^{1/3}. \tag{3.95}$$

## 9. SOME QUALITATIVE IMPLICATIONS

The following important qualitative results may be inferred from the foregoing discussion.

**1.** With increasing intensities $W^t$ there is a continuous transition from the growth rates given by (3.85), which vary as $W^t$, to those of (3.94) and (3.95), which vary as $\sqrt{W^t}$ (Fig. 13).

Therefore, the "hydrodynamic" instability, like the "kinetic" instability, is merely a limiting case of the instability described by the general relation (3.83). The growth (3.94) depends on the energy $W^t$ in the same way as in the case of interaction between fixed-phase waves. Equation (3.94) differs from the corresponding fixed-phase relation (3.30) by the additional factor $\sqrt{k/k_2}$. Its oc-

* The instability of beams of electromagnetic waves is discussed in more detail in [15]; self-focusing effects are associated with the nonlinearities of spatially inhomogeneous electromagnetic waves [16, 17, 18].

currence arises because, in addition to $N^t_{k_2}$, Eq. (3.83) includes the term $N^t_{k_2-k}$ with $k \ll k_2$, which is approximately equal to $N^t_{k_2} - k dN^t_{k_2}/dk_2$, i.e., $N^t_{k_2} - N^t_{k_2-k} \simeq k dN^t_{k_2}/dk_2$, which is of the same order of magnitude as $N^t_{k_2} k/k_2$. Had we included only the first term of (3.83) containing $N^t_{k_2}$, the result would have been of precisely the same order of magnitude as that obtained for the estimate of the nonlinear interaction time for fixed-phase waves. Moreover, if in estimating the instability of fixed-phase waves we had included interaction not only with the waves of the lower frequency $\Omega^t - \omega_{0e}$, but also with waves of frequency $\Omega^t + \omega_{0e}$, the result would have been the same as (3.92). Thus we have established a correspondence between nonlinear "hydrodynamic" instability and the interaction of fixed-phase waves.

2. Typical of nonlinear "hydrodynamic" instabilities is the fact that for strong linear damping $\gamma^l_k \gg \gamma^{nlr}_k$ growth of the oscillations is possible. This is immediately apparent from the expression (3.92) for the frequency $\omega$ in the presence of linear damping:

$$\omega = \frac{i\gamma^l_k}{2} \pm i\sqrt{\frac{(\gamma^{l_2}_k)^2}{4} + (\gamma^{nlr}_k)^2}. \tag{3.96}$$

For $\gamma^{nlr}_k \ll \gamma_k$, taking the plus sign, we have

$$\omega = i\frac{(\gamma^{nlr}_k)^2}{\gamma^l_k}, \tag{3.97}$$

i.e., oscillation growth takes place. In the case of the instability (3.85), which is proportional to $W^t$, growth occurs for $\gamma^{nlr}_k > \gamma^l_k$.

3. A few words are in order concerning the possible elimination of "hydrodynamic" instability. The longitudinal wave intensity grows at the expense of the transverse waves, therefore the energy of the latter must decrease. On the other hand, the reverse action of the excited longitudinal waves on the transverse waves tends to flatten the spectrum of the latter and to increase $\delta\Omega^t$. The mechanism of this flattening effect consists in the appearance of nonlinear growth of transverse waves with frequencies outside those of the original packet. These waves will extract energy from the waves in the initial packet, i.e., the spectrum will flatten. Both effects, flattening of the spectrum and the reduction of its intensity, cause the condition $\gamma \gg \Delta\Omega$ to be violated and transform the instability into the form described by (3.85). This subsequently leads to the quasistationary equilibrium states investigated earlier.

**4.** The "hydrodynamic" instabilities occurred through the interaction of Langmuir and transverse waves. Similar nonlinear instabilities are possible when random waves of another kind are present in the plasma. An instability similar to that described above appears in connection with the interaction of Langmuir and low-frequency waves such as, for example, sound waves, Alfvén waves, etc. This problem is of interest in that, first, intense random Langmuir waves are easily generated in a plasma, for example by the presence of beams in the plasma, and, second, the instability mechanism can bring about the transfer of vibrational energy to the plasma ions taking part in the low-frequency waves.

Nonlinear "hydrodynamic" instabilities associated with the interaction of Langmuir waves and sound waves in a plasma without a magnetic field were first investigated by Sagdeev and Oraevskii [3].* From the point of view of the effective transfer of energy from electrons to ions, the "kinetic" and "hydrodynamic" instabilities play a similar part. Moreover, during the development of instability the "hydrodynamic" generation of low-frequency waves changes into "kinetic" generation.

**5.** In addition to the imaginary correction to the Langmuir frequency $\omega_{0e}$ there is a real correction [as for the fixed-phase case; see Eq. (3.33)]. Since this correction is small compared with $\omega_{0e}$, it might appear at first sight scarcely worth considering. Actually it is important, for at least two reasons. First, in a number of plasma processes the frequency difference between two interacting Langmuir waves can be important. The frequency $\omega_{0e}$ is not involved in this difference. However, the small corrections associated with thermal motion can become comparable with the nonlinear corrections. Second, through the nonlinear corrections it is possible for the phase velocities to become amplitude dependent, which, though weak, can be significant. This effect makes it possible to regulate the phase velocity by means of the amplitude, a capability that gains importance in problems involving the acceleration of charged particles by waves [12]. In the limit $|\omega| \ll \Delta\Omega$ we have from (3.85)

$$\mathrm{Re}\,\omega = \frac{1}{2}\int \frac{w_{\mathbf{k}\mathbf{k}_2}}{\Delta\Omega}\left(N^t_{\mathbf{k}_2} - N^t_{\mathbf{k}_2-\mathbf{k}}\right) d\mathbf{k}_2. \tag{3.98}$$

* These ideas are extended to the interaction of several coupled waves in [19. 20].

Here the principal value of the integral has been taken.

As an example we estimate Re $\omega$ in terms of $W_t$ for the case of isotropic transverse waves for $k \ll k_2$, $k_2 \gg \omega_{0e}/c$ and $k \gg \omega_{0e}/c$:

$$\mathrm{Re}\,\omega \approx \left(\frac{k}{k_2}\right)^3 \omega_{0e} \left(\frac{\omega_{0e}}{\Omega^t}\right)^2 \frac{W^t}{n_0 mc^2}. \tag{3.99}$$

The existence of nonlinear corrections to the frequency of the interacting waves is a general characteristic of nonlinear interactions. They were first calculated for Langmuir waves by Sturrock [13].*

**6.** The similarity between the effects of the nonlinear hydrodynamic instabilities of intense random waves and the decay instabilities of fixed-phase waves is easily understood. During one period of the low-frequency wave into which the high-frequency wave decays there are several periods of the latter, hence the effect of the high-frequency wave is averaged over many periods. This frequency averaging is equivalent to averaging over the phases of the high-frequency waves.

## Literature Cited


1. J. A. Armstrong, N. Bloembergen, J. Ducuing, and P. S. Pershan, "Interactions between light waves in a nonlinear dielectric," Phys. Rev., 127:1918 (1962).
2. S. Akhmanov and R. V. Khokhlov, Problems in Nonlinear Optics. VINITI, Moscow (1965).
3. R. Z. Sagdeev and V. N. Oraevskii, "Stability of the steady-state longitudinal oscillations of a plasma," Zh. Tekh. Fiz., 32:1291 (1962); V. N. Oraevskii, "Stability of nonlinear steady-state oscillations," Yadernyi Sintez, 2:263 (1964); R. Z. Sagdeev, "Collective processes and shock waves in a rarefied plasma," Problems in Plasma Theory, Vol. 4. Atomizdat (1964), p. 20.
4. I. S. Danilkin, "Conversion of two transverse waves into a longitudinal wave in a collisionless plasma," Zh. Tekh. Fiz., 35:435 (1965).
5. L. D. Landau and E. M. Lifshits, Electrodynamics of Continuous Media. Addison-Wesley, Reading, Mass. (1960).
6. S. M. Rytov, Theory of Electrical Fluctuations and Thermal Radiation. Izd. Akad. Nauk SSSR (1959).
7. L. M. Kovrizhnykh and V. N. Tsytovich, "Effects of the decay of transverse waves in a plasma," Zh. Éksp. Teor. Fiz., 47:1454 (1964).


* The corrections to the real part of the frequency are discussed in more detail in Chap. X, as well as in the recent papers [21, 22].

8. V. N. Tsytovich, "Nonlinear generation of plasma waves by a beam of transverse waves (I)," Zh. Tekh. Fiz., 34:773 (1965).
9. L. M. Kovrizhnykh and V. N. Tsytovich, "Interaction of intense high-frequency radiation with a plasma," Dokl. Akad. Nauk SSSR, 158:1306 (1964).
10. V. A. Liperovskii and V. N. Tsytovich, "Oscillation spectra of a weakly turbulent plasma," Zh. Tekh. Fiz., 36:575 (1966).
11. V. A. Liperovskii, L. M. Kovrizhnykh, and V. N. Tsytovich, "Nonlinear generation of plasma waves by a beam of transverse waves (II)," Zh. Tekh. Fiz., 36:1339 (1966).
12. Ya. B. Fainberg, "Particle acceleration in a plasma," Atomnaya Énergiya, 6:431 (1959).
13. P. A. Sturrock, "Nonlinear effects in an electron plasma," Proc. Roy. Soc., 242A:277 (1957).
14. V. N. Tsytovich and A. B. Shvartsburg, "Nonlinear generation of plasma waves by a beam of transverse waves (III)," Zh. Tekh. Fiz., 37:589 (1967).
15. V. N. Tsytovich, Paper at the International Conference on Phenomena in Ionized Gases, Vienna (1967).
16. G. A. Askar'yan, "Effect of the field gradient of a high-intensity arc on the medium," Zh. Éksp. Teor. Fiz., 42:1567 (1962).
17. V. I. Talanov, "Self-focusing of electromagnetic waves," Izv. Vyssh. Uch. Zav., Radiofizika, 7:564 (1964).
18. S. A. Akhmanov, A. P. Sukhorukov, and R. V. Khokhlov, "Self-focusing and diffraction of light in a nonlinear medium," Uspekhi Fiz. Nauk, 93:19 (1967).
19. K. S. Karplyuk and V. N. Oraevskii, "Excitation of oscillations in the decay instability of steady-state waves in a plasma," ZhÉTF Pis. Red., 5:451 (1967).
20. V. N. Oraevskii and V. N. Tsytovich, "Multiplasma decay instabilities of a turbulent plasma," Zh. Éksp. Teor. Fiz., 53:1116 (1967).
21. L. M. Gorbunov and A. M. Timerbulatov, "The dispersion law and nonlinear interaction of Langmuir waves in a weakly turbulent plasma," Zh. Éksp. Teor. Fiz., 53:1492 (1967).
22. V. G. Makhan'kov and V. N. Tsytovich, "Coulomb collisions of particles in a turbulent plasma," Zh. Éksp. Teor. Fiz., 53:1789 (1967).

*Chapter IV*

# Induced Processes in Plasma

In the preceding chapter we made a detailed examination into the interaction of waves having random phases, making use of the nonlinear equations for the field in the plasma. We now show how to derive the nonlinear equations from simple physical considerations based on the concept of induced processes.

## 1. SPONTANEOUS AND INDUCED EMISSION PROCESSES IN A TWO-LEVEL SYSTEM

The induced process idea was first introduced by Einstein applied to a simple two-level system (Fig. 14). We now summarize briefly the basic idea of induced processes in a two-level system.

Suppose that the energy of level 1 is greater than that of level 2 ($\varepsilon_1 > \varepsilon_2$). If the system comprising the two levels is not acted upon by any external influence and it has an initial energy $\varepsilon_1$, i.e., if it is in state 1, it can then go to state 2 spontaneously. This spontaneous transition can be described by a certain probability of a

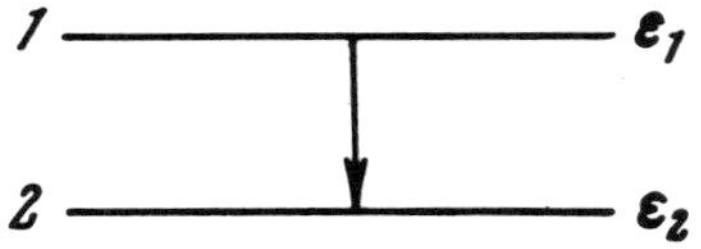


Fig. 14. Schematic representation of a two-level system.

spontaneous emission per second, $u_{12}$. This probability has the following physical interpretation: if a set of $n_1$ atoms (or, as we say, an ensemble of atoms) were initially in state 1, then after one second there will be $n_2$ atoms in state 2, where $n_2 = u_{12}n_1$, i.e., $u_{12}$ is the ratio $n_2/n_1$ after one second. Consequently, the rate of increase of $n_2$ per unit time $\partial n_2/\partial t$ is

$$\frac{\partial n_2}{\partial t} = u_{12} n_1 = - \frac{\partial n_1}{\partial t}. \tag{4.1}$$

The derivative $\partial n_2/\partial t$ is equal to the rate of decrease of the number of atoms in state 1.

Now let external radiation act on the two-level system, namely electromagnetic waves with a frequency near that of spontaneous emission: $\omega_{12} = \varepsilon_1 - \varepsilon_2/\hbar$, where $\hbar$ is Planck's constant. This radiation will induce transitions in the system, both from state 1 to state 2 and from state 2 to state 1. We call such transitions induced processes. The probability of the induced processes depends on the radiation intensity. Since it is zero at zero intensity, the probabilities of the induced processes must in the first approximation be proportional to the radiation intensity.

We denote the probability of induced emission, i.e., transition from state 1 to state 2, by $w_{12}$, and the probability of induced absorption by $w_{21}$. The intensity of the radiation inducing this transition is proportional to the square of the electric field $\mathbf{E}_k$ of the wave:

$$w_{12} \sim |\mathbf{E}_k|^2; \; w_{21} \sim |\mathbf{E}_k|^2. \tag{4.2}$$

Here $\mathbf{k}$ is the wave number corresponding to the frequency $\omega_{12}$.

Quantum notation can be used to replace $|\mathbf{E}_k|^2$ with a quantity that is uniquely related to it and is equal to the number of radiation quanta. From the quantum point of view the energy in the radiation field is the sum of the energies of the quanta, i.e., the radiation intensity is proportional to the number of quanta. We denote the latter by $N_k$. From the above $|\mathbf{E}_k|^2$ and $N_k$ are proportional to one another. We shall determine the proportionality coefficient a little later on, when we are ready to apply it to plasma. By virtue of the proportionality of $N_k$ and $|\mathbf{E}_k|^2$ the probabilities of the induced processes are proportional to $N_k$:

$$w_{12} = \widetilde{w}_{12} N_k; \; w_{21} = \widetilde{w}_{21} N_k, \tag{4.3}$$

where $\tilde{w}_{12}$ and $\tilde{w}_{21}$ are the proportionality coefficients.

In the presence of induced processes Eq. (4.1) must be altered to account for the decrease in the number of particles in state 2 resulting from induced emission and their increase caused by induced absorption from state 1:

$$\frac{\partial n_1}{\partial t} = -(u_{12} + \tilde{w}_{12}N_k)\,n_1 + \tilde{w}_{21}N_k n_2 = -\frac{\partial n_2}{\partial t}. \tag{4.4}$$

Einstein's simple arguments enable us to establish a relationship between $u_{12}$, $\tilde{w}_{21}$, and $\tilde{w}_{12}$. Equation (4.4) must be true in the special case of statistical equilibrium, when $\partial n/\partial t = 0$ and $n_2 = \text{const} \cdot e^{-\varepsilon_2/T}$; $n_1 = \text{const} \cdot e^{-\varepsilon_1/T}$($T$ is the temperature, and the statistical weights of the states are equal to one), and $N_k$ corresponds to equilibrium radiation, i.e., to Planck's formula

$$N_k = \frac{1}{e^{\frac{\hbar\omega}{T}} - 1}. \tag{4.5}$$

From (4.4) we obtain for $\partial n_1/\partial t = 0$

$$N_k = \frac{u_{12}}{\tilde{w}_{21}e^{\frac{\hbar\omega_{21}}{T}} - \tilde{w}_{12}}. \tag{4.6}$$

Since (4.5) and (4.6) must be the same for all temperatures,

$$\tilde{w}_{12} = u_{12} = \tilde{w}_{21}. \tag{4.7}$$

This has some very important consequences: 1) the total probability of spontaneous and induced emission is

$$u_{12}(1 + N_k); \tag{4.8}$$

2) the probability of absorption is

$$u_{12}N_k, \tag{4.9}$$

ie., in order to obtain the emission probability we must multiply the probability of spontaneous emission by $1 + N_k$, and for the absorption probability we must multiply it by $N_k$.

Although the above arguments apply to a two-level system, they naturally have a more general application. This follows from the fact that the energy levels in any system are, strictly speaking, discrete, and for any discrete spectrum it suffices to limit the discussion to two arbitrarily chosen levels. Statistical equilibrium, which we used to derive (4.8) and (4.9), has, as we know, the property that any two states chosen are in equilibrium with one another. Rules (4.8) and (4.9) also have a more profound electrodynamical foundation, which becomes clear in the quantum treatmeant of the radiation field [1].

## 2. INTERACTION OF ELECTROMAGNETIC WAVES WITH A TWO-LEVEL SYSTEM

We introduce the differential probability of spontaneous emission $u_{12}(\mathbf{k})$, referred to the interval $d\mathbf{k}$ of the radiated waves:

$$u_{12} = \int u_{12}(\mathbf{k}) \frac{d\mathbf{k}}{(2\pi)^3}. \tag{4.10}$$

The quantity $u_{12}(\mathbf{k})$ has a maximum about those values of $|\mathbf{k}|$ corresponding to $\omega_{12}/c$ and depends on the angles involved.

Consider the interaction of the wave $\mathbf{k}$ with a two-level system. If we are concerned with the variation of the total number of particles at levels 1 and 2 (they can have a finite width), we must include all possible transitions, i.e., write (4.4) for the differential probabilities and integrate over all $\mathbf{k}$. In place of (4.4) we have

$$\frac{\partial n_1}{\partial t} = -u_{12}n_1 + (n_2 - n_1)\int u_{12}(\mathbf{k}) N_{\mathbf{k}} \frac{d\mathbf{k}}{(2\pi)^3}. \tag{4.11}$$

Let us see how the radiation field varies when it interacts with a two-level system. It is clear that the number of quanta increases due to emission and decreases due to absorption. Therefore,

$$\frac{\partial N_{\mathbf{k}}}{\partial t} = u_{12}(\mathbf{k})(1 + N_{\mathbf{k}}) n_1 - u_{12}(\mathbf{k}) N_{\mathbf{k}} n_2 = \\ = u_{12}(\mathbf{k})(n_1 - n_2) N_{\mathbf{k}} + u_{12}(\mathbf{k}) n_1. \tag{4.12}$$

Here n is the number of particles per $cm^3$.

Consider a narrow wave packet $N_{\mathbf{k}}$ whose frequency is near $\omega_{12}$ and whose $\mathbf{k}$ is near $\mathbf{k}_0$. Introducing the total number of quanta in the packet, N, given by

$$N = \int N_{\mathbf{k}} \frac{d\mathbf{k}}{(2\pi)^3}, \tag{4.13}$$

and integrating over $\mathbf{k}$, we obtain in place of (4.11) and (4.12)

$$\frac{\partial n_1}{\partial t} = -u_{12}n_1 + (n_2 - n_1) u_{12}(\mathbf{k}_0) N = -\frac{\partial n_2}{\partial t}; \tag{4.14}$$

$$\frac{\partial N}{\partial t} = u_{12}(\mathbf{k}_0)(n_1 - n_2)N + u_{12}n_1. \tag{4.15}$$

If the intensity of the wave packet is sufficiently large, so that the induced processes are dominant, then $N \gg 1$, and

$$\frac{\partial N}{\partial t} = \gamma N; \quad \gamma = u_{12}^0(n_1 - n_2); \quad u_{12}^0 = u_{12}(k_0). \tag{4.16}$$

If $n_1 < n_2$, the wave is absorbed, and if $n_1 > n_2$, it is amplified. In the case $n_1 > n_2$ it is customarily said that the temperature is negative. This is entirely a matter of convention, as for $n_1 > n_2$ the system is in a nonequilibrium state, hence the whole concept of temperature, which characterizes equilibrium states, does not apply. If, however, we make the formal assumption that $n_2 = \text{const} \cdot e^{-\varepsilon_2/T}$ and $n_1 = \text{const} \cdot e^{-\varepsilon_1/T}$, then $n_1 > n_2$ is possible only for $T < 0$. Only in this limited sense is it proper to speak of negative temperature.

For $n_1 > n_2$ the number of quanta increases exponentially, a fact that is readily understandable. An exponential law implies that the process has an avalanche type of development. If an atom emits one quantum, then the latter, on striking another atom, induces emission from that atom; now two quanta appear, which induce emission from other atoms. This yields four quanta, and so on, i.e., an avalanche actually develops.

The growth rate of the instability may be regarded as independent of time, i.e., determined by the initial population $n_1(0)$, $n_2(0)$, only if the radiation intensity is sufficiently low, so that it can significantly alter the populations of the energy levels during the characteristic time of the process. If we neglect spontaneous processes, the population change is determined from (4.14):

$$\frac{\partial n_1}{\partial t} = -\gamma N; \tag{4.17}$$

$$\frac{\partial (n_1 - n_2)}{\partial t} = -2(n_1 - n_2)u_{12}^0 N = -2\gamma N. \tag{4.18}$$

In other words, the population difference decreases with time. This causes the so-called saturation effect.

It is a simple matter to find the simultaneous solution of the set of equations (4.16) and (4.17), taking account of the variations in the absorption and the populations. In fact, recognizing that

$$n_1 + n_2 = \text{const} = n_1(0) + n_2(0);$$
$$N + n_1 = \text{const} = N(0) + n_1(0),$$

we obtain the following equation for N:

$$\frac{\partial N}{\partial t} = u_{12}^0 (2N(0) + n_1(0) - n_2(0) - 2N)\, N. \tag{4.19}$$

It is evident from the structure of this equation that N does not grow indefinitely, as it would in the linear approximation when N is small and $n_1(0) > n_2(0)$. The number N, we note, cannot be greater than

$$N^{\text{equi}} = N(0) + \frac{n_1(0) - n_2(0)}{2},$$

because then the sign of the right-hand side changes. It is also clear that the equilibrium value of N is stable, since a small increase in N over the equilibrium value produces damping, whereas a small decrease produces growth of N.

The solution of (4.19) is

$$N = \frac{N^{\text{equi}}}{1 + \frac{\varepsilon}{2} e^{-2N^{\text{equi}} u_{12}^0 t}}; \quad \varepsilon = \frac{n_1(0) - n_2(0)}{N(0)}. \tag{4.20}$$

The characteristic time to reach an equilibrium value of N is determined by the initial growth rate [provided N(0) is small]. Moreover,

$$n_1 - n_2 = \frac{N^{\text{equi}} e^{-2N^{\text{equi}} u_{12}^0 t}}{\frac{1}{\varepsilon} + \frac{1}{2} e^{-2N^{\text{equi}} u_{12}^0 t}}. \tag{4.21}$$

This shows that $n_1 - n_2$ tends exponentially to zero, while the growth rate (4.16) tends to zero.

Consider the case of a normal population $n_2(0) > n_1(0)$. If $N(0) \ll n_2(0) - n_1(0)$, then, as implied by (4.21), $N^{\text{equi}} < 0$ and, consequently, the population is essentially unchanged. For large intensity $N(0) \gg n_2(0) - n_1(0)$ we have $N^{\text{equi}} \sim N(0) > 0$, and the populations quickly equalize. It is important to note that during equalization of the populations the spontaneous emission in (4.15) will not be small compared with the induced emission. This means that a small difference will occur in the populations through competition between the induced processes, which tend to equalize the populations, and the spontaneous processes, which tend to change all the atoms into the lowest state.

The saturation effect was first discussed in [2] for the case in which collisions compete with the induced processes of emission and absorption (see also [3, 4]).

## 3. PACKETS OF INTERACTING WAVES AS NEGATIVE-TEMPERATURE SYSTEMS

We have analyzed the example of the two-level system in detail because of its strong analogy with the nonlinear interaction of three random-phase wave packets. One is easily convinced of the analogy by comparing the solutions (4.21) with (3.69). Nonlinear interaction causes the intensities of the two transverse-wave packets to equalize by their interaction with plasma waves. This is analogous to the equalization of the populations of the two-level system. Consequently, the beam of high-frequency transverse waves plays the part of the atoms in the upper level, and the beam of low-frequency transverse waves is the analog of the atoms in the lower level.

From the quantum point of view, of course, the beam of transverse waves represents a set of particles, quanta, whose energy is $\hbar\omega$. One might say that quanta with energy $\hbar\omega_1$ are converted into quanta with energy $\hbar\omega_2 < \hbar\omega_1$, with the emission of longitudinal waves. In point of fact, then, we clearly have a physical correspondence with the case of the two-level system. Nonlinear effects lead to the interaction of different waves and the emission of longitudinal by transverse waves.

It is interesting that in the purely classical setting we can encounter discrete level effects. This happens when the spread in the quantum energies $\hbar\omega_2$ and $\hbar\omega_1$ are considerably smaller than their difference, which can be radiated in the form of longitudinal waves, $\delta\omega_1, \delta\omega_2 \ll \omega_{0e}$. In this case, on emitting a longitudinal wave, the transverse quantum can change its energy only by a discrete amount, $\hbar\omega_{0e}$. If we have an intense packet of quanta $\hbar\omega_1$, this corresponds to a negative-temperature state. This packet is unstable, because any longitudinal quantum emitted by a "particle" of the packet can induce the emission from another transverse quantum. The result is an avalanche effect. This continues to develop until the number of quanta $\hbar\omega_2$ becomes appreciable and the reverse process of induced absorption of longitudinal quanta occurs. As a result, the intensities of the packets $\hbar\omega_1$ and $\hbar\omega_2$ tend to equalize. Conversely, if there are intense longitudinal waves and a transverse-wave packet (by analogy with the atoms at the lower levels), the longitudinal waves are strongly absorbed. This continues until the absorption-produced beam $\hbar\omega_1$ is comparable in intensity with the original beam.

This last effect may be regarded as the nonlinear absorption of longitudinal waves in a plasma containing intense transverse waves, and the equalization of the intensities of the two transverse-wave packets as saturation of the nonlinear absorption of longitudinal waves. This saturation, of course, can be produced by linear absorption caused by collisions (as in the work of Karplus and Schwinger [2] for a two-level system).

## 4. DENSITY OF QUANTA IN A PLASMA

In order to present a mathematical equivalence between the two problems we need a rigorous definition of N (the number of quanta) for the various plasma wave modes. For this we analyze the energy of the electromagnetic field of waves in random phase. Consider first a longitudinal wave. Longitudinal waves are not accompanied by oscillating magnetic fields, hence the rate of change of energy density $W^l$ is

$$\frac{\partial W^l}{\partial t} = \frac{1}{4\pi} \mathbf{E}^l \frac{\partial \mathbf{D}^l}{\partial t},$$

or

$$W^l = \frac{1}{4\pi} \int_{-\infty}^{t} \mathbf{E}^l \frac{\partial \mathbf{D}^l}{\partial t} dt. \tag{4.22}$$

We represent the field in (4.22) as a plane wave expansion:

$$\mathbf{E}^l = \int \mathbf{E}^l_{\mathbf{k}\omega} e^{-i(\omega t - \mathbf{k}\mathbf{r})} d\mathbf{k} d\omega;$$
$$\mathbf{D}^l = \varepsilon^l \mathbf{E}^l = \int \varepsilon^l(\omega', \mathbf{k}') \mathbf{E}^l_{\mathbf{k}'\omega'} e^{-i(\omega' t - \mathbf{k}'\mathbf{r})} d\mathbf{k}' d\omega'$$

and average over the phases. We obtain

$$\langle W^l \rangle = \frac{1}{8\pi}\int d\mathbf{k}d\omega d\mathbf{k}'d\omega' \langle \mathbf{E}^l_{\mathbf{k}\omega}\mathbf{E}^l_{\mathbf{k}'\omega'}\rangle e^{-i(\omega+\omega')t+i(\mathbf{k}+\mathbf{k}')\mathbf{r}} \times \times \frac{\varepsilon^l(\omega', \mathbf{k}')\omega' + \omega\varepsilon^l(\omega, \mathbf{k})}{\omega + \omega} . \tag{4.23}$$

This result has been symmetrized with respect to $\omega$ and $\omega'$.

We recall, moreover, that for random propagating waves

$$\mathbf{E}^l_{\mathbf{k}\omega} = \mathbf{E}^l_{\mathbf{k}}\delta(\omega - \Omega^l(\mathbf{k})),$$

where $\Omega^l(\mathbf{k})$ is the solution of the linear dispersion equation. Since

$$\langle \mathbf{E}^l_{\mathbf{k}}\mathbf{E}^l_{\mathbf{k}'}\rangle = |\mathbf{E}^l_{\mathbf{k}}|^2\delta(\mathbf{k}+\mathbf{k}') \text{ and } \Omega^l(-\mathbf{k}) = -\Omega^l(\mathbf{k}), \tag{4.24}$$

we have

$$\langle \mathbf{E}^l_{\mathbf{k}\omega}\mathbf{E}^l_{\mathbf{k}'\omega'}\rangle = |\mathbf{E}^l_{\mathbf{k}}|^2\delta(\mathbf{k}+\mathbf{k}')\,\delta(\omega - \Omega^l(\mathbf{k}))\,\delta(\omega' - \Omega^l(\mathbf{k}')) = \\ = |\mathbf{E}^l_{\mathbf{k}}|^2\delta(\mathbf{k}+\mathbf{k}')\,\delta(\omega+\omega')\,\delta(\omega - \Omega^l(\mathbf{k})).$$

Hence the numerator and denominator of the expression

$$\frac{\varepsilon^l(\omega', \mathbf{k}')\omega' + \omega\varepsilon^l(\omega, \mathbf{k})}{\omega + \omega'},$$

which occurs in (4.23), both tend to zero. To resolve the indeterminacy, we introduce $(\partial/\partial\omega)\omega\varepsilon^l$. Thus (henceforth we drop the average sign for $W^l$),

$$W^l = \frac{1}{8\pi}\int d\mathbf{k}\,|\mathbf{E}^l_{\mathbf{k}}|^2 \frac{\partial}{\partial\omega}\omega\varepsilon^l\Big|_{\omega=\Omega^l(\mathbf{k})} \tag{4.25}$$

Since $\varepsilon^l(\Omega^l, \mathbf{k}) = 0$, we have

$$\frac{\partial}{\partial\omega}\omega\varepsilon^l = \frac{1}{\omega}\frac{\partial}{\partial\omega}\omega^2\varepsilon^l.$$

We now write the expression for the energy of the field as the sum of the energies of the individual quanta. Introducing the number of quanta, we write

$$W^l = \int \frac{d\mathbf{k}}{(2\pi)^3}\hbar\Omega^l(\mathbf{k})N^l_{\mathbf{k}}. \tag{4.26}$$

Here $\hbar\Omega^l(\mathbf{k})$ is the energy of an individual quantum, $d\mathbf{k}/(2\pi)^3$ is the number of possible quantum states with momentum $\hbar\mathbf{k}$. Comparing (4.26) with (4.25), we have

$$N^l_{\mathbf{k}} = \frac{\pi^2}{\hbar}\left(\frac{1}{(\omega)^2}\frac{\partial}{\partial\omega}\omega^2\varepsilon^l\right)\Big|_{\omega=\Omega^l(\mathbf{k})}|\mathbf{E}^l_{\mathbf{k}}|^2. \tag{4.27}$$

Comparing this expression with that for $N_{\mathbf{k}}^{l}$, introduced earlier, we see that they differ only by Planck's constant. It is also easy to verify that the $N_{\mathbf{k}}^{t}$ introduced before corresponds to the number of transverse-mode quanta (if the factor $1/\hbar$ is included). In fact, the transverse-wave energy consists of the electric energy (4.22), which gives the result (4.25) with the replacement $t \to l$:

$$W_1^t = \frac{1}{8\pi}\int d\mathbf{k}\,|\,\mathbf{E}_{\mathbf{k}}^t\,|^2\,\frac{\partial}{\partial\omega}\,\omega\varepsilon^t\Big|_{\omega=\Omega^t(\mathbf{k})},$$

and the magnetic energy, which is equal to

$$W_2^t = \left\langle\frac{H^2}{8\pi}\right\rangle = \frac{1}{8\pi}\int|\,\mathbf{H}_{\mathbf{k}}\,|^2 d\mathbf{k};$$

$$\frac{1}{c}\,\mathbf{H}_{\mathbf{k}} = \left[\frac{\mathbf{k}}{\omega}\,\mathbf{E}_{\mathbf{k}}^t\right];\quad |\,\mathbf{H}_{\mathbf{k}}\,|^2 = \frac{k^2c^2}{\omega^2}\,|\,\mathbf{E}_{\mathbf{k}}^t\,|^2 = \varepsilon^t\,|\,\mathbf{E}_{\mathbf{k}}^t\,|^2,$$

because $k^2c^2 = \omega^2\varepsilon^t$. Consequently,

$$W^t = W_1^t + W_2^t = \frac{1}{8\pi}\int d\mathbf{k}\,|\,\mathbf{E}_{\mathbf{k}}^t\,|^2\left(\frac{\partial}{\partial\omega}\,\omega\varepsilon^t + \varepsilon^t\right) =$$

$$= \frac{1}{8\pi}\int d\mathbf{k}\,|\,\mathbf{E}_{\mathbf{k}}^t\,|^2\,\frac{1}{\omega}\,\frac{\partial}{\partial\omega}\,\omega^2\varepsilon^t\Big|_{\omega=\Omega^t(\mathbf{k})}$$

Comparing this result with the quantum expression

$$W^t = \int\frac{2\Omega^t N_{\mathbf{k}}^t d\mathbf{k}^t\hbar}{(2\pi)^3}$$

(the coefficient 2 accounts for the two polarizations), we obtain

$$N_{\mathbf{k}}^t = \frac{\pi^2}{2\hbar\,(\Omega^t(\mathbf{k}))^2}\,\frac{\partial}{\partial\omega}\,\omega^2\varepsilon^t(\omega,\mathbf{k})\Big|_{\omega=\Omega^t(\mathbf{k})}\,|\,\mathbf{E}_{\mathbf{k}}^t\,|^2. \tag{4.28}$$

## 5. NONLINEAR WAVE INTERACTION RESULTING FROM THE BALANCE OF INDUCED EMISSION AND ABSORPTION

The interaction of random-phase waves may be interpreted explicitly as effects involving the induced emission of certain waves by others. Consider the process in which a transverse wave emits a longitudinal wave. This may be depicted graphically as shown in Fig. 15. The shaded circle represents effective nonlinear interaction, which depends in general on the momenta of the waves $\mathbf{k}_1^t$, $\mathbf{k}^l$, and $\mathbf{k}_2^t$. This process must comply with the laws of energy

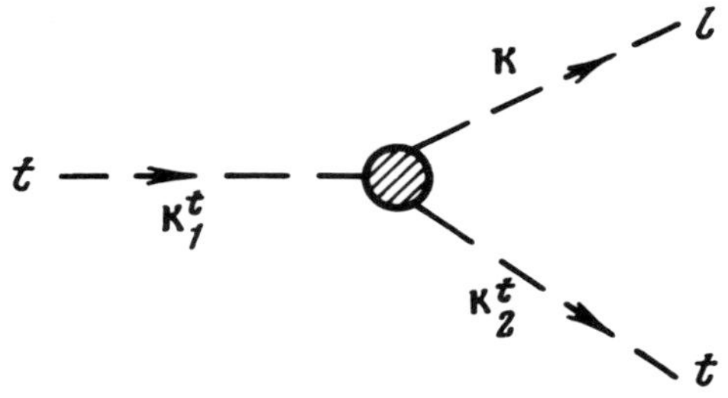


Fig. 15. Decay of a transverse wave into a Langmuir and a transverse wave.

and momentum conservation:*

$$\hbar \mathbf{k}_1^t = \hbar \mathbf{k}_2^t + \hbar \mathbf{k}^l; \tag{4.29}$$

$$\hbar \Omega_1^t(\mathbf{k}_1) = \hbar \Omega_2^t(\mathbf{k}_2) + \hbar \Omega^l(\mathbf{k}). \tag{4.30}$$

Planck's constant is retained in the conservation laws. These laws correspond to the conservation laws that were obtained earlier for nonlinear wave interaction.

From the conservation laws (4.29) and (4.30) we can easily find the angle $\Theta$ between $\mathbf{k}_1$ and $\mathbf{k}$ at which a longitudinal quantum can be emitted:

$$\cos\theta = \frac{\Omega^l(\mathbf{k})\,\Omega^t(\mathbf{k}_1)}{kk_1c^2} + \frac{k^2 - (\Omega^l(k))^2c^{-2}}{2kk_1}. \tag{4.31}$$

The expression for cos $\Theta$ is particularly simple for $k \ll k_1$, when the second term of (4.31) is small:

$$(\mathbf{k}\mathbf{v}_{gr}^t) = \Omega^l(\mathbf{k}); \quad \cos\theta = \frac{v_\varphi^l}{v_{gr}^t}. \tag{4.32}$$

Consequently, in this limit a longitudinal wave can be emitted by a transverse wave only if the phase velocity of the longitudinal wave is less than the group velocity of the transverse wave. The longitudinal-wave field in this case is essentially detached from the transverse-wave field, in the same way that in hydrodynamics sound is radiated by a body moving at a speed greater than that of sound (Mach effect).

* The fruitfulness of the quantum approach, especially regarding the interpretation of the Vavilov-Cerenkov effect, has been argued in detail in [5]. It is shown here and in the ensuing paragraphs that the quantum approach affords a very straightforward interpretation of nonlinear interactions.

Let us now consider the induced emission and absorption of longitudinal by transverse waves. We begin with the case for which the density of the emitting quanta $N_{\mathbf{k}}^{t}$ is much smaller than that of the emitted quanta $N_{\mathbf{k}}^{l}$. Since one emitted quantum is ascribed to each emitting quantum, $N_{\mathbf{k}}^{t}\,\Delta\mathbf{k}^{t} \sim N_{\mathbf{k}}^{l}\,\Delta\mathbf{k}^{l}$, which is possible for $|\Delta\mathbf{k}^{t}| \gg |\Delta\mathbf{k}^{l}|$ ($|\Delta\mathbf{k}| = \Delta k_x \Delta k_y \Delta k_z$), i.e., when the phase volume occupied by the transverse waves is much larger than that of the longitudinal waves. As we saw above, this requires that the transverse waves have frequencies much higher than the plasma frequency. The ratio of the phase volumes is then of order $(\Omega^{t}/\omega_{0e})^{3}$.

Notice that the nonlinear wave interactions are analogous to the two-level system specifically in the approximation $N_{\mathbf{k}}^{t} \ll N_{\mathbf{k}}^{l}$. The analogous situation for a two-level system is tantamount to neglect of the Pauli principle, which does not allow particles to enter a state in which there is already another particle. In other words, it is assumed that the atomic gas is very rarefied.

Let the quantum $\Omega_1^{t}$, $\mathbf{k}_1^{t}$ emit a longitudinal wave $\Omega^{l}$, $\mathbf{k}^{l}$, changing to another quantum $\Omega_2^{t}$, $\mathbf{k}_2^{t}$. The differential probability of this process per unit time, referred to an interval $d\mathbf{k}^{l}$ of the emitted quanta, is denoted by $w_{\mathbf{k}^{t}}(\mathbf{k}^{l})$. We write the equation for the balance of the induced processes, which is the analog of that for the two-level system. The number of longitudinal quanta $N_{\mathbf{k}}^{l}$ is increased by induced emission at the expense of quanta $\Omega_1^{t}$ and is decreased by induced absorption of quanta $\Omega_2^{t}$ [6]:

$$\frac{\partial N_{\mathbf{k}}^{l}}{\partial t} = \int w_{\mathbf{k}_1^{t}}(\mathbf{k})\, N_{\mathbf{k}}^{l} N_{\mathbf{k}_1}^{t} \frac{d\mathbf{k}_1}{(2\pi)^3} - \int w_{\mathbf{k}_1^{t}}(\mathbf{k})\, N_{\mathbf{k}}^{l} N_{\mathbf{k}_2}^{t} \frac{d\mathbf{k}_2}{(2\pi)^3}. \tag{4.33}$$

Here $w_{\mathbf{k}_1}(\mathbf{k}) N_{\mathbf{k}}^{l}$ is the probability of induced emission, and $N_{\mathbf{k}_1}^{t}\frac{d\mathbf{k}_1}{(2\pi)^3}$ is the number of transverse quanta in the interval $d\mathbf{k}_1$.

By the conservation of momentum $\mathbf{k}_2 = \mathbf{k}_1 - \mathbf{k}$, i.e.,

$$\frac{\partial N_{\mathbf{k}}^{l}}{\partial t} = N_{\mathbf{k}}^{l} \int w_{\mathbf{k}_1}^{t}(\mathbf{k})\,(N_{\mathbf{k}_1}^{t} - N_{\mathbf{k}_1-\mathbf{k}}^{t}) \frac{d\mathbf{k}_1}{(2\pi)^3}. \tag{4.34}$$

The earlier equation (3.50) has the following form for $N^{t} \ll N^{l}$:

$$\frac{\partial N_{\mathbf{k}}^{l}}{\partial t} = N_{\mathbf{k}}^{l} \int w_{\mathbf{k},\mathbf{k}_1,\mathbf{k}_2} d\mathbf{k}_1 d\mathbf{k}_2\,(N_{\mathbf{k}_1}^{t} - N_{\mathbf{k}_2}^{t})\,(2\pi)^{-6}. \tag{4.35}$$

Consequently, by comparison with $w_{\mathbf{k},\mathbf{k}_1,\mathbf{k}_2}$ we can write the exact expression for the emission probability:

$$\frac{w_{\mathbf{k}_1}(\mathbf{k})}{(2\pi)^3} = \int w_{\mathbf{k},\,\mathbf{k}_1,\,\mathbf{k}_2} d\mathbf{k}_2\,(2\pi)^{-6} =$$

$$= \frac{4\pi^2 e^2 \omega_{0e}^2 k^2 \hbar \left(1 + \frac{(\mathbf{k}-\mathbf{k}_1,\ \mathbf{k}_1)^2}{k^2(\mathbf{k}-\mathbf{k}_1)^2}\right)}{m_e^2 (\Omega^l(\mathbf{k}))^2 \left(\frac{\partial}{\partial\omega}\varepsilon^l(\omega,\ \mathbf{k})\right)\Big|_{\omega=\Omega^l(\mathbf{k})}} \times$$

$$\times \frac{\delta(\Omega^t(\mathbf{k}_1-\mathbf{k}) - \Omega^t(\mathbf{k}_1) + \Omega^l(\mathbf{k}))(2\pi)^{-3}}{\left(\frac{\partial}{\partial\omega}\omega^2\varepsilon^t(\omega,\ \mathbf{k}_1)\right)_{\omega=\Omega^t(\mathbf{k}_1)}\left(\frac{\partial}{\partial\omega}\omega^2\varepsilon^t(\omega,\ \mathbf{k}_1-\mathbf{k})\right)_{\omega=\Omega^t(\mathbf{k}_1-\mathbf{k})}}. \tag{4.36}$$

We thus arrive at a specific expression for the probability of emission of longitudinal waves by transverse waves (decay probability).

We now show how equation (3.50) can be deduced from elementary physical considerations [7]. For this we introduce the probability $\widetilde{w}$, referred to the interval $d\mathbf{k}_1 d\mathbf{k}_2/(2\pi)^6$.

This process of emission of a longitudinal by a transverse wave may be treated as follows: the transverse wave $\mathbf{k}_1^t$, on emitting $\mathbf{k}^l$, changes into a wave $\mathbf{k}_2^t$; consequently, the wave $\mathbf{k}_1^t$ is absorbed, and the waves $k^l$, $\mathbf{k}_2^t$ are emitted. The total probability of the process, allowing for spontaneous and induced emission and absorption, is obtained from $\widetilde{w}$, as explained earlier, by multiplying it by $N_\mathbf{k}+1$ for the emitted waves and by $N_\mathbf{k}$ for the absorbed waves. It has the form

$$\widetilde{w}\,(N_\mathbf{k}^l + 1)(N_{\mathbf{k}_2}^t + 1)\,N_{\mathbf{k}_1}^t. \tag{4.37}$$

In addition to the emission process there is the inverse process, in which a longitudinal wave $\mathbf{k}^l$ is absorbed and a transverse wave $\mathbf{k}_2^t$ is converted to $\mathbf{k}_1^t$, i.e., $\mathbf{k}_2^t$ is absorbed and $\mathbf{k}_1^t$ is emitted. The probability of this process is

$$\widetilde{w} N_\mathbf{k}^l N_{\mathbf{k}_2}^t (N_{\mathbf{k}_1}^t + 1). \tag{4.38}$$

In order to determine the rate of change of $N_\mathbf{k}^l$ we must subtract (4.38) from (4.37) and multiply the result by the number of possible states with a given $\mathbf{k}$, i.e., by $d\mathbf{k}_1/(2\pi)^3$, then to integrate (sum) over $\mathbf{k}_1$ and $\mathbf{k}_2$:

$$\frac{\partial N_\mathbf{k}^l}{\partial t} = \int \widetilde{w}\frac{d\mathbf{k}_1 d\mathbf{k}_2}{(2\pi)^6}((N_\mathbf{k}^l+1)(N_{\mathbf{k}_2}^t+1)\,N_{\mathbf{k}_1}^t - N_\mathbf{k}^l N_{\mathbf{k}_2}^t (N_{\mathbf{k}_1}^t + 1)) =$$

$$= \int \widetilde{w}\frac{d\mathbf{k}_1 d\mathbf{k}_2}{(2\pi)^6}(N_{\mathbf{k}_2}^t N_{\mathbf{k}_1}^t + N_\mathbf{k}^l N_{\mathbf{k}_1}^t - N_\mathbf{k}^l N_{\mathbf{k}_2}^t) + \int \widetilde{w}\frac{d\mathbf{k}_1 d\mathbf{k}_2}{(2\pi)^6} N_{\mathbf{k}_1}^t. \tag{4.39}$$

The first term of (4.39) describes the induced processes and corresponds to the earlier result (3.50). By comparison with the latter we have

$$\widetilde{w} = \hbar\, w_{\mathbf{k}\mathbf{k}_1\mathbf{k}_2} \text{ and } w_{\mathbf{k}_1}(\mathbf{k}) = \frac{1}{(2\pi)^3}\int \widetilde{w} d\mathbf{k}_2.$$

The other equations, corresponding to (3.51) and (3.52), are obtained similarly. It is important to remember here that the sign of the change of $N^t_{\mathbf{k}_1}$ is opposite to the sign of (4.39), because in this process $\mathbf{k}_1$ is absorbed; moreover, the integration must be carried out over all wave numbers for the given $\mathbf{k}_1$ (or $\mathbf{k}_2$ for $N^t_{\mathbf{k}_2}$).

## 6. SPONTANEOUS EMISSION OF WAVES BY WAVES

It is appropriate at this point to discuss the spontaneous emission of plasma waves by transverse waves. This effect is described by the last term of (4.39). The power emitted from 1 $cm^3$ of plasma is

$$Q^l = \int \frac{\hbar\Omega^l(\mathbf{k})}{(2\pi)^3} \cdot \frac{\partial N^l_{\mathbf{k}}}{\partial t} d\mathbf{k}.$$

Recognizing that for spontaneous emission

$$\frac{\partial N^l_{\mathbf{k}}}{\partial t} = \int w_{\mathbf{k}_1}(\mathbf{k})\, N^t_{\mathbf{k}_1} \frac{d\mathbf{k}_1}{(2\pi)^3},$$

we obtain

$$Q^l = \int \frac{\hbar\Omega^l(\mathbf{k})}{(2\pi)^6} w_{\mathbf{k}_1}(\mathbf{k})\, N^t_{\mathbf{k}_1} d\mathbf{k} d\mathbf{k}_1.$$

The power $Q^l$ is the sum of the emission intensities of the individual quanta:

$$Q^l = \int Q^l(\mathbf{k}_1)\, N^t_{\mathbf{k}_1} \frac{d\mathbf{k}_1}{(2\pi)^3};$$
$$Q^l(\mathbf{k}_1) = \int \hbar\Omega^l(\mathbf{k})\, w_{\mathbf{k}_1}(\mathbf{k}) \frac{d\mathbf{k}}{(2\pi)^3}. \quad (4.40)$$

Using the probability (4.36), it is easy to estimate the emitted power $Q(\mathbf{k}_1)$, for example, for high-frequency transverse quanta, $\Omega^t(\mathbf{k}_1) \gg \omega_{0e}$:

$$Q^l(\mathbf{k}_1) = \frac{\hbar^2 e^2 \omega_{0e}^2}{2m_e^2 (\Omega^t(\mathbf{k}_1))^2} \int\limits_{\omega_{0e}}^{k_{max}} k^3 dk \simeq \frac{\hbar^2 e^2 \omega_{0e}^2}{8m_e^2 \Omega^2(\mathbf{k}_1)} k_{max}^4 \simeq \frac{\hbar^2 e^2 \omega_{0e}^6}{8m_e^2 (\Omega(\mathbf{k}_1))^2} \frac{1}{v_{\varphi_{min}}^4} \qquad (4.41)$$

For the smallest possible $v_\varphi \sim v_{Te}$ (4.41) is to order of magnitude

$$Q^l \simeq \left(\frac{\hbar\omega_{0e}}{m_e c^2}\right)^2 \frac{e^2\omega_{0e}^2}{8c} \frac{\omega_{0e}^2}{\Omega_1^2} \frac{c^4}{v_{Te}^4}.$$

Notice that the quantum effect of spontaneous emission of a longitudinal by a transverse wave has been deduced by a classical calculation.*

## 7. ANALOGY WITH THE VAVILOV – CERENKOV EFFECT

The emission of waves by waves bears a strong analogy with the Cerenkov radiation of waves by particles. Thus, the role of the particles in this case is taken by the transverse waves. It is well known that Cerenkov radiation by particles is possible only if the phase velocity of the waves is smaller than the particle velocity. In this case the waves, in a manner of speaking, cannot keep up with the particles and become "detached" from them, i.e., they are radiated. The intensity of this radiation was first calculated by Tamm and Frank [8]. The Cerenkov radiation condition is easily deduced from the conservation of energy and momentum for the emission of a wave by a particle, for which the laws are entirely analogous to (4.29) and (4.30):

$$\mathbf{p}_1 = \mathbf{p} + \hbar\mathbf{k};$$

$$\varepsilon_{\mathbf{p}_1} = \varepsilon_{\mathbf{p}} + \hbar\omega, \qquad (4.42)$$

or

$$\varepsilon_{\mathbf{p}+\hbar\mathbf{k}} = \varepsilon_{\mathbf{p}} + \hbar\omega.$$

Since the momentum of the radiated wave is normally small compared with that of the particle $\hbar\mathbf{k} \ll \mathbf{p}$, we have

$$\varepsilon_{\mathbf{p}+\hbar\mathbf{k}} \simeq \varepsilon_{\mathbf{p}} + \hbar\mathbf{k}\frac{d\varepsilon_{\mathbf{p}}}{d\mathbf{p}} = \varepsilon_{\mathbf{p}} + \hbar\mathbf{k}\mathbf{v}, \qquad (4.43)$$

* The emission of waves by waves has also been treated in [9].

where $\mathbf{v}$ is the particle velocity. The Cerenkov condition is analogous to (4.32), where the role of the particle velocity is taken by the group velocity of the waves $d\omega/d\mathbf{k}$.

The particle velocity cannot exceed the velocity of light. Therefore, only those waves whose phase velocities are smaller than the velocity of light can be radiated. The transverse-wave phase velocity in an isotropic plasma is always larger than the velocity of light:

$$\frac{\omega}{k} = \frac{\sqrt{k^2c^2 + \omega_{0e}^2}}{k} > c,$$

hence only longitudinal waves can be radiated.

The transverse-wave group velocity in a plasma, as opposed to the phase velocity, is smaller than the velocity of light:

$$\mathbf{v}_{\text{gr}}^t = \frac{d\Omega^t}{d\mathbf{k}} = \frac{d\sqrt{k^2c^2 + \omega_{0e}^2}}{d\mathbf{k}} = \frac{\mathbf{k}}{k}\frac{kc}{\sqrt{k^2 + \frac{\omega_{0e}^2}{c^2}}}. \tag{4.44}$$

Consequently, transverse waves, like the particles in the analogous case, can emit only longitudinal waves. Although this result was deduced for $|\mathbf{k}| \ll |\mathbf{k}^t|$, it is equally valid for $|\mathbf{k}|$ comparable with $|\mathbf{k}^t|$.

## 8. THE KINETIC EQUATION FOR PLASMONS

Plasmon is the name given to any random plasma wave.* The kinetic equation for plasmons can be written using simple, straightforward considerations. The distribution function of the plasmons (quanta $N_{\mathbf{k}}^l$ and $N_{\mathbf{k}_1}^t$, $N_{\mathbf{k}_2}^t$ in the above example) is aptly represented by the quantum density $N_{\mathbf{k}}(\mathbf{r}, t)$. This quantity is caused to vary for two reasons: first, the quanta can move from one point in space to another, so that the number of quanta at a given point at a given time changes; secondly, on "colliding" with one another or gener-

* From now on we refer to any quanta in a plasma, including high-frequency transverse quanta, as plasmons. Even though in the linear approximation the plasma has scarcely any effect on high-frequency transverse quanta, $\Omega^t \gg \omega_{0e}$, we shall see in the nonlinear approximation (at large intensities) that their interaction with the plasma is no longer slight. In a roundabout way this justifies the name plasmon for any plasma quantum.

ating one another inductively, they can locally alter the direction and magnitude of their momenta, as well as the quantum density. The variation in the number of quanta due to generation is described precisely by expressions of the type (4.39). The variation due to translation, on the other hand, is easily determined by allowing for the fact that the quanta move with the wave group velocities, i.e.,

$$\frac{dN_{\mathbf{k}}}{dt}=\frac{\partial N_{\mathbf{k}}}{\partial t}+\mathbf{v}_{\mathrm{gr}}\frac{\partial N_{\mathbf{k}}}{\partial \mathbf{r}};\ \mathbf{v}_{\mathrm{gr}}=\frac{d\omega(\mathbf{k})}{d\mathbf{k}}. \tag{4.45}$$

Thus, if we neglect the spontaneous process [7], we obtain for longitudinal waves

$$\frac{dN_{\mathbf{k}}^{l}}{dt}=\frac{\partial N_{\mathbf{k}}^{l}}{\partial t}+\mathbf{v}_{\mathrm{gr}}^{l}\frac{\partial N_{\mathbf{k}}^{l}}{\partial \mathbf{r}}=$$
$$=\int\widetilde{w}(\mathbf{k},\mathbf{k}_1,\mathbf{k}_2)\frac{d\mathbf{k}_1 d\mathbf{k}_2}{(2\pi)^6}(N_{\mathbf{k}_1}^{t}N_{\mathbf{k}_2}^{t}+N_{\mathbf{k}}^{l}N_{\mathbf{k}_1}^{t}-N_{\mathbf{k}}^{l}N_{\mathbf{k}_2}^{t}). \tag{4.46}$$

This is precisely the equation (with the left-hand side including the change in the number of quanta as they move in space) we obtained by direct averaging of the nonlinear equations over the random phases.

An additional comment is required, of course, concerning the form of the right-hand side of (4.46). We dealt above by and large with the case when three wave packets were involved. However, the method used to derive (4.46) shows that this equation is not restricted to this case. For example, if $\Omega_1$ is the frequency of some initial transverse wave, we have considered only the effects of induced emission of a wave $\omega_{0e}$ with transition to $\Omega_2=\Omega_1-\omega_{0e}$. If, however, there were waves in the plasma capable of absorbing the wave $\Omega_1$, this would lead to the generation of waves with frequency $\Omega_3=\Omega_1+\omega_{0e}$. The new wave, in turn, could emit and absorb plasma waves, i.e., in general it is necessary to consider the interaction of a great many wave packets. In order to analyze the entire process we need know only how each of the frequencies varies, i.e., account for both emission and absorption for the frequency $\Omega_1$.

It is not difficult to find the probability of absorption by the wave $\mathbf{k}_1$. Thus, we know that the probability $\widetilde{w}(\mathbf{k},\mathbf{k}_1,\mathbf{k}_2)$ of emission of the wave $\mathbf{k}$ by $\mathbf{k}_1$ with transition to $\mathbf{k}_2$ is equal to the probability of the inverse process of absorption of $\mathbf{k}$ by $\mathbf{k}_2$ with transition to $\mathbf{k}_1$. In other words, the probability of absorption by $\mathbf{k}_1$ with transition to $\mathbf{k}_3$ is $\widetilde{w}(\mathbf{k}, \mathbf{k}_3, \mathbf{k}_1)$. Along with processes involving transitions from $\mathbf{k}_1$ to $\mathbf{k}_2$, we must include the processes of transition from $\mathbf{k}_1$ to $\mathbf{k}_3$:

$$\frac{dN_{\mathbf{k}}^l}{dt} = \int \widetilde{w}(\mathbf{k}, \mathbf{k}_3, \mathbf{k}_1)\,(N_{\mathbf{k}_1}^t N_{\mathbf{k}_3}^t + N_{\mathbf{k}}^l N_{\mathbf{k}_3}^t - N_{\mathbf{k}}^l N_{\mathbf{k}_1}^t)\,\frac{d\mathbf{k}_1\,d\mathbf{k}_3}{(2\pi)^6}. \tag{4.47}$$

It is easily seen that the substitution $3 \to 1$, $1 \to 2$ reduces (4.47) to (4.46). In other words, if we denote by $N_{\mathbf{k}}^t$ the total distribution function accounting for all wave beams (this function being equal to the corresponding values of $N_{\mathbf{k}_1}$, $N_{\mathbf{k}_2}$, and $N_{\mathbf{k}_3}$ inside the beams), then Eq. (4.46), which includes integration over all possible $\mathbf{k}_1$ and $\mathbf{k}_2$, automatically takes account of the contribution of (4.47), i.e., both the emission and the absorption of longitudinal waves by transverse waves.

But if we try to apply the same reasoning to the equations for $N_{\mathbf{k}_1}^t$ and $N_{\mathbf{k}_2}^t$, we do not succeed, because these equations do not include integration either over $\mathbf{k}_1$ or over $\mathbf{k}_2$. In other words, the wave interaction is not analogous to induced processes in a two-level system, but in a multilevel system.

Let us see how the equations for transverse waves are subsequently modified. We write the result for three wave packets:

$$\frac{dN_{\mathbf{k}_1}^t}{dt} = -\int \widetilde{w}(\mathbf{k}, \mathbf{k}_1, \mathbf{k}_2)\,\frac{d\mathbf{k}\,d\mathbf{k}_2}{(2\pi)^6}\,(N_{\mathbf{k}_1}^t N_{\mathbf{k}_2}^t - N_{\mathbf{k}_2}^t N_{\mathbf{k}}^l + N_{\mathbf{k}_1}^t N_{\mathbf{k}}^l);$$

$$\frac{dN_{\mathbf{k}_2}^t}{dt} = \int \widetilde{w}(\mathbf{k}, \mathbf{k}_1, \mathbf{k}_2)\,\frac{d\mathbf{k}\,d\mathbf{k}_1}{(2\pi)^6}\,(N_{\mathbf{k}_1}^t N_{\mathbf{k}_2}^t - N_{\mathbf{k}_2}^t N_{\mathbf{k}}^l + N_{\mathbf{k}_1}^t N_{\mathbf{k}}^l).$$

Both equations account only for the emission of $\mathbf{k}_1$ and absorption of $\mathbf{k}_2$, whereas the absorption of $\mathbf{k}_1$ is described by the relation

$$\frac{dN_{\mathbf{k}_1}^t}{dt} = \int \widetilde{w}(\mathbf{k}, \mathbf{k}_3, \mathbf{k}_1)\,\frac{d\mathbf{k}\,d\mathbf{k}_3}{(2\pi)^6}\,(N_{\mathbf{k}_1}^t N_{\mathbf{k}_3}^t + N_{\mathbf{k}}^l N_{\mathbf{k}_3}^t - N_{\mathbf{k}}^l N_{\mathbf{k}_1}^t).$$

The general equation for the total $N_{\mathbf{k}}^t$, taking account of both absorption and emission, becomes

$$\begin{aligned}\frac{dN_{\mathbf{k}_1}^t}{dt} = \int [&\widetilde{w}(\mathbf{k}, \mathbf{k}_2, \mathbf{k}_1)\,(N_{\mathbf{k}_1}^t N_{\mathbf{k}_2}^t + N_{\mathbf{k}}^l N_{\mathbf{k}_2}^t - N_{\mathbf{k}}^l N_{\mathbf{k}_1}^t) - \\ &- \widetilde{w}(\mathbf{k}, \mathbf{k}_1, \mathbf{k}_2)\,(N_{\mathbf{k}_1}^t N_{\mathbf{k}_2}^t + N_{\mathbf{k}}^l N_{\mathbf{k}_1}^t - N_{\mathbf{k}}^l N_{\mathbf{k}_2}^t)]\,\frac{d\mathbf{k}\,d\mathbf{k}_2}{(2\pi)^6}.\end{aligned} \tag{4.48}$$

## 9. INDUCED COMBINATION SCATTERING

The physical meaning of the nonlinear interactions discussed above can also be elucidated by invoking the well-known "combination" scattering effect discovered by Mandel'shtam, Landsberg, and Raman. In this situation we are concerned with combination scattering by plasma waves, particularly by longitudinal waves. Since the plasma-wave frequency is $\omega_{0e}$, the combination frequencies of the scattered satellite waves must be

$$\Omega^t \pm \omega_{0e}. \tag{4.49}$$

We have seen that when the absorption and emission of plasma waves are considered, each wave can generate satellite waves at the frequencies (4.49). We should recall that in Raman scattering the satellite intensities are roughly equal. This can happen only in special cases.

Let the intensity of a scattered wave of momentum $\mathbf{k}_2$ be so small that the intensities of the satellites are much smaller than the incident intensity. Then in (4.48) we can drop terms containing $N^t_{\mathbf{k}_1}$ :

$$\frac{dN^t_{\mathbf{k}_1}}{dt} = \int \{\widetilde{w}(\mathbf{k}, \mathbf{k}_2, \mathbf{k}_1) N^l_{\mathbf{k}} N^t_{\mathbf{k}_2} + \widetilde{w}(\mathbf{k}, \mathbf{k}_1, \mathbf{k}_2) N^l_{\mathbf{k}} N^t_{\mathbf{k}_2}\} \frac{d\mathbf{k}\, d\mathbf{k}_2}{(2\pi)^6}. \tag{4.50}$$

The first term describes the appearance of the red satellite, and the second describes the blue satellite. For thermal noise $N^l_{\mathbf{k}} \simeq T/\hbar\omega_{0e}$ is practically independent of $\mathbf{k}$, and the intensities of the two satellites are equal.

With an increase in the intensity of the scattered wave the situation is radically altered. The scattered waves begin to generate plasma waves, and the scattering process is no longer determined by the thermal noise, but by the plasma oscillations. It must be recognized here that an intense wave is analogous to a negative-temperature system.

In this case, as a rule, asymmetry develops in the satellite intensities. This is easily understood. The mere act of the generation of oscillations shows that the induced emission of longitudinal waves, resulting in the formation of the red satellite, prevails over induced absorption, which results in the formation of the blue satellite. Consequently, the red satellite turns out to be stronger than the blue satellite. Moreover, when $N^l$ becomes large due to generation we get stronger scattering. The first red satellite generates a second one, and so on; instability develops with the formation of a large number of satellites.

The process in which waves are scattered by the oscillations generated by the scattered waves is properly called induced combination scattering. This concept is identical with that represented by the term "decay process" often used in plasma physics.

## Literature Cited

1. A.I. Akhiezer and V.B. Berestetskii, Quantum Electrodynamics. Wiley, New York (1965).
2. R. Karplus and J. Schwinger, "A note on saturation in microwave spectroscopy," Phys. Rev., 73:1020 (1948).
3. N.G. Basov and A.M. Prokhorov, "The molecular oscillator and amplifier," Uspekhi Fiz. Nauk, 57:485 (1955).
4. A.N. Oraevskii, Molecular Oscillators. Izd. "Nauka" (1964).
5. V.L. Ginzburg, "Aspects of the theory of radiation at velocities above the velocity of light," Uspekhi Fiz. Nauk, 69:537 (1959).
6. V.N. Tsytovich, "Resonance induced scattering and emission in a medium," Dokl. Akad. Nauk SSSR, 154:76 (1964).
7. M. Camac, et al., "Shock waves in collision-free plasmas," Nucl. Fusion Suppl., 2:423 (1962).
8. I.E. Tamm and I.M. Frank, "Coherent radiation of a fast electron in a medium," Dokl. Akad. Nauk SSSR, 14:107 (1937).
9. G.A. Askar'yan, "Cerenkov and transition radiation from electromagnetic waves," Zh. Éksp. Teor. Fiz., 42:1360 (1962).

*Chapter V*

# General Characteristics and Applications of Decay Processes in Plasma

## 1. DECAY PROCESSES

Processes of induced emission of waves by other waves belong to the general category of what are known as decay processes. This term is used to denote nonlinear interactions of waves in which one wave decays into other waves. In particular, the emission of a longitudinal wave by a transverse wave may be regarded as the decay of a transverse wave into two waves, one longitudinal and one transverse.

The kinetic equation describing this type of interaction accounts for both this process and its inverse, in which a transverse wave coalesces with a longitudinal wave to form a new transverse

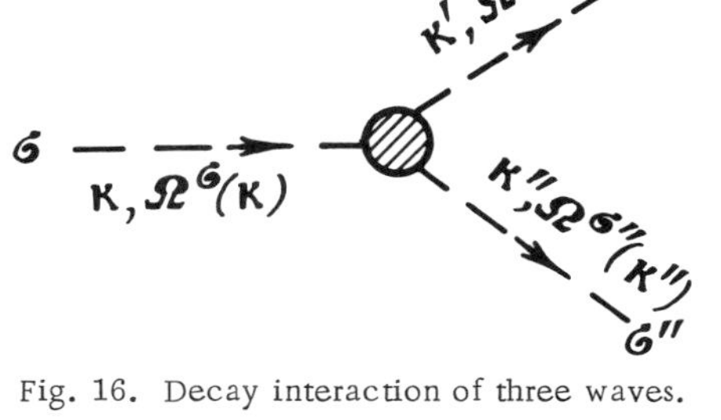


Fig. 16. Decay interaction of three waves.

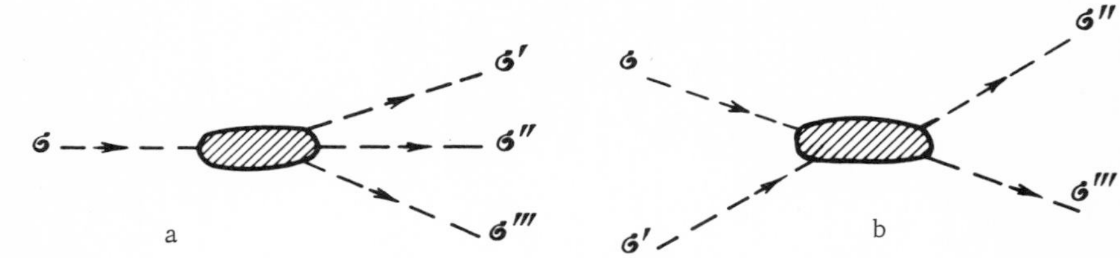


Fig. 17. Four-plasmon decay interactions: a) Emission of three waves from one; b) scattering of waves by waves.

wave. In general, we shall think of decay processes in terms of both the direct and inverse forms (except when the problem specifically states that only processes in one direction are to be considered).

An elementary decay process is illustrated graphically in Fig. 16, which also shows the values of the momenta and energies of the interacting waves; the index $\sigma$ refers to the wave type. In particular, the problem can involve two transverse and one longitudinal wave, as analyzed above, but other combinations are possible. In, for example, a plasma in an external magnetic field $\Omega^{\sigma}(\mathbf{k})$ is known to have a great many branches, describing different waves. The remainder of the discussion refers to any such wave in plasma and in other media.

Figure 16 illustrates the simplest wave interaction process. In principle more complex processes are possible, in which one of the waves decays into three (Fig. 17a) or two waves are transformed into two other (Fig. 17b). Naturally, the only way the second process differs from the first is that one of the waves (for example, $\sigma'$) is absorbed rather than emitted. If $\sigma'' = \sigma$ and $\sigma''' = \sigma'$, the process shown in Fig. 17b may be interpreted as the effect of nonlinear scattering of the waves $\sigma$ by $\sigma'$. In general this kind of scattering can be accompanied by the conversion of one kind of wave into others. Sometimes the processes illustrated in Fig. 16 are called three-plasmon decays, those in Fig. 17 four-plasmon decays.

In elementary decay processes the conservation of energy and momentum must be observed. Thus, for three-plasmon decays it is required that

$$\Omega^{\sigma}(\mathbf{k}) = \Omega^{\sigma'}(\mathbf{k}') + \Omega^{\sigma''}(\mathbf{k}''); \tag{5.1}$$

$$\mathbf{k} = \mathbf{k}' + \mathbf{k}'', \tag{5.2}$$

and for the four-plasmon decay processes shown in Fig. 17a

$$\Omega^{\sigma}(\mathbf{k}) = \Omega^{\sigma'}(\mathbf{k}') + \Omega^{\sigma''}(\mathbf{k}'') + \Omega^{\sigma'''}(\mathbf{k}'''); \quad \mathbf{k} = \mathbf{k}' + \mathbf{k}'' + \mathbf{k}'''. \tag{5.3}$$

These conservation laws cannot always be satisfied for one particular type. For example, it is seen immediately that a longitudinal Langmuir wave cannot decay into two Langmuir waves. This follows from the fact that the frequency of a Langmuir wave is near $\omega_{0e}$, hence the energy of the system before decay is $\omega_{0e}$, and after decay it would be near $2\omega_{0e}$. Consequently, the conservation of energy would be violated.

Three-plasmon decay processes are described by equations whose structure is precisely the same as those derived in the preceding chapter. In fact, the probabilistic approach used in the derivation of the equations is equally well suited to any quanta, and any differences in the interactions are attributable simply to different values of the decay probabilities, as well as to the different dispersion characteristics of the oscillations, which forbid or allow decays for certain regions of the wave spectra.

For instance, let us write the equations for the emittèd wave $\sigma'$. We introduce the probability $w_{\sigma}^{\sigma', \sigma''}(\mathbf{k}, \mathbf{k}', \mathbf{k}'')$ for decay of the wave $\sigma$ into $\sigma'$ and $\sigma''$. Taking account of the induced emission of $\sigma'$ and $\sigma''$, the induced absorption of $\sigma$, and their inverse processes, we obtain

$$\frac{dN_{\mathbf{k}'}^{\sigma'}}{dt} = \int w_{\sigma}^{\sigma', \sigma''}(\mathbf{k}, \mathbf{k}', \mathbf{k}'') \frac{d\mathbf{k}\, d\mathbf{k}''}{(2\pi)^6} (N_{\mathbf{k}}^{\sigma} N_{\mathbf{k}''}^{\sigma''} + N_{\mathbf{k}'}^{\sigma'} N_{\mathbf{k}}^{\sigma} - N_{\mathbf{k}'}^{\sigma'} N_{\mathbf{k}''}^{\sigma''}). \tag{5.4}$$

This equation differs from (4.46) only in notation.

The equations for four-plasmon processes are deduced similarly. We can show, for example, how the equation is obtained for the decay process in Fig. 17b. Let the probability of this process be $w_{\sigma,\sigma'}^{\sigma'',\sigma'''}(\mathbf{k}, \mathbf{k}', \mathbf{k}'', \mathbf{k}''')$. We write the equations for the quanta $\sigma$. In this process the waves $\sigma$ and $\sigma'$ are absorbed, while $\sigma''$ and $\sigma'''$ are emitted. The decrease in the number of waves $\sigma$ is

$$\int w_{\sigma,\sigma'}^{\sigma'',\sigma'''}(\mathbf{k}, \mathbf{k}', \mathbf{k}'', \mathbf{k}''') N_{\mathbf{k}}^{\sigma} N_{\mathbf{k}'}^{\sigma'} \times (N_{\mathbf{k}''}^{\sigma''} + 1)(N_{\mathbf{k}'''}^{\sigma'''} + 1) \frac{d\mathbf{k}'\, d\mathbf{k}''\, d\mathbf{k}'''}{(2\pi)^9}, \tag{5.5}$$

and their increase due to the inverse process is

$$\int w_{\sigma,\sigma'}^{\sigma'',\sigma'''}(\mathbf{k},\mathbf{k}',\mathbf{k}'',\mathbf{k}''')N_{\mathbf{k}''}^{\sigma''}N_{\mathbf{k}'''}^{\sigma'''}(N_{\mathbf{k}}^{\sigma}+1)(N_{\mathbf{k}'}^{\sigma'}+1)\frac{d\mathbf{k}'\,d\mathbf{k}''\,d\mathbf{k}'''}{(2\pi)^9}. \tag{5.6}$$

The total change in $N_{\mathbf{k}}^{\sigma}$ is determined by the difference between (5.6) and (5.5):

$$\frac{dN_{\mathbf{k}}^{\sigma}}{dt}=\int w_{\sigma,\sigma'}^{\sigma'',\sigma'''}(\mathbf{k},\mathbf{k}',\mathbf{k}'',\mathbf{k}''')(N_{\mathbf{k}}^{\sigma}N_{\mathbf{k}''}^{\sigma''}N_{\mathbf{k}'''}^{\sigma'''}+N_{\mathbf{k}'}^{\sigma'}N_{\mathbf{k}''}^{\sigma''}N_{\mathbf{k}'''}^{\sigma'''}-$$
$$-N_{\mathbf{k}}^{\sigma}N_{\mathbf{k}'}^{\sigma'}N_{\mathbf{k}''}^{\sigma''}-N_{\mathbf{k}}^{\sigma}N_{\mathbf{k}'}^{\sigma'}N_{\mathbf{k}'''}^{\sigma'''})\frac{d\mathbf{k}'\,d\mathbf{k}''\,d\mathbf{k}'''}{(2\pi)^9}. \tag{5.7}$$

The equations for $N_{\mathbf{k}'}^{\sigma'}$, etc., are found in a similar way.

## 2. THE DIFFUSION APPROXIMATION

The whole picture of decay processes becomes very simple when the energy and momentum (frequency and wave number) of certain quanta are much smaller than the energy and momentum of other quanta. Consider the three-plasmon decay of a high-frequency wave $\lambda$ into a low-frequency wave $\sigma$. The conservation of energy requires the existence of another high-frequency wave; let this wave be of the same type as the emitting wave $\lambda$.

As we explained in the example of the decay process $t \to l + t$, terms containing the product of the decaying wave quanta $(N^t N^t)$ can be neglected. For the waves $\sigma$, therefore, we can write

$$\frac{dN_{\mathbf{k}}^{\sigma}}{dt}=N_{\mathbf{k}}^{\sigma}\int w_{\lambda}^{\sigma}(\mathbf{k}_1)(N_{\mathbf{k}_1}^{\lambda}-N_{\mathbf{k}_1-\mathbf{k}}^{\lambda})\frac{d\mathbf{k}_1}{(2\pi)^3}. \tag{5.8}$$

Here we have made use of the following relation, dictated by the conservation law (5.2), between the momentum $\mathbf{k}_2$ of the quantum $\lambda$ after the emission of $\sigma$ and the original momentum $\mathbf{k}_1$ of $\lambda$:

$$\mathbf{k}_2=\mathbf{k}_1-\mathbf{k}. \tag{5.9}$$

This allows us to replace $\mathbf{k}_2$ by $\mathbf{k}_1 - \mathbf{k}$ in the second term of (5.8) and to integrate over $\mathbf{k}_1$. In (5.8) we have introduced the notation

$$w_{\lambda}^{\sigma}(\mathbf{k}_1)=\int w_{\lambda}^{\sigma\lambda}(\mathbf{k}_1,\mathbf{k},\mathbf{k}_2)\frac{d\mathbf{k}_2}{(2\pi)^3}. \tag{5.10}$$

If the width $\delta\omega^{\lambda}$ of the spectrum of the emitting waves is much larger than the frequency of the emitted waves $\sigma$, then, provided $\mathbf{k} \ll \mathbf{k}_1$, we can assume that

$$N^{\lambda}_{\mathbf{k}_1-\mathbf{k}} \simeq N^{\lambda}_{\mathbf{k}_1} - \mathbf{k}\frac{\partial N^{\lambda}_{\mathbf{k}_1}}{\partial \mathbf{k}_1}, \tag{5.11}$$

i.e.

$$\frac{dN^{\sigma}_{\mathbf{k}}}{dt} = N^{\sigma}_{\mathbf{k}} \int w^{\sigma}_{\lambda}(\mathbf{k}_1)\left(\mathbf{k}\frac{\partial N^{\lambda}_{\mathbf{k}_1}}{\partial \mathbf{k}_1}\right)\frac{d\mathbf{k}_1}{(2\pi)^3}. \tag{5.12}$$

Consequently, the nonlinear growth rate for wave generation depends on the derivative of the wave number (momentum) distribution function of the emitting waves. The interesting thing is that for a directional beam of emitting waves the growth rate is simply proportional to $dN_{\mathbf{k}_1}/d\mathbf{k}_1$, where $\mathbf{k}_1$ is the wave number in the direction of propagation of the beam. This is apparent from the fact, stipulated by the law of conservation of $\delta(\Omega^{\sigma} - \mathrm{k}v^{\lambda}_{\mathrm{gr}})$, that in the direction of the beam the only $\sigma$ waves generated will be those whose phase velocity coincides with the group velocity of the waves $\lambda$, so that the only term left from the sum (5.12) is the one satisfying this condition. Owing to the unique relation between $k_1$ and $\omega_1$ the derivative $dN_1/dk_1$ is proportional to $dN_1/d\omega_1$ (the signs of the derivatives are the same for normal dispersion $d\omega/dk > 0$). Hence, for example, an intense spectral line whose width is much greater than $\omega_{0e}$ can, in general, generate plasma waves from that part of the spectrum in which the derivative of the intensity with respect to the frequency is positive (Fig. 18).

Let us now see how the distribution of the emitted waves changes. From an equation analogous to (4.48), neglecting the terms $N^{\lambda} N^{\lambda}$, we obtain (omitting the indices $\sigma$ and $\lambda$ in w from now on)

$$\frac{dN^{\lambda}_{\mathbf{k}_1}}{dt} = \int [\widetilde{w}(\mathbf{k}_2, \mathbf{k}, \mathbf{k}_1)(N^{\sigma}_{\mathbf{k}}N^{\lambda}_{\mathbf{k}_2} - N^{\sigma}_{\mathbf{k}}N^{\lambda}_{\mathbf{k}_1}) -$$
$$- \widetilde{w}(\mathbf{k}_1, \mathbf{k}, \mathbf{k}_2)(N^{\sigma}_{\mathbf{k}}N^{\lambda}_{\mathbf{k}_1} - N^{\sigma}_{\mathbf{k}}N^{\lambda}_{\mathbf{k}_2})]\frac{d\mathbf{k}d\mathbf{k}_2}{(2\pi)^6}. \tag{5.13}$$

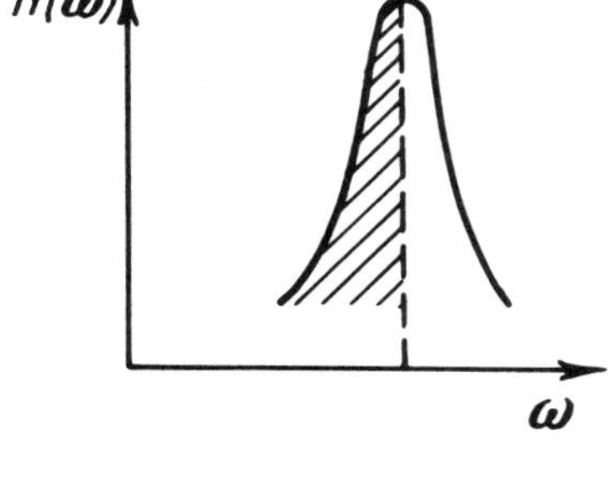


Fig. 18. Instability of transverse waves in a plasma. The hatched region indicates the part of the wave packet spectrum in which instability is possible.

Equation (5.13) is easily modified, assuming $k \ll k_1$. We note that $\tilde{w}(\mathbf{k}, \mathbf{k}_2, \mathbf{k}_1)$ is proportional to $\delta(\mathbf{k} - \mathbf{k}_2 + \mathbf{k}_1)$, i.e.,

$$\tilde{w}(\mathbf{k}_1, \mathbf{k}, \mathbf{k}_2) = (2\pi)^3 w_{\mathbf{k}\mathbf{k}_1\mathbf{k}_2} \delta(\mathbf{k} - \mathbf{k}_1 + \mathbf{k}_2).$$

After integration over $\mathbf{k}_2$, then, we write

$$\frac{\partial N^{\lambda}_{\mathbf{k}_1}}{dt} = \frac{1}{(2\pi)^3} \int [w_{\mathbf{k}, \mathbf{k}_1+\mathbf{k}, \mathbf{k}_1} (N^{\lambda}_{\mathbf{k}_1+\mathbf{k}} - N^{\lambda}_{\mathbf{k}_1}) - - w_{\mathbf{k}, \mathbf{k}_1, \mathbf{k}_1-\mathbf{k}} (N^{\lambda}_{\mathbf{k}_1} - N^{\lambda}_{\mathbf{k}_1-\mathbf{k}})] N^{\sigma}_{\mathbf{k}} d\mathbf{k}. \quad (5.14)$$

We recognize $w_{\mathbf{k}, \mathbf{k}_1, \mathbf{k}_1-\mathbf{k}}$ as none other than the quantity introduced above

$$w_{\mathbf{k}_1}(\mathbf{k}) = \int \tilde{w} \frac{d\mathbf{k}_2}{(2\pi)^3} = \int w_{\mathbf{k}, \mathbf{k}_1, \mathbf{k}_2} \delta(\mathbf{k} - \mathbf{k}_1 + \mathbf{k}_2) d\mathbf{k}_2.$$

It is also seen at once that $w_{\mathbf{k}, \mathbf{k}_1+\mathbf{k}, \mathbf{k}_1} = w_{\mathbf{k}_1+\mathbf{k}}(\mathbf{k})$. Consequently, Eq. (5.14) reduces to a form amenable to expansion in $\mathbf{k}$:

$$\frac{\partial N^{\lambda}_{\mathbf{k}_1}}{\partial t} = \int [w_{\mathbf{k}_1+\mathbf{k}}(\mathbf{k}) (N^{\lambda}_{\mathbf{k}_1+\mathbf{k}} - N^{\lambda}_{\mathbf{k}_1}) - - w_{\mathbf{k}_1}(\mathbf{k}) (N^{\lambda}_{\mathbf{k}_1} - N^{\lambda}_{\mathbf{k}_1-\mathbf{k}})] N^{\sigma}_{\mathbf{k}} \frac{d\mathbf{k}}{(2\pi)^3}. \quad (5.15)$$

Substituting into (5.15) the expansions

$$N^{\lambda}_{\mathbf{k}_1 \pm \mathbf{k}} = N^{\lambda}_{\mathbf{k}_1} \pm k_i \frac{\partial N^{\lambda}_{\mathbf{k}_1}}{\partial k_{1i}} + \frac{1}{2} k_i k_j \frac{\partial^2 N^{\lambda}_{\mathbf{k}_1}}{\partial k_{1i} \partial k_{1j}};$$

$$w_{\mathbf{k}_1+\mathbf{k}}(\mathbf{k}) = w_{\mathbf{k}_1}(\mathbf{k}) + k_i \frac{\partial w_{\mathbf{k}_1}(\mathbf{k})}{\partial k_{1i}}$$

and including only the first nonvanishing second-order terms in $k_i$, we obtain

$$\frac{dN^{\lambda}_{\mathbf{k}_1}}{dt} = \int \left( k_i \frac{\partial w_{\mathbf{k}_1}(\mathbf{k})}{dk_{1i}} k_j \frac{\partial N^{\lambda}_{\mathbf{k}_1}}{\partial k_{1j}} + w_{\mathbf{k}_1}(\mathbf{k}) k_i k_j \frac{\partial^2 N^{\lambda}_{\mathbf{k}_1}}{\partial k_{1i} \partial k_{1j}} \right) N^{\sigma}_{\mathbf{k}} \frac{d\mathbf{k}}{(2\pi)^3} =$$

$$= \frac{\partial}{\partial k_{1i}} D_{ij} \frac{\partial N^{\lambda}_{\mathbf{k}_1}}{\partial k_{1j}}; \qquad D_{ij} = \int k_i k_j N^{\sigma}_{\mathbf{k}} w_{\mathbf{k}_1}(\mathbf{k}) \frac{d\mathbf{k}}{(2\pi)^3}. \quad (5.16)$$

Equation (5.16) is a diffusion equation. The diffusion coefficient $D_{ij}$ depends on the wave intensity. Therefore, to put it the other way around, the quanta $\lambda$ diffuse into the field of the quanta $\sigma$. This diffusion leads, on the one hand, to a change in the direction of the momentum of the quanta $\lambda$, i.e., to scattering, and, on the other hand, to a change in the magnitude of the momentum and, hence, of the energy (frequency) of $\lambda$. The generation or absorption of the waves $\sigma$ by $\lambda$, as described by (5.12), changes the energy density of the waves by which the quanta $\lambda$ are scattered.

The resulting form of the equations for the decay interactions provides a basis for qualitative judgements and estimates without actually having to solve the equations. Thus, for example, it may be inferred from the form of (5.12) that an anisotropic distribution of the waves $N^{\lambda}_{\mathbf{k}_1}$ is liable to become unstable. Let $N^{\lambda}_{\mathbf{k}_1}$ depend on the modulus of $\mathbf{k}_1$ and on the angle $\Theta$ of the vector $\mathbf{k}_1$ relative to some direction: $N^{\lambda}_{\mathbf{k}_1} = N^{\lambda}(k_1, \theta)$. The growth rate (5.12) contains $dN^{\lambda}/d\mathbf{k}$ and in the given instance also $dN^{\lambda}/d\Theta$, i.e., the nonlinear growth rate depends on the degree of anisotropy of the waves $N^{\lambda}$. The characteristic buildup time of the instability is determined by the nonlinear growth rate (5.12). Moreover, the diffusion (5.16) leads to scattering and broadening of the spectrum of the waves $\lambda$, and the scattering, in particular, can render the distribution of $\lambda$ isotropic. For example, a transverse wave transmitted through a plasma in which longitudinal waves have been excited is subjected to scattering, and its spectrum is broadened. The characteristic scattering time may be estimated from the relation

$$\tau \simeq \frac{(\Delta k)^2}{D}, \tag{5.17}$$

where $\Delta k$ is the change in the wave number.

It should also be noted that the diffusion approximation can be used for more complex decay interactions, for example, four-plasmon processes. In the diffusion approximation we can write an expression for the scattering of longitudinal by transverse waves:

$$\frac{dN^{l}_{\mathbf{k}_1}}{\partial t} = N^{l}_{\mathbf{k}_1} \int w^{t,\,l}_{t,\,l}(\mathbf{k}_1, \mathbf{k}_2, \mathbf{k})(\mathbf{k}_1 - \mathbf{k}_2) \frac{\partial N^{t}_{\mathbf{k}}}{\partial \mathbf{k}} N^{l}_{\mathbf{k}_2} \frac{d\mathbf{k}\, d\mathbf{k}_2}{(2\pi)^6}. \tag{5.18}$$

Here

$$w^{t,\,l}_{t,\,l}(\mathbf{k}_1, \mathbf{k}_2, \mathbf{k}) = \int w^{t,\,l}_{t,\,l}(\mathbf{k}_1, \mathbf{k}_2, \mathbf{k}, \mathbf{k}') \frac{d\mathbf{k}'}{(2\pi)^3};$$

$w^{t,l}_{t,l}(\mathbf{k}_1, \mathbf{k}_2, \mathbf{k}, \mathbf{k}')$ is the scattering probability, $\mathbf{k}_1$ and $\mathbf{k}_2$ are the longitudinal-wave vectors, and $\mathbf{k}$ and $\mathbf{k}'$ are the transverse-wave vectors.

## 3. CONSERVATION LAWS FOR DECAY PROCESSES

The conservation laws for elementary emission and absorption events lead to the conservation of total energy and momentum of the quanta. Introducing the energy of the quanta

$$W^{\sigma}=\int\frac{\hbar\omega^{\sigma}(\mathbf{k})\,N_{\mathbf{k}}^{\sigma}}{(2\pi)^{3}}\,d\mathbf{k}, \tag{5.19}$$

it is found at once that for the process illustrated in Fig. 16

$$W^{\sigma}+W^{\sigma'}+W^{\sigma''}=\text{const.} \tag{5.20}$$

Introducing the momentum of the quanta

$$\mathbf{P}^{\sigma}=\int\frac{\hbar\mathbf{k}N_{\mathbf{k}}^{\sigma}}{(2\pi)^{3}}\,d\mathbf{k}, \tag{5.21}$$

we obtain for the same process

$$\mathbf{P}^{\sigma''}+\mathbf{P}^{\sigma'}+\mathbf{P}^{\sigma}=\text{const.} \tag{5.22}$$

These laws show that the interaction in decay processes is such that the quanta exchange energy and momentum only with each other.

Of course, the conservation laws (5.20) and (5.22) are also true for the diffusion approximation discussed above. However, when a quantum of the same type as the original one ($\lambda$-quantum) appears after decay, another conservation law is satisfied, namely that of the total number of quanta of type $\lambda$:

$$N^{\lambda}=\int N_{\mathbf{k}}^{\lambda}\,\frac{d\mathbf{k}}{(2\pi)^{3}}=\text{const.} \tag{5.23}$$

This conservation law is derived easily either from the exact equation or from the approximate diffusion relation (5.16). Since the spectrum $N_{\mathbf{k}}^{\lambda}$, as seen earlier, is broadened, the quantum density must decrease, according to (5.23). The conservation law (5.23) is a simple consequence of the fact that in every emission or absorption of a quantum $\sigma$ a quantum $\lambda$ is neither created nor destroyed. This law has powerful physical implications.

Let us consider, for example, the decay of a Langmuir wave into another Langmuir wave and a low-frequency wave, where the frequency of the latter is much lower than the Langmuir-wave frequency $\omega_{0e}$. According to (5.23) the total number of Langmuir waves in this process does not change. Since the frequency of those waves depends only very slightly on the wave number, $\Omega^l(\mathbf{k}) \approx \omega_{0e}$, it can be verified by multiplying (5.23) by $\omega_{0e}$ that the energy of the Langmuir waves is approximately conserved. This implies that this particular decay process cannot produce a significant change in the Langmuir-wave energy.

## 4. PROBABILITIES OF DECAY PROCESSES

The number of branches for the various oscillation modes in a plasma is very large. Their nonlinear decay interaction is completely determined by the appropriate probabilities. There is a straightforward general method for finding them (Appendix 3), based on the correspondence principle [1]. The general decay equations can be used in this case to calculate the power radiated in the emission of a third wave from two primaries in the limit of small intensity of the third wave, when the decay interaction equation takes the form

$$\frac{\partial N_{\mathbf{k}}^{\sigma}}{\partial t} = \int w_{\sigma}^{\sigma',\sigma''}(\mathbf{k}, \mathbf{k}_1, \mathbf{k}_2)\, N_{\mathbf{k}_1}^{\sigma'} N_{\mathbf{k}_2}^{\sigma''} \frac{d\mathbf{k}_1\, d\mathbf{k}_2}{(2\pi)^6}, \tag{5.24}$$

and the radiation intensity is

$$I^{\sigma} = \frac{\partial W^{\sigma}}{dt} = \int \hbar\omega^{\sigma} \frac{\partial N_{\mathbf{k}}^{\sigma}}{\partial t} \frac{d\mathbf{k}}{(2\pi)^6} = \int w_{\sigma}^{\sigma'\sigma''} \hbar\omega^{\sigma} N_{\mathbf{k}_1}^{\sigma'} N_{\mathbf{k}_2}^{\sigma''} \frac{d\mathbf{k}\, d\mathbf{k}_1\, d\mathbf{k}_2}{(2\pi)^9}. \tag{5.25}$$

Alternatively, the radiation intensity can be found if the nonlinear plasma current is known:

$$j_{k,i}^{(2)} = \int S_{i,j,l}(k, k_1, k_2)\, E_{k_1 j} E_{k_2 l}\, dk_1\, dk_2 \delta(k - k_1 - k_2). \tag{5.26}$$

This current was found for a cold plasma in Chap. II. In any dynamical description of the medium, however (including thermal motion, neutral atoms, quantum effects, etc.), its determination seldom presents any difficulties, because all that is required is a simple expansion of the variables in terms of the electromagnetic field amplitudes. Also important is the fact that in the limit of small intensity of the excited $\sigma$-waves their influence on the fields of the $\sigma'$- and $\sigma''$-waves may be neglected and the latter regarded as fixed. Replacing the fields $E_j$ and $E_l$ in (5.26) by those of the indicated waves, we can then regard the current $j^{(2)}$ as known.

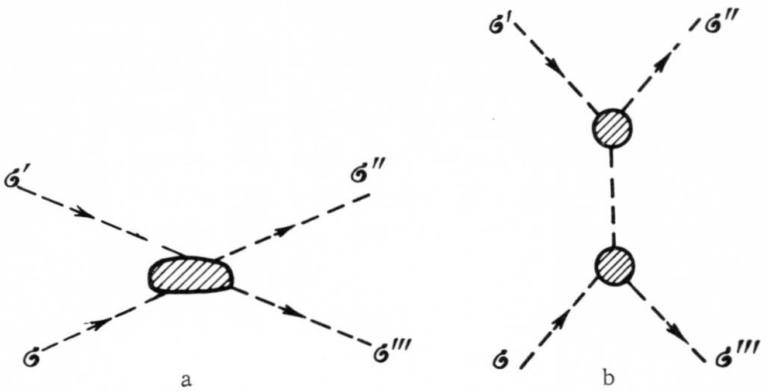


Fig. 19. Possible types of scattering of waves by other waves. a) Direct interaction of four waves; b) interaction of four waves through an intermediate wave and direct interaction of three waves. Diagram b cancels the effect arising from diagram a. This cancellation is analogous to that which occurs in the scattering of waves by particles.

Now all that remains is to solve a simple electrodynamical problem, namely to calculate the radiation intensity of the wave $\sigma$ from the known exciting current $\mathbf{j}^{(2)}$. Averaging the result over the phases of the waves $\sigma'$ and $\sigma''$ and comparing with (5.25), we readily find the probability for the corresponding process (Appendix 3) [1]:

$$w_{\sigma}^{\sigma'\sigma''}(\mathbf{k}, \mathbf{k}', \mathbf{k}'') = 8(2\pi)^7\delta(\mathbf{k}-\mathbf{k}'-\mathbf{k}'')\,\delta(\Omega^{\sigma}(\mathbf{k})-\Omega^{\sigma'}(\mathbf{k}')-\Omega^{\sigma''}(\mathbf{k}''))\times$$
$$\times\frac{(\Omega^{\sigma'}(\mathbf{k}')^2)(\Omega^{\sigma''}(\mathbf{k}''))^2\,|S_{\sigma,\sigma',\sigma''}(\Omega^{\sigma}(\mathbf{k}),\mathbf{k},\Omega^{\sigma'}(\mathbf{k}'),\mathbf{k}',\Omega^{\sigma''}(\mathbf{k}''),\mathbf{k}'')|^2}{\left(\frac{\partial}{\partial\omega}\omega^2\varepsilon^{\sigma}(\omega,\mathbf{k})\right)_{\omega=\Omega^{\sigma}(\mathbf{k})}\left(\frac{\partial}{\partial\omega}\omega^2\varepsilon^{\sigma'}(\omega,\mathbf{k}')\right)_{\omega=\Omega^{\sigma'}(\mathbf{k}')}\left(\frac{\partial}{\partial\omega}\omega^2\varepsilon^{\sigma''}(\omega,\mathbf{k}'')\right)_{\omega=\Omega^{\sigma''}(\mathbf{k}'')}},\tag{5.27}$$

where

$$S_{\sigma,\sigma',\sigma''}=e_i^{\sigma*}(\mathbf{k})\,S_{ijl}e_j^{\sigma'}(\mathbf{k}')\,e_l^{\sigma''}(\mathbf{k}'');$$
$$\varepsilon^{\sigma}(\omega_1,\mathbf{k})=\varepsilon_{ij}(\omega_1,\mathbf{k})\,e_i^{\sigma*}(\mathbf{k})\,e_j^{\sigma}(\mathbf{k})+\omega^{-2}(\mathbf{k}e^{\sigma}(\mathbf{k}))(\mathbf{k}e^{\sigma*}(\mathbf{k})).\tag{5.28}$$

$\varepsilon_{ij}$ is the tensor dielectric constant, and $e_j^{\sigma}(\mathbf{k})$ are the unit vectors of the corresponding waves.

These relations can be used to solve the problem stated.

The probabilities for four-plasmon decays are similarly found if the nonlinear current $\mathbf{j}^{(3)}$, which is of third order with respect to the wave field, is known. Here, as a rule, a significant role is played by the current $\mathbf{j}^{(3)}$ derived from $\mathbf{j}^{(2)}$ when the first powers of the field are expressed in terms of the second powers by Maxwell's equations containing $\mathbf{j}^{(2)}$ in the right-hand side. The contribution of these processes is of the same order as from those derived directly from $\mathbf{j}^{(3)}$. Moreover, mutual cancellation of these two currents can also occur.

The best visual interpretation of the two processes is afforded by analogy with the scattering of particles. We merely point out that if the first process is illustrated graphically as in Fig. 19a, the second must be represented as in Fig. 19b, containing only nonlinear interactions of three waves.

Some special relations for the probabilities of four-plasmon interactions are presented in Appendix 3.

## 5. DECAY PROCESSES IN ISOTROPIC PLASMA

Let us briefly describe the waves of an isotropic plasma.

In a cold isotropic plasma only two wave types exist, transverse and longitudinal, whose interaction we have discussed above. In a hot isotropic plasma these waves can be accompanied by ion-acoustic modes, whose frequencies do not exceed $\sim \omega_{0i} = \sqrt{\frac{m_e}{m_i}}\, \omega_{0e}$, where $m_e$ and $m_i$ are the masses of the electron and ion, respectively. The dependence of the frequencies $\Omega$ on k is illustrated in Fig. 20 for the hot plasma three branches, t, $l$, and s. These branches are described analytically by the following approximate formulas.

1. Transverse waves:

$$\Omega^t(k) = \sqrt{k^2c^2 + \omega_{0e}^2}. \tag{5.29}$$

The phase velocities of the transverse waves are always greater and the group velocities $v_{\text{gr}}^t = c^2k/\sqrt{k^2c^2 + \omega_{0e}^2}$ less than the velocity of light. For transverse waves spatial dispersion has little effect on their propagation.

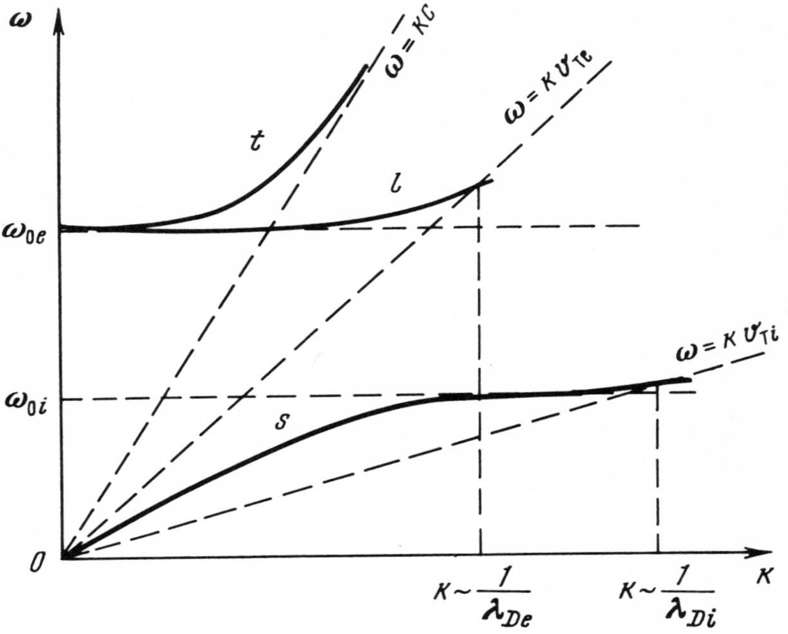


Fig. 20. Dispersion curves for weakly damped waves in isotropic plasma: t) transverse branch; *l*) longitudinal branch; s) ion-acoustic branch.

2. Langmuir waves:

$$\Omega^{l}(k) \simeq \omega_{0e} + \frac{3}{2}\frac{k^2 v_{Te}^2}{\omega_{0e}}, \tag{5.30}$$

where $v_{Te}$ is the mean thermal velocity of the plasma electrons. The phase velocities of the Langmuir waves lie in the range $v_{Te} < v_{\varphi}^{l} < \infty$, their group velocities in the range $0 < v_{gr}^{l} < v_{Te}$. For $v_{\varphi} < v_{Te}$ the waves are strongly damped.

3. Ion-acoustic waves ($T_e \gg T_i$)

$$\Omega^{s}(k) = \frac{k v_s}{\sqrt{1 + \frac{k^2 v_{Te}^2}{\omega_{0e}^2}}} + \frac{3}{2}\frac{k^2 v_{Ti}^2}{\omega_{0i}}, \tag{5.31}$$

where $v_{Ti}$ is the mean thermal velocity of the ions:

$$v_s = \sqrt{\frac{m_e}{m_i}} v_{Te}.$$

For $k \ll \omega_{0e}/v_{Te} = \lambda_{De}^{-1}$ the spectrum is acoustic, $\Omega^s(k) = kv_s$, and for $k \gg \lambda_{De}$

$$\Omega^s \approx \omega_{0i}. \tag{5.32}$$

The phase velocities of the waves lie in the range $v_{Ti} < v_\varphi^s < v_s$, the group velocities in the range $v_{Ti}(T_i/T_e)^{1/4} < v_{gr}^s < v_s$.

Let us now examine how three-plasmon decays can occur in isotropic plasma [2-8].

Any process is possible in which a high-frequency wave emits a low-frequency wave, changing to a wave of the same type as the primary. This allows the following processes:

$$t \rightleftarrows t + l; \tag{5.33}$$

$$t \rightleftarrows t + s; \tag{5.34}$$

$$l \rightleftarrows l + s. \tag{5.35}$$

The arrows indicate the directions that the process can take. It is apparent from the law of energy conservation that the first process $t \rightleftarrows t + l$ is possible not only for waves with $\Omega^t \gg \Omega^l$, but also for waves with $\Omega^t$ of the same order as $\Omega^l$, but whereas the coalescence process $t + l \rightarrow t$ can occur for transverse waves with frequencies near $\omega_{0e}$, the decay process $t \rightarrow t + l$ is possible only for $\Omega^t > 2\omega_{0e}$ (the decay produces two waves having frequencies greater than $\omega_{0e}$).

The conservation laws strictly forbid wave decays in which all three waves are of the same type. In other words, the following decay processes are not allowed:

$$t \rightleftarrows t + t, \quad l \rightleftarrows l + l, \quad s \rightleftarrows s + s.$$

The following processes are permitted:

$$l + l \rightleftarrows t \tag{5.36}$$

and

$$l + s \rightleftarrows t; \tag{5.37}$$

$$t + s \rightleftarrows l. \tag{5.38}$$

The processes (5.36) are possible only for transverse waves whose frequencies are near $2\omega_{0e}$, and (5.37) and (5.38) are possible only for t-waves whose frequencies are near $\omega_{0e}$. The probabilities for the processes (5.37) and (5.38) differ only in the signs of $\mathbf{k}^s$ and $\Omega^s(\mathbf{k}^s)$, according to the conservation laws for the decay processes, occurring in the $\delta$-functions.

Proceeding to four-plasmon processes, we note that processes involving the decay of an $l$-wave into three $l$-waves, as well as $t \rightleftarrows 3t$, $s \rightleftarrows 3s$, are also forbidden by the conservation laws. Similar plasmons, however, can interact with one another in processes of the type $l + l \rightleftarrows l + l$, which correspond to the scattering of plasmons by each other. Another interesting process is $t + l \rightleftarrows l + t$, resulting in the scattering of transverse waves in a plasma by strong Langmuir waves. This can also lead to the scattering of Langmuir waves in a plasma in which there is strong electromagnetic radiation (transverse waves). Objects in which the intensity of the transverse waves is high are often met in astrophysical situations.

The decay interactions discussed here are important in such problems as nonlinear absorption, stabilization of instabilities, nonthermal radiation from plasma, plasma diagnostics, energy transfer from electrons to ions, etc. These examples will illustrate how the theory of decay interaction can be used very simply to obtain rough estimates.

## 6. THREE-PLASMON DECAY PROCESSES AND NONLINEAR WAVE ABSORPTION

Consider the decay effect $\boldsymbol{\lambda} \rightleftarrows \lambda + \sigma$, where $\sigma$ is a wave whose frequency is much lower than that of $\lambda$. We pose the following problem: can the decay processes discussed result in the effective absorption of the waves $\boldsymbol{\lambda}$ or $\sigma$, which does not occur in the linear approximation?

Let us direct our attention first to the $\lambda$-waves. Since the number of $\lambda$-waves in this decay process remains constant, they can transfer energy only from one part of the spectrum to another. How can spectral energy transfer produce effective absorption of $\lambda$-waves? Two possibilities present themselves.

First, spectral transfer can move the $\lambda$-waves from that part of the spectrum where their absorption is weak to the part where it is appreciable. This is one possible mechanism of non-linear absorption.

Second, transformation can transfer the $\lambda$-waves to the part of the spectrum where the frequencies of the quanta are lower. Clearly, with the total number of quanta held constant, their energy can decrease significantly only if the frequencies of the quanta are strongly dependent on their wave number. This is another possible mechanism for nonlinear wave absorption.

Both mechanisms are typified by the fact that the wave energy cannot vanish altogether, because the total number of quanta remains the same. We should note that this property is typical of many nonlinear absorption mechanisms, since nonlinear interactions often become negligible at low intensities. Neither is a true nonlinear absorption mechanism, because in the first case the number of quanta vanishes through linear absorption, while in the second case only the quantum energy decreases. However, absorption of this type would not occur without nonlinear effects.

We now examine the affect on the $\sigma$-waves. These can be created or destroyed. Consequently, for them true nonlinear absorption is possible, but then so is nonlinear growth.

Note that the same waves can be either radiating or absorbing ($\lambda$) in one process, while in another they can be radiated and absorbed ($\sigma$). An example is the case of Langmuir waves in the multistage decay process

$$t \rightleftarrows t + l; \quad l \rightleftarrows l + s. \tag{5.39}$$

## 7. DECAY INTERACTIONS OF LANGMUIR AND ION-ACOUSTIC WAVES

Consider the decay $l \rightleftarrows l + s$ [3, 4, 9] of a Langmuir wave $l$

$$\Omega^l = \omega_{0e} + \frac{3}{2}\frac{k^2 v_{Te}^2}{\omega_{0e}} \tag{5.40}$$

into a Langmuir and an ion-acoustic wave

$$\Omega^s = \frac{kv_s}{\sqrt{1 + k^2\lambda_{De}^2}}. \tag{5.41}$$

This process is important not only as an example of a mechanism resulting in nonlinear wave absorption, but also as a mechanism for transferring oscillatory energy from electrons to ions [3]. In addition, it is very important since the energy converted into the s-modes can be fairly rapidly transformed into thermal energy of the plasma owing to their considerable damping.

In the case of one-dimensional decay processes, even if $\Omega^l \gg \Omega^s$, the diffusion approximation may prove inapplicable because $k^l \sim k^s$. This follows from the conservation of energy and momentum of the wave in the elementary decay event, which we now propose to discuss.

Langmuir oscillations are only weakly damped for $k^l \ll \lambda_{De}^{-1}$. The maximum momentum that the s-quantum can acquire in emission is $2k^l$. In this case the emitting $l$-quantum reverses its direction of motion, so that $k^s \ll \lambda_{De}^{-1}$. In other words, only acoustic quanta for which Eq. (5.41) has the form

$$\Omega^s = kv_s. \tag{5.42}$$

can take part in decay interaction with Langmuir quanta. The laws of energy and momentum conservation for the quanta during decay have the form

$$\omega^{l'} = \omega^l + |\mathbf{k}^s| v_s; \tag{5.43}$$

$$\mathbf{k}^{l'} = \mathbf{k}^l + \mathbf{k}^s, \tag{5.44}$$

or

$$(k^{l'})^2 = (k^l)^2 + \frac{2\omega_{0e}}{3v_{Te}} \sqrt{\frac{m_e}{m_i}} |\mathbf{k}^s|; \tag{5.45}$$

$$(k^{l'})^2 = (k^l)^2 + (k^s)^2 + 2k^l k^s \cos\theta_+, \tag{5.46}$$

where $\Theta_+$ is the angle between the s-wave and the emitting $l$-wave. We have

$$\cos\theta_+ = \frac{|\mathbf{k}^s| + 2k_0}{2|\mathbf{k}^l|}; \quad k_0 = \frac{1}{3}\frac{\omega_{0e}}{v_{Te}} \sqrt{\frac{m_e}{m_i}}. \tag{5.47}$$

Since $\cos\Theta_+ < 1$,

$$0 < |\mathbf{k}^s| < 2(k^l - k), \tag{5.48}$$

i.e.,

$$k^l > k_0. \tag{5.49}$$

If we introduce the phase velocity of the Langmuir waves, $\Omega^l/k^l$, which is approximately given by $v_\varphi^l = \omega_{0e}/k^l$, condition (5.49) becomes

$$v_\varphi^l < 3v_{Te}\sqrt{\frac{m_i}{m_e}}. \tag{5.50}$$

Consequently, decay processes are possible only for $l$-waves whose phase velocities do not exceed certain maximum values (5.50). On the other hand, we have seen that the transfer of energy for intense Langmuir waves is such as to reduce k, i.e., to increase $v_\varphi^l$. The decay process continues until condition (5.50) no longer holds. Since the Langmuir-wave frequency depends only very slightly on the wave number, the change in the total wave energy in this transfer is small.

We also know that Landau damping of Langmuir waves is appreciable only if $v_\varphi^l$ is comparable with $v_{Te}$. The nonlinear interaction in question therefore tends to reduce the Landau damping. This causes a slight reduction in the energy as a result of the drop in frequency of the Langmuir waves.

In order to estimate the characteristic energy transfer time we need to know the decay probability. Making use of the representative probabilities listed in Appendix 3, we write

$$w_{\mathbf{k}^l}(\mathbf{k}^s) = \frac{\hbar e^2 k^s \sqrt{\dfrac{m_e}{m_i}}(2\pi)^3}{8\pi m_e^2 v_{Te}} \frac{(\mathbf{k}^l - \mathbf{k}^s, \mathbf{k}^l)^2}{(\mathbf{k}^l - \mathbf{k}^s)^2 (k^l)^2} \times$$

$$\times \delta\left(\frac{3}{2}\frac{v_{Te}^2}{\omega_{0e}}[(\mathbf{k}^l - \mathbf{k}^s)^2 - (k^l)^2] + k^s v_s\right). \tag{5.51}$$

The corresponding estimates differ somewhat, depending on whether the problem is one-dimensional or not. A particular direction might be chosen because the waves considered have the same initial direction. In the one-dimensional case $\cos\vartheta_+ = 0$, hence from (5.47)

$$|k^s| = 2|k^l| - 2k_0. \tag{5.52}$$

If $k^l$ is of the same order as $k_0$, transfer has in effect already been realized, so that it is more sensible to examine the transfer process for $k^l \gg k_0$. But then, according to (5.52), $k^s \sim 2k^l$, i.e., the longitudinal wave changes direction on emitting an acoustic wave. The momentum of the $l$-wave after decay is

$$|k^l| - |k^s| = -(k^l - k_0) + k_0.$$

It has the same direction as the initial momentum if $k^l - k_0 < k_0$. If this inequality holds, a single decay terminates the entire process, because immediately after the emission of the s-wave the momentum of the $l$-wave has a magnitude smaller than $k_0$.

Consequently, in the one-dimensional case the change in momentum of the $l$-wave for $k^l \gg k_0$ is not small, and the diffusion approximation is no longer valid. As a result of two decays, however, the momentum changes by a small amount, and $|k^l|$ decreases relatively slowly. The characteristic time for one decay can be estimated by assuming that the initial value of $N^{\lambda}_{k_1-k}$ is small:

$$\frac{1}{N^s}\frac{\partial N^s}{\partial t} = \gamma = \int w_{k^l}(k^s)\, N^l_k \frac{dk^l}{(2\pi)^3}. \qquad (5.53)$$

Letting $N^l_k$ be one-dimensional and inserting the probability (5.51) into (5.53), we have

$$\gamma = \frac{e^2k^s\sqrt{\dfrac{m_e}{m_i}}}{8\pi m_e^2 v_{Te}} \int \hbar N^l_{k^l}\, dk^l \delta\left(\frac{3}{2}\frac{v_{Te}^2}{\omega_{0e}}\left((k^s)^2 - 2k^l k^s\right) + k^s v_{Te}\sqrt{\frac{m_e}{m_i}}\right) = \frac{e^2 k_0}{8\pi m_e^2 v_{Te}^2} N^l_{\frac{k^s+2k_0}{2}}\hbar. \qquad (5.54)$$

On the other hand, the longitudinal-wave energy is of order

$$W^l = \int \frac{\hbar\omega_{0e} N^l_k\, dk}{(2\pi)^3} = \frac{\omega_{0e} N^l \Delta k \hbar}{8\pi^3}. \qquad (5.55)$$

Hence

$$\gamma = \frac{\pi}{4}\frac{k_0}{\Delta k}\frac{W^l}{n m_e v_{Te}^2}\,\omega_{0e}. \qquad (5.56)$$

Here $\Delta k$ is the initial width of the wave spectrum. The time (5.56) characterizes the change of $k^l$ by an amount of order $k_0$.

To find the time for $k^l$ to change by an amount on the order of $k^l$, this being the time of spectral transfer from $k^l$ to $k_0$ for

$k^l \gg k_0$, we need to recognize that the above process has to be repeated $k^l/k_0$ times, the process rate reducing with each transition, because the initial energy (which is approximately conserved by the number of quanta remaining constant) is redistributed between a large number of wave packets. For this reason the total transfer time is in effect determined by the time of the slowest final stage, when each packet holds only $k_0/k^l$ times the initial energy. Moreover, for a large number of wave packets the increment is determined by the intensity difference between two consecutive packets, which must not be larger than $k_0/k^l$.

Consequently, we obtain the following estimate for the transfer time:

$$\frac{1}{\tau} \simeq \frac{\pi}{4}\left(\frac{k_0}{k^l}\right)^2 \frac{k_0}{\Delta k} \frac{W^l}{nm_e v_{Te}^2}\,\omega_{0e} = \frac{\pi}{36}\left(\frac{v_\Phi^l}{v_{Te}}\right)^2 \frac{m_e}{m_i} \frac{k_0}{\Delta k} \frac{W^l}{nm_e v_{Te}^2}.$$

Notice that if $\Delta k < 2k_0$, energy transfer corresponds to the formation of satellite spectra, because in this case $\Delta\omega^l < \omega^s$. If, on the other hand, $\Delta k \gg 2k_0$, a diffusion type of picture appears, although the small change in the momentum of the $l$-quantum is brought on by a twofold reversal of direction. This is enough to show that decay processes in which an s-quantum is emitted at an angle with respect to the $l$-quantum can play a very important role. It is also clear that this causes the Langmuir oscillations to tend to become isotropic, because the acoustic waves carry a momentum component perpendicular to the primary packet of $l$-waves. If $k^l \gg k_0$ and $k^l \gg k^s$, then according to (5.47) $\cos \vartheta_+ \ll 1$, i.e., the diffusion approximation is applicable only to the generation of s-waves propagating at an angle nearly perpendicular to the direction of the $l$-wave packet. In this case

$$\frac{1}{\tau} = \frac{1}{N_s^s}\frac{dN_s^s}{dt} \simeq \int w_l^s \mathbf{k}^s \frac{dN_{\mathbf{k}^l}^l}{d\mathbf{k}^l}\frac{d\mathbf{k}^l}{(2\pi)^3}; \tag{5.58}$$

$$w_l^s(\mathbf{k}^s, \mathbf{k}^l) = \frac{\hbar e^2 (\Omega^s)^3 (2\pi)^3 m_i}{16\pi m_e^3 v_{Te}^4 (k^s)^2}\,\delta(\Omega^s - \mathbf{k}^s \mathbf{v}_{gr}^l. \tag{5.59}$$

If the packet of $l$-waves subtends angles $\Delta\vartheta \sim 1$, then

$$\frac{1}{\tau} \simeq \omega_{0e} \frac{W^l}{nT_e} \frac{k^s k_0}{k^l \Delta k^l}, \tag{5.60}$$

where $k^s$ and $k^l$ are the characteristic wave numbers of the s- and $l$-waves.

Let us now consider the case in which a relatively weak Langmuir wave is transmitted through a plasma in which strong acoustic oscillations are excited. We wish to estimate the time for this wave to become isotropic in a field of isotropic acoustic waves. We assume that the diffusion approximation applies, in other words, that $k^s \ll k^l$ and $k^l \gg k_0$. The time to become isotropic is

$$\tau \simeq \frac{(\Delta k)^2}{D} \simeq \frac{(k^l)^2}{D};$$

$$D \sim \int (2\pi)^{-3}(k^s)^2 N_{\mathbf{k}^s}\, d\mathbf{k}^s w_l^s \sim \frac{\omega_{0e}^3 k^s}{k^l v_{Te}^2} \frac{W^s}{n m_e v_{Te}^2},$$

i.e.,

$$\tau \simeq \frac{1}{\omega_{0e}} \frac{n m_e v_{Te}^2}{W^s} \frac{v_{Te}^2}{(v_\varphi^l)^2} \frac{k^l}{k^s}. \tag{5.61}$$

In addition to becoming isotropic, the Langmuir waves suffer a change in energy associated with the change of the modulus of $k^l$. This change is best estimated by using the exact equation for the $l$-waves

$$\frac{\partial N_{\mathbf{k}_1}^l}{\partial t} = \int \Big\{ w_l^{l,\,s}(k_1, k_2, k_s) \frac{d\mathbf{k}_2 d\mathbf{k}_s}{(2\pi)^6} (N_{\mathbf{k}_1}^l N_{\mathbf{k}_2}^l - N_{\mathbf{k}_1}^l N_{\mathbf{k}_s}^s + N_{\mathbf{k}_2}^l N_{\mathbf{k}_s}^s) - \text{terms in which } 1 \rightleftarrows 2 \Big\}. \tag{5.62}$$

From this we obtain the following expression for the change in the total $l$-wave energy:

$$\frac{\partial W^l}{\partial t} = \int \frac{d\mathbf{k}_1^l d\mathbf{k}_2^l d\mathbf{k}_s}{(2\pi)^9} \Omega^s(\mathbf{k}_s) \left( w_l^{l,\,s}(k_1^l, k_2^l, k_s) - w_l^{l,\,s}(k_2^l, k_1^l, k_s) \right) N_{\mathbf{k}_1}^l N_{\mathbf{k}_s}^s. \tag{5.63}$$

Putting in the expression for the probabilities, we obtain for isotropic s-waves

$$\frac{\partial}{\partial t} W^l = \int N_{\mathbf{k}}^l N_{\mathbf{k}_s}^s d\mathbf{k}_s \frac{e^2 m_i \omega_{0e}^2 \Omega_s^5 \left( k^2 k_s^2 - \frac{k_s^4}{2} - \frac{2}{9} \frac{\omega_{0e}^2 \Omega_s^2}{v_{Te}^4} \right)}{4\pi m_e^3 k^3 k_s^3 9 v_{Te}^8 \left( k^4 - \frac{4\omega_{0e}^2 \Omega_s^2}{9 v_{Te}^4} \right)} \frac{dk\hbar^2}{(2\pi)^3}. \tag{5.64}$$

Recognizing that $W^l$ is almost constant, we readily estimate the rate of change of the wave energy:

$$\frac{1}{\tau} \simeq \frac{1}{4}\sqrt{\frac{m_e}{m_i}}\,\omega_{0i}\left(\frac{v_\Phi^l}{v_{Te}}\right)^2\left(\frac{k_s}{k^l}\right)^3\frac{W^s}{nm_e v_{Te}^2}. \tag{5.65}$$

For $k^l \gg k_0$ this time is approximately $(k^l/k_0)^2(m_e/m_i)$ times the characteristic scattering time. In other words, the scattering of the Langmuir waves in the field of the acoustic waves is quasi-elastic, i.e., the direction of the $l$-waves suffers a greater change than does their energy.

The actual diffusion results from balancing the induced decay and induced coalescence processes. In the case of isotropic s-waves (for the incident spectrum) the situation can arise in which decay predominates. Then the wave energy decreases, and $v_\varphi^l$ increases. Otherwise there is only a reduction in $v_\varphi^l$. It is important to note in this connection that if $k^s$ is comparable with $k^l$, a situation can occur in which the conservation laws allow only the coalescence processes. If $k_1 > k_0$, it is then required that

$$2(k_1 - k_0) < k^s < 2(k_0 + k_1^l). \tag{5.66}$$

But if $k_1 < k_0$, then

$$2(k_0 - k_1) < k^s < 2(k_0 + k_1^l). \tag{5.67}$$

Consequently, if $k_1 \gg k_0$, $k^s$ must lie within a small interval of $4k_0$ about $k_1$, but if $k_1 < k_0$, it must lie in an interval smaller than $4k_0$ about $k_0$. For example, let the initial conditions satisfy (5.66), and suppose that only coalescence occurs. This process causes the $l$-waves to spread beyong the interval defined by (5.66). Moreover, it is unlikely that the s-waves will be grouped for $k_1 \gg k_0$ in a small wave number interval $\Delta k_1 \sim 4k_0$, because their different mutual nonlinear interactions tend to alter their spectrum. We observe that the case in which nonlinear interactions are significant is that in which the intensity of the s-waves is fairly high, as assumed above. Under conditions (5.67) a decrease in $v_\varphi^l$ cannot yield additional absorption, because

$$k_1 < k_0 \text{ and } \frac{v_\varphi}{v_{Te}} > 3\sqrt{\frac{m_i}{m_e}},$$

and linear absorption occurs for $v_\varphi$ comparable with $v_{Te}$.

It is obvious from the above how the direction and energy of the $l$-waves change and when this change decreases their phase velocities, producing additional absorption. For an isotropic plasma this absorption occurs near $v_\varphi \sim v_{Te}$, so that in a given case it is very important to determine the direction of energy transfer of the waves in terms of the phase velocities. In a magnetoactive plasma absorption is possible at frequencies close to the gyrofrequencies of the electrons and ions and their harmonics. For a magnetoactive plasma, therefore, the energy transfer of waves into the region of gyroresonance absorption $\sim \omega_{H\alpha}$ can also be important. Another point to consider is that in magnetic fields the frequency of plasma waves can be strongly dependent on their wave number. For example, in a strong magnetic field $\omega_{He} \gg \omega_{0e}$ ($\omega_{He} = eH/m_e c$) the frequency of plasma waves perpendicular to the field is small, while parallel to the field it is near $\omega_{0e}$. Consequently, the energy transfer of the waves in different directions can strongly affect the total energy of the plasma waves while the number of such waves remains constant.

We note in conclusion that the condition for the one-dimensional random phase decay interaction of Langmuir and ion-acoustic waves to apply is

$$\gamma < \frac{3k_s v_{Te}^2}{v_\varphi^l} \frac{\Delta k^l}{k^l}.$$

## 8. FOUR-PLASMON DECAY PROCESSES AND NONLINEAR WAVE ABSORPTION

We now consider briefly the contributions of four-plasmon interactions, which can also lead to energy transfer of the waves in the region of the spectrum where absorption is pronounced. We let a single example suffice. Suppose that a directional packet of high-intensity, high-frequency transverse waves is present in a plasma. We wish to see what effects result from the scattering of Langmuir waves by transverse waves. We use the equation of the diffusion approximation for the change in the number of Langmuir waves:

$$\frac{\partial N_{\mathbf{k}_1}^l}{\partial t} = N_{\mathbf{k}_1}^l \int u_l^t(\mathbf{k}_1, \mathbf{k}_2, \mathbf{k}') N_{\mathbf{k}_2}^l(\mathbf{k}_1 - \mathbf{k}_2) \frac{\partial N_{\mathbf{k}'}^t}{\partial \mathbf{k}'} \frac{1}{(2\pi)^6} d\mathbf{k}' \, d\mathbf{k}_2. \qquad (5.68)$$

It is readily seen that

$$\int N^l_{\mathbf{k}_1} \frac{d\mathbf{k}_1}{(2\pi)^3} = \text{const}, \tag{5.69}$$

i.e., in this process the Langmuir waves undergo spectral transfer [8]. The validity of (5.69) is confirmed by the realization that the probability $u^t_l$ in (5.68) is symmetrical with respect to $k_1$ and $k_2$, whereas the factor $(\mathbf{k}_1 - \mathbf{k}_2)$ is antisymmetric.

Let the transverse-wave distribution function have the form

$$N^t_{\mathbf{k}'} = \text{const}\,\exp\left[-\frac{(\mathbf{k}' - \mathbf{k}_0)^2}{2(\Delta k)^2}\right], \tag{5.70}$$

where $\mathbf{k}_0$ is the mean directional momentum of the waves, and $\Delta k$ is the width of the packet. Substituting (5.70) into (5.68), we obtain:

$$(\mathbf{k}_1 - \mathbf{k}_2)\frac{\partial N^t_{\mathbf{k}'}}{\partial \mathbf{k}'} = -\frac{(\mathbf{k}_1\mathbf{k}') - (\mathbf{k}_2\mathbf{k}') - (\mathbf{k}_1 - \mathbf{k}_2)\,\mathbf{k}_0}{(\Delta k)^2} N^t_{\mathbf{k}'}. \tag{5.71}$$

The law of energy conservation for the scattering event has the following form in the diffusion approximation:

$$(\mathbf{k}_1 - \mathbf{k}_2)\,\mathbf{v}^t_{\text{gr}} = (\mathbf{k}_1 - \mathbf{k}_2)\frac{\mathbf{k}'}{k'}c = \omega_1 - \omega_2 = \frac{3}{2\omega_{0e}}(k_1^2 - k_2^2)\,v^2_{Te}.$$

The direction of transfer is determined by the sign of (5.71), i.e., by the sign of

$$-(\mathbf{k}_1 - \mathbf{k}_2)(\mathbf{k}' - \mathbf{k}_0) = -\frac{\frac{3}{2c\omega_{0e}}(k_1^2 - k_2^2)\,k' v^2_{Te} - (\mathbf{k}_1 - \mathbf{k}_2)\,\mathbf{k}_0}{(\Delta k)^2}. \tag{5.72}$$

If $\mathbf{k}_1 - \mathbf{k}_2$ is parallel to $\mathbf{k}_0$ and $k_1 > k_2$, while $k'$ is of the same order as $k_0$, then (5.72) is positive for

$$\frac{k_1 + k_2}{\omega_{0e}} v^2_{Te} < c,$$

which is always satisfied. Only if $\mathbf{k}_1 - \mathbf{k}_2$ is almost perpendicular to $\mathbf{k}_0$ does the expression (5.72) become negative for $k_1 > k_2$. In the case of positive (5.72) the waves having the larger wave numbers grow, i.e., transfer of these waves is in the direction of smaller phase velocities.

## 9. DECAY INTERACTIONS AND NONLINEAR STABILIZATION OF A PLASMA INSTABILITY

At the outset we note that the general problems of nonlinear stabilization discussed in this section are valid not only for decay interactions, but also for other types of nonlinear interactions, as we shall show later on, especially for induced scattering effects (Chap. 9).

Consider an unstable plasma. As a consequence of the instability a definite mode of oscillation is excited in the plasma. What possible mechanism can limit the amplitudes of the unstable oscillations?

As a mechanism for energy depletion, nonlinear absorption could tend to stabilize the instability. It must be realized, however, that the stabilization of instabilities can occur entirely from causes other than nonlinear absorption. Remember that instabilities can arise from induced processes which lead to an avalanche type of wave growth. As a rule, the range of generation of oscillations has an upper and lower frequency limit, i.e., the wave numbers are bounded. Suppose the oscillations generated have values of k in an interval $\Delta k$. Outside this interval there is neither generation nor absorption. If some nonlinear mechanism transfers the oscillations from the interval $\Delta k$ into a region outside $\Delta k$, this inhibits the avalanche buildup of instability, i.e., it can result in the cessation of instability. In order for such a cessation to occur the rate of transfer out of the interval $\Delta k$ must exceed the rate of generation. What happens to the removed waves subsequently is immaterial; it is not important, in particular, whether or not they are absorbed.

How then do we visualize the development of an instability?

Initially, when the wave intensity is small, nonlinear decay interactions of course play no part. However, small fluctuation fields are present both in the interval $\Delta k$ and outside it. With the development of instability the fluctuation waves within $\Delta k$ grow. At a certain intensity of these waves nonlinear effects set in, resulting in their diffusion out of the interval $\Delta k$. If the value of $\Delta\omega$ corresponding to $\Delta k$ is smaller than the frequency $\omega^{\sigma}$ of the wave emitted in decay, the escape of oscillations from the vicinity of k does not correspond to diffusion through the boundaries of $\Delta k$, but results in the formation of satellites, i.e., an effectively discontinuous change in k. If only one satellite is allowed by the conservation laws to be excited, a well-defined distribution of energy is set up between these satellites (for example, if the instability of Langmuir waves is involved, the energy in the first satellite will be three times that in the fundamental line). But if the formation of many satellites is allowed, the energy distribution is almost diffuse; then the energy in the interval $\Delta k$ will decrease since it must be distributed among an ever-increasing number of satellites.

In both the diffusion case and in the formation of satellites oscillation energy is withdrawn from $\Delta k$. This withdrawal increases with the energy of the oscillations in $\Delta k$. Finally, at some definite value of the energy it becomes comparable with the rate of energy increase from the instability. Does this imply stabilization of the instability? Of course it does not, because even though the oscillation energy inside the interval $\Delta k$ does not increase, it does so outside $\Delta k$. However, the energy increase outside $\Delta k$ does not go on indefinitely.

The fact is that the rate of energy diffusion from $\Delta k$ is determined by the energy in $\Delta k$ only so long as this energy exceeds that outside $\Delta k$. But the energy within $\Delta k$ stops increasing at some time during the development of the instability, and the energy outside that interval continues to grow. Therefore, the process reaches a stage at which the withdrawal of energy from $\Delta k$ is regulated by the intensity of the oscillations outside $\Delta k$. The behavior of the oscillations outside $\Delta k$ is to a certain extent independent of the generation inside $\Delta k$ (for example, when the energy flux to the region outside $\Delta k$ is much smaller than the energy inside this region).

As mentioned, the distribution of energy among an ever-increasing number of waves outside $\Delta k$ can cause a reduction in the intensity inside $\Delta k$. In other words, the intensity of the oscillations in $\Delta k$ diminishes, becoming less than the value which occurred when the rates of generation and loss were equal. As a matter of fact, in this case the loss is greater than the generation, i.e., the generation of oscillations outside $\Delta k$ has resulted in an equally large loss from $\Delta k$, and this is equivalent to the presence of a decay rate within $\Delta k$ greater than the growth rate of the instability. Under these conditions the oscillations in $\Delta k$ could tend to zero.

Consequently, the temporal development of instability corresponds to a buildup of oscillations in $\Delta k$ up to a certain maximum and then a decrease to zero.

We now determine the nonlinear state into which the plasma has gone after the development of instability. In this state there is a distribution of oscillations which is zero inside $\Delta k$ and a certain wave packet outside $\Delta k$. The packet generated outside $\Delta k$ results from the nonlinear effect that stabilizes the instability in $\Delta k$. It is also clear that if such a packet is initially present, instability will not, in general, set in.

Consider the next stage of evolution of this packet. We saw that the transfer of energy in k space can produce absorption of the packet's energy. When this vanishes, however, instability occurs. Therefore, nonlinear absorption could result in the onset of instability and a repetition of the entire sequence. In other words, a relaxation behavior is possible in the variation of the oscillation intensity in $\Delta k$. In this case, even on the average, a certain oscillatory energy is present in the entire range, both inside and outside $\Delta k$.

The above line of reasoning suffers from one deficiency. We have postulated rather vaguely that the oscillations generated in the interval $\Delta k$ do not significantly affect the growth rate $\gamma$. It is clear that this is not true in general. The generated oscillations can affect the growth when the oscillatory energy reaches large values. This places considerable importance on two values of the oscillatory energy density inside $\Delta k$, which are characteristic of the particular instability: a) the energy density for which the outflow of energy is comparable with the generation; b) the energy density for which the instability growth rate begins to change. Depending on the ratio of these two values, it is possible to have either the stabilization described above caused by the withdrawal of energy from $\Delta k$, or stabilization due to the decrease in the growth rate under the influence of the oscillations generated.

We illustrate both possible situations in two examples of decay interactions. These mechanisms are also important in the stabilization of a beam instability, which we discuss later.

## 10. PROPAGATION OF A STRONG ELECTROMAGNETIC WAVE IN A PLASMA

As an illustration of the stabilization of an instability through the change in growth rate by the generated oscillations we examine the interaction between a plasma and a high-frequency transverse wave having a broad initial spectrum, $\Delta\omega \gg \omega_{0e}$ and $\Delta\omega \ll \omega$[10]. In the one-dimensional case the diffusion equation becomes

$$\frac{\partial N^l}{\partial t} = \gamma N^l; \quad \gamma = \omega_{0e} \frac{e^2 \hbar}{8\pi m_e^2 c^2} k^t \frac{\partial N^t}{\partial k^t}; \tag{5.73}$$

$$\frac{\partial N^t}{\partial t} = \frac{\partial}{\partial k^t} D \frac{\partial N^t}{\partial k^t}; \quad D = \frac{e^2 \omega_{0e}^5 \hbar}{8\pi m_e^2 (\Omega^t)^2 c^5} N^l. \tag{5.74}$$

The initial spectrum is illustrated by the solid curve in Fig. 21. Instability occurs in the interval for which $\partial N^t / \partial\omega > 0$. Oscillations occur here which force the transverse waves to diffuse toward smaller $\omega$. As a result oscillations with lower phase velocities begin to be excited, and the transverse waves are subjected to further diffusion. The dashed curves in Fig. 21 represent the successive stages 1, 2, and 3 in the variation of the transverse-wave distribution. In the final stage $\partial N^t / \partial\omega = 0$ or $\partial N^t / \partial\omega < 0$, and the instability disappears. Moreover, the energy of the majority of the t-quanta decreases, i.e., nonlinear absorption takes place in the transverse modes, and in the longitudinal modes the instability stabilizes through the reduction in the growth rate.

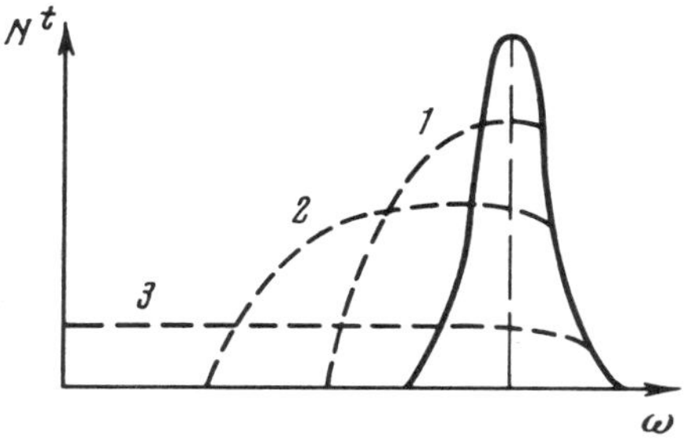


Fig. 21. The change in the transverse wave spectrum which produces stabilization of the wave instability.

The loss of transverse-wave energy is quickly deduced from elementary considerations. The initial energy of the transverse wave is

$$W_0^t = \int N_{0\omega}^t \frac{\hbar\omega \, dk}{(2\pi)^3} = \bar{\omega} \int_{\Delta\omega} \frac{N_0^t \, dk\hbar}{(2\pi)^3}, \tag{5.75}$$

where $\bar{\omega}$ is the mean wave frequency. The integration in (5.75) is taken over the initial distribution (solid curve in Fig. 21).

On the other hand, by conservation of the number of quanta

$$\int N^t\,dk = \text{const} = \int_{\Delta\omega} N_0^t dk = N_\infty^t k = N_\infty^t \bar{\omega}/c, \tag{5.76}$$

where $N_\infty^t$ is the final stable distribution corresponding to curve 3 in Fig. 21. In the final state

$$W_\infty^t = \int N_\infty^t \frac{\hbar\omega\, d\omega}{(2\pi)^3} = N_\infty^t \frac{\hbar\bar{\omega}^2}{2\,(2\pi)^3} = \frac{\bar{\omega}}{2}\int N_0^t \frac{\hbar dk}{(2\pi)^3} = \frac{W_0^t}{2}, \tag{5.77}$$

i.e., the transverse waves lose half their energy. By the conservation of energy it is transferred to longitudinal waves.

To find the distribution of the longitudinal waves generated at the onset of instability, we substitute (5.73) into (5.74):

$$\frac{\partial N^t}{\partial t} = \frac{\partial}{\partial k^t}\frac{\omega_{0e}^4}{(k^t)^3}\frac{\partial N^l}{\partial t}\frac{1}{c^4}. \tag{5.78}$$

We then obtain the following conservation relation:

$$N^t - \frac{\partial}{\partial k^t}\frac{\omega_{0e}^4}{(k^t)^3 c^4} N^l = \text{const}. \tag{5.79}$$

Integrating over k in the interval from zero to k, which does not contain $\Delta\omega$, we neglect $N_0^t$. However, inserting $N_\infty^t$ from (5.76), we have

$$N^l = N_\infty^t \frac{(k^t)^4 c^4}{\omega_{0e}^4} = \frac{W_0^t (2\pi)^3 (k^t)^4}{\hbar\bar{\omega}^2 c^{-4}}. \tag{5.80}$$

## 11. STABILIZATION OF LONGITUDINAL WAVES GENERATED BY A BEAM OF TRANSVERSE WAVES

In the preceding example $l$-waves were stabilized by a reduction of $\gamma$. In this case half the energy of the t-waves was transferred, i.e., generation throughout the entire unstable stage led to a large energy density in the longitudinal waves.

We now examine the stabilization of the $l$-wave instability in the decay process $t \rightleftarrows t + l$ caused by the withdrawal of oscillatory energy from the generation interval $\Delta\Omega$. We assume that this withdrawal is effected by the decay of a Langmuir wave into an acoustic wave. It is important to understand that the decay process $l \rightleftarrows l + s$ is not always possible; for example, it is precluded if $T_i = T_e$, or if $v_\varphi^l > 3v_{Te}\sqrt{m_i/m_e}$. The longitudinal waves generated by transverse waves have phase velocities near the velocity of light, i.e., it is required that $v_{Te} > (c/3)\sqrt{m_e/m_i}$. For a hydrogen plasma this means that the electron temperature must be greater than 30-20 eV. Let us suppose that the temperature only slightly exceeds this value. Let us also assume that $\Delta k < 2k_0$, so that the development of instability leads to the generation of Langmuir satellites and only one satellite can occur.

Since $\Delta k^l c \simeq \left(\frac{\omega_0}{\Omega^t}\right)^3 \Delta\Omega^t$, $\Omega^t \gg \omega_{0e}$, the conditions $\Delta k < 2k_0$, $\Delta\Omega^t\left(\frac{\omega_0}{\Omega^t}\right)^3 < \frac{\omega_{0e}}{3v_{Te}}\sqrt{\frac{m_e}{m_i}}c$ are normally fulfilled either in the case $\Delta\Omega_t < \omega_{0e}$ or in the case $\Delta\Omega^t > \omega_{0e}$, when a diffusion pattern occurs for decay processes $t \rightleftarrows l + t$.

The characteristic time for a single decay in the case $\Delta\Omega^t < \omega_{0e}$ is

$$\frac{1}{\tau^{t,\,l}} \approx \frac{W^t}{nm_e}\omega_{0e}\frac{\omega_{0e}}{\Delta\Omega^t}\frac{\pi}{2c^2}, \tag{5.81}$$

where $W^t$ is the energy of the primary packet of t-waves. During a time $\tau$ an $l$-wave energy of order $W^t\omega_{0e}/\Omega^t$ is produced. The time for the transfer of this energy into the first Langmuir satellite is

$$\frac{1}{\tau^{l,\,s}} \sim \frac{W^l}{nm_e v_{Te}^2}\omega_e\frac{\pi}{4}\frac{k_0}{\Delta k} \simeq \frac{\omega_{0e}}{\Omega^t v_{Te}}\frac{cW^t}{nm_e v_{Te}^2}\frac{\omega_{0e}\pi\sqrt{\frac{m_e}{m_i}}\omega_{0e}}{\Delta\Omega^t\left(\frac{\omega_{0e}}{\Omega^t}\right)^3 12} \sim$$

$$\sim \left(\frac{\Omega^t}{\omega_{0e}}\right)^2\frac{\omega_{0e}^2}{\Delta\Omega^t}\frac{W^t}{nm_e}\sqrt{\frac{m_e}{m_i}}\frac{c}{v_{Te}^3}\frac{\pi}{12}. \tag{5.82}$$

Comparison of (5.81) and (5.82) reveals that $\tau^{l,s}$ is in general much smaller than $\tau^{t,l}$ $\Bigl(\Omega^t \gg \omega_{0e}$, $v_{Te} \ll c$, and the factor $\left(\frac{\Omega^t}{\omega_0}\right)^2\left(\frac{c}{v_{Te}}\right)^3$ is very large and cannot cancel $\frac{\pi}{12}\sqrt{\frac{m_e}{m_i}}\Bigr)$. This estimate shows that transfer occurs before $W^l$ can attain the value $W^t\omega_{0e}/\Omega^t$. Equating $\tau^{t,l}$ and $\tau^{l,s}$, we infer that the transfer loss is comparable with the generation for very small values of $W^l$:

$$W^l \simeq W^t \frac{v_{Te}^3}{c^3}\left(\frac{\omega_{0e}}{\Omega^t}\right)^3 \sqrt{\frac{m_i}{m_e}} = W^l_*. \tag{5.83}$$

If $W^l$ is greater than the value determined by (5.83), the transfer loss greatly affects the process.

The withdrawal of energy leads to different results, depending on whether or not the acoustic oscillations can be absorbed during the transfer time. The rate of transfer at the time of equal generation and loss is determined by (5.81), and the damping rate of the acoustic waves is

$$\frac{1}{\tau} \approx \sqrt{\frac{\pi}{8}\frac{m_e}{m_i}}\, k^s v_s \sim 2k^l \sqrt{\frac{\pi}{8}\frac{m_e}{m_i}}\, v_{Te} \sqrt{\frac{m_e}{m_i}} \sim \sim \omega_{0e}\frac{m_e}{m_i}\sqrt{\frac{\pi}{2}}\frac{v_{Te}}{c}. \tag{5.84}$$

Setting (5.81) equal to (5.84), we obtain

$$W^t = W^t_* \simeq n m_e v_{Te} \frac{m_e}{m_i}\sqrt{\frac{2}{\pi}}\frac{\Delta\Omega^t}{\omega_{0e}} 4c. \tag{5.85}$$

If $W^t \gg W^t_*$, the absorption of sound is unimportant. In this case for $W^l \gg W^l_*$ the transfer processes are so rapid as to permit equilibrium to be established between the energy densities inside and outside $\Delta k^l$ at every instant of time (for the first $l$-satellite). This energy ratio is 1:3, i.e., in the first satellite the energy is three times that inside $\Delta k^l$. Consequently, the transverse waves must generate four times as many waves in order to perpetuate the growth of waves inside $\Delta k$. In other words, the instability growth rate is reduced to one fourth if $W^l$ begins to exceed $W^l_*$, i.e., the transfer effect causes only a reduction of the growth rate. The instability continues to develop and can turn out to be similar to that in the preceding section, in which event the time for half the transverse-wave energy to be lost is four times its value in the absence of the decay process $l \rightleftarrows l + s$.

If $W^t \ll W^t_*$, the absorption of sound is very strong. It causes the equilibrium value of the $l$-waves for relatively fast $l$-s decay to correspond to zero $l$-wave intensity inside $\Delta k^l$. This is easily verified. By virtue of the strong absorption of s-waves $N^s \ll N^l$, and the equation describing the transfer of energy from $\Delta k$ to the outside becomes

$$\frac{\partial N_1^l}{\partial t} = -\alpha N_1^l N_2^l; \quad \frac{\partial N_2^l}{\partial t} = \alpha N_1^l N_2^l, \tag{5.86}$$

where $N_1^l$ and $N_2^l$ are the numbers of $l$-quanta inside and outside $\Delta k^l$, respectively, and $\alpha$ is a quantity determined from the probability of $l-s$ decay:

$$\alpha = \frac{\hbar\pi}{48} \frac{e^2 \sqrt{m_e/m_i}}{m_e^2 v_{Te}^3} \omega_{0e}.$$

By the conservation of the number of $l$-quanta

$$N_1^l + N_2^l = \text{const} = N_{10}^l + N_{20}^l \simeq N_{10}^l \tag{5.87}$$

it is a simple matter to solve (5.86):

$$N_1^l = N_{10} \frac{\exp\left[-\alpha t (N_{10} + N_{20})\right]}{\frac{N_{20}}{N_{10}} + \exp\left[-\alpha t (N_{10} + N_{20})\right]}. \tag{5.88}$$

For $N_1$ to tend to zero means the complete removal of instability. The $l$-wave energy increases to $W_*^l$, then is transferred outside $\Delta k$. This terminates the process. The transverse waves lose only a minute fraction of their total energy in this case.

We see that, depending on the specific oscillation conditions, either a large or small fraction of the transverse-wave energy can be transferred. Accordingly, the nonlinear effects in one case yield large values of the steady-state energy, in others they yield small values. It is necessary, of course, that the divergence angle of the transverse-wave beam be sufficiently small, $\Delta\Theta \ll (\omega_{0e}/\Omega)^{3/2}$.

In presenting these examples we have merely attempted to clarify the essential points covered in Section 9. The examples also help us understand how low-frequency plasma oscillations are generated by high-frequency plasma oscillation. The former can produce additional diffusion of the plasma, a result that is important to plasma confinement. We stress the fact that both nonlinear absorption and nonlinear stabilization effects can turn off the excitation of low-frequency oscillations. The excitation of low-frequency by high-frequency oscillations has been observed experimentally [11-13].*

## 12. DECAY PROCESSES AND PLASMA DIAGNOSTICS

Nonlinear effects can assume considerable importance when one of the interacting waves is fairly weak. When this is so, as we have seen, a weak wave can be amplified at the expense of energy from a large-amplitude wave. This offers real possibilities for plasma diagnostics.

For example, suppose that a relatively weak transverse wave is transmitted through a plasma. If longitudinal waves are present in the plasma, acoustic or Langmuir modes in particular, or any other modes of sufficiently high intensity, then the transmitted wave is scattered by them, its frequency being altered at the same time.† If the plasma waves have a sufficiently narrow spectrum, satellite waves are generated. The intensity of the transmitted wave must be such as not to change the energy of the observed waves. From the intensity profile of the satellite and the angle of scattering it is possible to find the spectral and angular distribution of the waves, the plasma density, temperature, and other parameters [18, 19].‡

This method of plasma diagnostics possesses several advantages over others. First, no instruments (such as probes) are introduced into the plasma where they might seriously perturb it. The intensity of the probing wave must be sufficiently small:

$$W^t \ll W^\sigma \frac{\Omega^\sigma}{\Omega^t}, \tag{5.89}$$

where $W^\sigma$ is the energy density of the observed waves, and $\Omega^\sigma$ is their frequency. If (5.89) is satisfied, the measuring wave cannot alter the distribution of the observed $\sigma$-waves. Second, and this is emphasized, this method can be used for high plasma densities.

* See also two papers presented at the International Conference on Phenomena in Ionized Gases, Vienna (1967) [32, 33].

† With regard to scattering by weak fluctuation waves, see, e.g., [14-17].

‡ See [20] for a discussion of the measurement of the parameters of cosmic plasma.

This is particularly important for the diagnostics of very dense plasmas, for which we must use probing waves in the optical range.

We now show how to estimate the energy density of plasma oscillations from the intensity of the scattered radiation. Suppose we are to detect Langmuir waves. We send into the plasma a transverse wave of frequency $\Omega^t$ with $\delta\Omega^t \ll \omega_{0e}$. Since the wave is weak, the intensity of the first satellite will be much smaller than that of the transmitted wave. Consequently, its intensity may be estimated from the decay interaction equation using only those terms which are proportional to the intensity of the incident transverse and the detected Langmuir waves:

$$\frac{dN^t_{\mathbf{k}_2}}{dt} = \int w_t^{l,\,t}(\mathbf{k}_1, \mathbf{k}_2, \mathbf{k})\, N^t_{0\mathbf{k}_1} N^l_{\mathbf{k}} \frac{d\mathbf{k}\, d\mathbf{k}_1}{(2\pi)^6}. \tag{5.90}$$

The right-hand side of (5.90) is known. Hence the radiation intensity of the satellite is

$$\frac{dW^t}{dt} \simeq S_0 W^l \alpha; \quad \alpha = \frac{\pi}{16} \frac{\omega_{0e}^3}{n_0 m_e c^2 v^l_\varphi (\Omega^t)^2}, \tag{5.91}$$

where $S_0 = W_0^l c$ is the energy flux of the incident radiation.

Note that (5.91) is suitable only for a rough estimate of $W^l$, whereas it is possible in principle to determine not only the average energy $W^l$ of the oscillations, but also their spectral distribution [18].

We next show how to find the plasma concentration. This requires that the difference between the frequencies of the fundamental line and satellite be measured:

$$\Delta\Omega = \omega_{0e} = \sqrt{\frac{4\pi n e^2}{m_e}}. \tag{5.92}$$

The temperature of a plasma can be determined from the scattering of electromagnetic waves by acoustic waves from the dependence of the frequency difference between the scattered and transmitted waves on the scattering angle:

$$\begin{aligned} \Omega_1^t - \Omega_0^t &= \Delta\Omega = \Omega^s = k^s v_s = |\mathbf{k}_1 - \mathbf{k}_0|\, v_s \simeq \\ &\simeq 2 \frac{\Omega_0^t}{c} v_{Te} \sqrt{\frac{m_e}{m_i}} \sin^2 \frac{\theta}{2}, \end{aligned} \tag{5.93}$$

where $\Theta$ is the scattering angle.

An equation analogous to (5.91) can be found from the probability of the decay process $t \rightleftarrows t + s$ and used for measuring the sound-wave intensity in a plasma. Finally, in the presence of a magnetic field it is possible also to measure the ion temperature from the combination scattering by ion-cyclotron waves.

If the diffusion approximation is valid, the oscillation energy density can be estimated from the scattering angle:

$$\frac{dN_{\mathbf{k}}^t}{dt} = \frac{\partial}{\partial k_i} D_{i,j} \frac{\partial N_{\mathbf{k}}^t}{\partial k_j} . \tag{5.94}$$

Over a path length L the scattering at $\Delta\Theta = \Delta k/k$ may be estimated as

$$(\Delta\theta)^2 = \frac{LD}{k^2 c}, \tag{5.95}$$

where D is the larger of the components $D_{i,j}$.

The energy of Langmuir oscillations has been found experimentally in this way by Demidov and Fanchenko [21] (see also [22-25]).

We conclude the present section with the observation that the above equations are valid as long as the appropriate plasma dimension is much greater than the wavelengths of the interacting waves. The opposite inequality is discussed in [26].

## 13. DECAY PROCESSES AND NONTHERMAL PLASMA RADIATION

Suppose that fairly high-intensity oscillations have been generated in a plasma by an instability. The following question arises: what will be the radiation from the plasma and, in particular, what will be its radiative energy losses? It is assumed that the plasma is confined, and we investigate the external radiation. In addition to decay processes, processes associated with the conversion of waves at the nonuniform boundary of the plasma will contribute to the radiation output (Tidman). As only transverse waves can exist outside the plasma, nonlinear processes which generate transverse waves are important when considering nonthermal radiation. Nonlinear effects occur throughout the entire volume of the plasma, boundary-related effects playing only a relatively small role. For small plasma dimensions the contribution of surface effects, surface waves in particular, increases. In addition to decay interactions, induced scattering effects can also play a major part in the conversion of plasma oscillations into transverse waves. Nonthermal radiation is important not only through its role in cooling the plasma, but also because it is useful, in conjunction with scattering measurements, for plasma diagnostics.

Let us summarize the decay processes that can lead to the generation of transverse waves in isotropic plasma. They include, for example, the coalescence of two Langmuir waves into a transverse wave, $l \rightleftarrows l + t$, which results in specific radiation from a plasma at frequencies near $2\omega_{0e}$ [6]. Other processes are the coalescence of a Langmuir and an ion-acoustic wave, $l + s \rightarrow t$ and the decay of a Langmuir into a transverse and an ion-acoustic wave, $l \rightarrow t + s$, resulting in radiation at frequencies near $\omega_{0e}$. These are the only processes that can generate waves with frequencies greater than $\omega_{0e}$ in isotropic plasma without involving relativistic beams. There are obviously a great many more such processes possible in an anisotropic plasma in a magnetic field.

In every case, however, it is required that at least one wave of sufficiently high frequency participate in the interaction. For t-waves of low intensity the equation describing the increase in the number of t-waves by the process $l + l \rightarrow t$ has the form

$$\frac{\partial N^t_{\mathbf{k}}}{\partial t} = \int w^t_{ll}(\mathbf{k},\ \mathbf{k}_1,\ \mathbf{k}_2)\, N^l_{\mathbf{k}_1} N^l_{\mathbf{k}_2} \frac{d\mathbf{k}_1 d\mathbf{k}_2}{(2\pi)^6}\,. \tag{5.96}$$

From this we obtain an order-of-magnitude estimate of the energy increase of the t-waves per unit time for isotropic $l$-waves:

$$\frac{dW^t}{dt} \simeq 10\, \frac{W^l W^l \omega_{0e}}{nm_e c^2} \left( \frac{\omega_{0e}}{k^l c} \right)^3. \tag{5.97}$$

Here $k^l$ is the characteristic wave number of the $l$-waves.

For the process $l + s \rightarrow t$ we similarly obtain an estimate of the radiation intensity [7] from a plasma ($k^s \ll k^l$):

$$\frac{dW^t}{dt} \simeq \frac{W^l W^s \omega_{0e}}{nm_e v^2_{Te}} \left( \frac{\omega_{0e}}{k^t c} \right)^2. \tag{5.98}$$

The process $l \to t + s$ is solved by the conservation laws for the case when the phase velocity of the Langmuir waves satisfies

$$\frac{v_\varphi^l}{v_{Te}} < \sqrt{\frac{9m_i}{4m_e}}. \qquad (5.99)$$

In the equation

$$\frac{\partial N_{\mathbf{k}}^t}{\partial t} = \int w_l^{ts}(\mathbf{k}, \mathbf{k}_1, \mathbf{k}_2)(N_{\mathbf{k}}^t N_{\mathbf{k}_1}^l + N_{\mathbf{k}_1}^l N_{\mathbf{k}_2}^s - N_{\mathbf{k}_2}^s N_{\mathbf{k}}^t)\frac{d\mathbf{k}_1 d\mathbf{k}_2}{(2\pi)^6} \qquad (5.100)$$

we neglect the terms $N_{\mathbf{k}_1}^l N_{\mathbf{k}_2}^s$ and $N_{\mathbf{k}_2}^s N_{\mathbf{k}}^t$, which are small if $N_{\mathbf{k}_2}^s$ is small. Then the growth of the t-waves is exponential:

$$\frac{\partial N_{\mathbf{k}}^t}{\partial t} = \gamma_{\text{nlr}}^t N_{\mathbf{k}}^t; \quad \gamma_{\text{nlr}}^t = \int w_l^{ts}(\mathbf{k}, \mathbf{k}_1, \mathbf{k}_2) N_{\mathbf{k}_1}^l \frac{d\mathbf{k}_1 d\mathbf{k}_2}{(2\pi)^6}. \qquad (5.101)$$

From this we estimate $\gamma_{\mathbf{k}}^t$ for the strong inequality (5.99):

$$\gamma_{\text{nlr}}^t = \frac{\pi}{32}\frac{W^l}{nm_e v_{Te}^2}\left(\frac{4m_e}{3m_i}\right)^{1/2}\omega_{0e}\frac{v_\varphi}{v_{Te}}\frac{k_1}{\Delta k_1}. \qquad (5.102)$$

Here $\Delta k_1 < k_1$ is the width of the $l$-wave spectrum.

The processes investigated of nonlinear conversion of longitudinal into transverse waves are important for the analysis of beam instabilities in plasma, when strong Langmuir oscillations can develop.

## 14. DECAY PROCESSES FOR NEGATIVE-ENERGY WAVES; MECHANISM OF NONLINEAR WAVE GENERATION

If the medium is not in equilibrium, its dispersion properties can change so that the quantity $(\partial/\partial\omega)\omega^2\varepsilon^\sigma|_{\omega>0}$ becomes negative.* Waves for which this condition is fulfilled are called negative-energy waves. The reason for the name stems from the fact that their interaction occurs as if each wave had an energy of $-\hbar\Omega^\sigma$

* For a medium in equilibrium $\text{Im}\,\varepsilon^\sigma > 0$ for $\omega > 0$ and $\gamma_{\mathbf{k}} < 0$, i.e., waves are absorbed, and as

$$\gamma_{\mathbf{k}} = -\frac{\omega^2\,\text{Im}\,\varepsilon^\sigma}{\frac{\partial}{\partial\omega}\omega^2\varepsilon^\sigma}, \text{ therefore } \frac{\partial}{\partial\omega}\omega^2\varepsilon^\sigma > 0.$$

[30]. To verify this we refer to the equations describing the nonlinear decay interaction of two transverse and one longitudinal random wave [see Eqs. (3.50) and (3.52)]. Let one of the transverse waves, say $\mathbf{k}_1$, be a negative-energy wave, i.e., let

$$\frac{\partial}{\partial\omega}\,\omega^2\varepsilon^t\,|_{\omega=\Omega^t(\mathbf{k}_1)>0}<0. \tag{5.103}$$

Refer to Eq. (3.50). By the definition (3.49) the quantity $w_{\mathbf{k}}(\mathbf{k}_1,\ \mathbf{k}_2)$ is negative in this case and cannot be interpreted as a probability. We replace it with $w_-(\mathbf{k}_1\mathbf{k}_2)=-w_{\mathbf{k}}(\mathbf{k}_1,\ \mathbf{k}_2)$. It also turns out that $N^t_{\mathbf{k}_1}$ is negative. We introduce in place of it $\widetilde{N}^t_{\mathbf{k}_1}=-N^t_{\mathbf{k}_1}$. Now the decay equations become

$$\frac{\partial}{\partial t}\ N^l_{\mathbf{k}}=\int w_-(\widetilde{N}^t_{\mathbf{k}_1}N^t_{\mathbf{k}_2}+N^t_{\mathbf{k}_2}N^l_{\mathbf{k}}+\widetilde{N}^t_{\mathbf{k}_1}N^l_{\mathbf{k}})\,\frac{d\mathbf{k}_1\,d\mathbf{k}_2}{(2\pi)^6}; \tag{5.104}$$

$$\frac{\partial}{\partial t}\ \widetilde{N}^t_{\mathbf{k}_1}=\int w_-(\widetilde{N}^t_{\mathbf{k}_1}N^t_{\mathbf{k}_2}+N^t_{\mathbf{k}_2}N^l_{\mathbf{k}}+\widetilde{N}^t_{\mathbf{k}_1}N^l_{\mathbf{k}_2})\,\frac{d\mathbf{k}\,d\mathbf{k}_2}{(2\pi)^6}; \tag{5.105}$$

$$\frac{\partial}{\partial t}\ N^t_{\mathbf{k}_2}=\int w_-(\widetilde{N}^t_{\mathbf{k}_1}N^t_{\mathbf{k}_2}+N^t_{\mathbf{k}_2}N^l_{\mathbf{k}}+\widetilde{N}^t_{\mathbf{k}_1}N^l_{\mathbf{k}})\,\frac{d\mathbf{k}\,d\mathbf{k}_1}{(2\pi)^6}. \tag{5.106}$$

Let us now show how probability arguments can be used to derive equations describing decay processes of this type. Assume that $\widetilde{N}^t_{\mathbf{k}_1}$ is the number of quanta, where the energy of each quantum is $-\hbar\Omega^t(\mathbf{k}_1)$ and the momentum is $-\hbar\mathbf{k}_1$. The total energy of the quanta is

$$W^t=-\int\frac{2\hbar\Omega^t(\mathbf{k}_1)}{(2\pi)^3}\ \widetilde{N}^t_{\mathbf{k}_1}\,d\mathbf{k}_1=$$

$$=\frac{1}{8\pi}\int\frac{|\,\mathbf{E}^t_{\mathbf{k}_1}|^2}{\Omega^t(\mathbf{k}_1)}\,\frac{\partial}{\partial\omega}\,\omega^2\varepsilon^t(\omega,\ \mathbf{k}_1)\,|_{\omega=\Omega^t(\mathbf{k}_1)}\,d\mathbf{k}_1. \tag{5.107}$$

Here $\widetilde{N}^t_{\mathbf{k}_1}$ is expressed in terms of $|\,\mathbf{E}^t_{\mathbf{k}_1}|^2$ according to (3.48). We note that the resulting expression for the energy of the field agrees with (4.28), which was obtained by averaging the classical expression for $W^t$ over the phases of the random waves. Equation (4.28) is a general expression obtained for a given choice of constant; the field energy can be found with an accuracy up to this constant.

The quantity $(\partial/\partial\omega)\omega^2\varepsilon^t$ cannot be negative for the entire frequency range, for example as $\omega\to\infty$ we find $\varepsilon^t\to 1$ and $(\partial/\partial\omega)\omega^2\varepsilon^t\to 2\omega>0$. In other words, it is possible for the energy

$$W^t=\frac{1}{8\pi}\int\frac{|\,\mathbf{E}^t_{\mathbf{k}_1}|^2}{\Omega^t(\mathbf{k}_1)}\,\frac{\partial}{\partial\omega}\,\omega^2\varepsilon^t(\omega,\ \mathbf{k}_1)\,|_{\omega=\Omega^t(\mathbf{k}_1)}\,d\mathbf{k}_1 \tag{5.108}$$

to have, in general, parts of its spectrum where the energy density is positive and parts where it is negative. The important thing is that if $|\mathbf{E}^t_{\mathbf{k}_1}|^2 \neq 0$ where $(\partial/\partial\omega)\omega^2\varepsilon^t < 0$, then $W^t < 0$. If there are two wave packets, one of which has $(\partial/\partial\omega)\omega^2\varepsilon^t < 0$, the other $(\partial/\partial\omega)\omega^2\varepsilon^t > 0$, then (5.108) determines their energy difference, which has a real physical meaning and is of course independent of the choice of constant in (5.108).

The total energy of the system decreases if a negative-energy wave is excited. A situation in which the generation of waves decreases the energy of a system is possible only for a nonequilibrium situation, because if the system is in equilibrium its energy is minimum and every wave excitation must increase the energy of the system.

It is reasonable to expect that if the energy of the system decreases through wave excitation, those waves will grow in the system. Since the energy difference of the waves is only manifested in the presence of both positive-energy and negative-energy waves, this growth can occur through nonlinear interaction between positive- and negative-energy waves. In fact, it is clear from (5.104)-(5.106) that the right-hand sides of the equations are positive, i.e., all three interacting waves grow in amplitude.

Equations (5.104)-(5.106) can be derived using probability arguments if we analyze the simultaneous emission of three waves, one of which has energy $-\Omega^t(\mathbf{k}_1)$ and momentum $-\mathbf{k}_1$. Let $w_-$ be the probability of the process. The rate of change of the number of waves due to induced emission is

$$w_-(\widetilde{N}^t_{\mathbf{k}_1}+1)(N^t_{\mathbf{k}_2}+1)(N^l_{\mathbf{k}}+1), \tag{5.109}$$

and that due to the inverse process of absorption of the three waves is

$$-w_-\widetilde{N}^t_{\mathbf{k}_1}N^t_{\mathbf{k}_2}N^l_{\mathbf{k}}. \tag{5.110}$$

The difference between (5.109) and (5.110) is

$$w_-(\widetilde{N}^t_{\mathbf{k}_1}N^t_{\mathbf{k}_2}+\widetilde{N}^t_{\mathbf{k}_1}N^l_{\mathbf{k}}+N^t_{\mathbf{k}_2}N^l_{\mathbf{k}}),$$

which corresponds to the right-hand side of (5.104)-(5.106).

It follows from these equations that the difference between the number of positive- and negative-energy quanta remains constant in this case:

$$\widetilde{N}_1^t - N_2^t = \text{const}; \quad \widetilde{N}_1^t - N^l = \text{const};$$
$$\widetilde{N}_1^t = \int \widetilde{N}_{\mathbf{k}_1}^t \frac{d\mathbf{k}_1}{(2\pi)^3}; \quad N_2^t = \int N_{\mathbf{k}_2}^t \frac{d\mathbf{k}_2}{(2\pi)^3}; \quad N^l = \int N_{\mathbf{k}}^l \frac{d\mathbf{k}}{(2\pi)^3}. \tag{5.111}$$

This conservation of only the difference between the numbers of quanta shows that the absolute value of the "number of quanta" of each type can grow without limit, but of course only up to intensities for which the equations are still valid.

It is readily seen that these equations lose their validity at a definite stage of the development of instability in that the growth rate of the waves does not always increase, i.e., the characteristic time $t_0$ for the nonlinear interaction decreases. As soon as $t_0$ becomes of order $1/\delta\Omega$, the random-phase approximation no longer applies. Consider the example when the transverse waves have frequencies much higher than the longitudinal-wave frequencies. Then the terms in $N_{\mathbf{k}_1}^t N_{\mathbf{k}_2}^t$ in (5.104) can be neglected, so that

$$\frac{\partial}{\partial t} N_{\mathbf{k}}^l = N_{\mathbf{k}}^l \int w_- (N_{\mathbf{k}_2}^t + \widetilde{N}_{\mathbf{k}_1}^t) \frac{d\mathbf{k}_1 \, d\mathbf{k}_2}{(2\pi)^6}. \tag{5.112}$$

Since $N_{\mathbf{k}}^t$ grows with time, $N_{\mathbf{k}}^l$ grows faster than exponentially, i.e., the effective growth rate characterizing the time $t_0$ is also a growing function of time.

Finally, we should point out that nonequilibrium systems are in general unstable in the linear approximation. If the nonequilibrium state is eliminated during the characteristic buildup time of nonlinear instability, the $\Omega$-interval in which the wave energy is negative also vanishes. Since the nonlinear instability time depends on the initial wave intensity, it is possible to arrive at a situation corresponding to a low initial wave intensity, when nonlinear instability does not occur.

Let us now explore what types of negative-energy wave interaction can exist within the context of the approximation of three interacting waves. If the wave with highest frequency has negative energy, then (5.104)-(5.106) tell us that all the waves grow. If the negative-energy wave has the lowest frequency, the equations retain the same form as when the energy of all the waves is positive, except that two of the waves exchange roles. For example, a strong wave with a narrow spectrum generates blue, rather than red, satellites in the plasma, or, in other words, the direction of spectral shift of the transverse waves reverses.

If the middle-frequency wave $(\Omega_2)$ has negative energy, the sum $N_1^t + \tilde{N}_2^t$ and the difference $N_1^t - \tilde{N}_2^t$ remain constant. Finally, two quanta can have negative energy, which in some measure is equivalent to the case when one quantum has negative energy, because only the relative signs of the wave energies are important.

The general aspects of the theory of unstable decay of waves with negative energy have been investigated by Dikasov, Rudakov, and Ryutov [30], who have shown that the process of instability development is accompanied by increasing entropy.

## 15. NONLINEAR DECAY INSTABILITIES OF TRANSVERSE AND LONGITUDINAL WAVES IN PARTIALLY IONIZED PLASMA

The above is illustrated by the example of transverse- and longitudinal-wave interaction in a partially ionized plasma containing neutral atoms with inverted population levels [31]. The nonequilibrium of the plasma in this case results from the fact that the number of atoms in a higher level is greater than the number in a lower level. We assume that one of the interacting waves has a frequency near $\omega_{21} = (\varepsilon_2 - \varepsilon_1)/\hbar$, i.e., the transition frequency between the inverted population levels. The dielectric constant of the plasma at frequencies $\omega$ near $\omega_{12}$ is described by the formula

$$\varepsilon^t(\omega) = 1 - \frac{\omega_{0*}^2(\nu_1 - \nu_2)}{\omega^2 - \omega_{12}^2}. \tag{5.113}$$

Here $\omega_{0*}^2 = 4\pi n_* e^2/m_e |f_{12}|$; $n_*$ is the concentration of neutral atoms, $\nu_1 = n_1/n_*$ and $\nu_2 = n_2/n_*$ are the relative populations of the levels $\varepsilon_1$ and $\varepsilon_2$, $|f_{12}|$ is the oscillator strength, and $n_1$ and $n_2$ are the concentrations of atoms with energy $\varepsilon_1$ and $\varepsilon_2$. By stipulation $\nu_2 > \nu_1$. We assume that $\omega - \omega_{12} \gg \gamma$, where $\gamma$ characterizes the damping associated with the width of the line in question. Let us say that $\omega_{12} \gg \omega_{0e} = \sqrt{4\pi n e^2/m_e}$ (n is the electron concentration) and $\omega_{12} \gg \omega_{0*}$. Then in the vicinity of $\omega \sim \omega_{0e}$ the effect of the neutral atoms is negligible, and the plasma waves are essentially unaltered by their presence. From (5.113) we have

$$\frac{\partial}{\partial\omega}\omega^2\varepsilon^t(\omega) = 2\omega\left(1 + \frac{\omega_{12}^2(\nu_1 - \nu_2)\,\omega_{0*}^2}{(\omega^2 - \omega_{12}^2)^2}\right). \tag{5.114}$$

This expression can be negative only if $\nu_2 > \nu_1$, i.e., for an inverted population. It is also obvious that it is negative if $(\omega - \omega_{12})/\omega_{12} \ll 1$, i.e., if the wave frequency is near $\omega_{12}$. We infer from (5.114) that the wave has negative energy if

$$\Delta\omega < \frac{\omega_{0*}}{2}\sqrt{\nu_2 - \nu_1}, \quad \Delta\omega = \omega - \omega_{12}. \tag{5.115}$$

Since $\Delta\omega \gg \gamma$ condition (5.115) is fulfilled for $\gamma \ll \omega_{0*}$, i.e., for a sufficiently large concentration of neutral atoms.

Let (5.115) be satisfied for the investigated wave but not satisfied for the wave of frequency $\omega - \omega_{0e}$, i.e., let the second transverse wave have positive energy. It is clear that if we are dealing with two waves for which (5.115) is valid and whose frequencies differ by $\omega_{0e}$ (which is necessary in order for decay interaction to be possible), it turns out that their interaction is consistent with the above considerations.*

If we vary the frequencies of the two waves properly, we arrive at a frequency such that one of the waves lies outside the interval $\Delta\omega$ defined by (5.115). In this case the nonlinear instability described above sets in. As for the temporal development of the system,† in the linear approximation the negative-energy wave decays, while the positive-energy wave grows. This follows from the expression for the decay rate $\gamma = -\dfrac{\omega^2 \operatorname{Im} \varepsilon^t}{\dfrac{\partial}{\partial\omega}\omega^2\varepsilon^t}$ and the fact that $\operatorname{Im}\varepsilon$ is negative for inverted populations, as well as the change in sign of $(\partial/\partial\omega)\omega^2\varepsilon^t$ for positive and negative energy. Let us estimate the value of this decay rate, allowing for a small imaginary part in $\varepsilon^t$:

$$\varepsilon^t(\omega) = 1 - \frac{\omega_{0*}^2(\nu_1 - \nu_2)}{\omega^2 - \omega_{12}^2 + 2\omega_{12}\gamma i}; \quad \operatorname{Im}\varepsilon^t(\omega) = \frac{-\omega_{0*}^2(\nu_2 - \nu_1)\,2\omega_{12}\gamma}{(\omega^2 - \omega_{12}^2)^2}. \tag{5.116}$$

From (5.114) and (5.116) we obtain

$$\gamma_k \simeq \gamma\frac{\alpha}{1-\alpha}; \quad \alpha = \frac{\omega_{12}^2(\nu_2 - \nu_1)\,\omega_{0*}^2}{(\omega^2 - \omega_{12}^2)^2}, \tag{5.117}$$

* As mentioned, spectral transfer will occur in the direction of higher frequencies. If the frequencies of both lines are lower than $\omega_{12}$, absorption takes place.

† The system is homogeneous in space, and at the initial time the variables are independent of the coordinates.

i.e., $\gamma_k$ is of the same order as the line width $\gamma$. The dispersion relation $\omega^2\varepsilon^t = k^2c^2$ with allowance for (5.113) leads to a biquadratic equation in the frequency, which has the solution

$$\omega^2 = \Omega^2(\mathbf{k}) = \frac{\omega_{12}^2 + k^2c^2 + \omega_{0_*}^2(\nu_1 - \nu_2)}{2} \pm$$

$$\pm \sqrt{\frac{(\omega_{12}^2 + k^2c^2 + \omega_{0_*}^2(\nu_1 - \nu_2))^2}{4} - k^2\omega_{12}^2c^2}. \quad (5.118)$$

The quantity $\Omega(\mathbf{k})$ is close to $\omega_{12}$ only if kc differs only slightly from $\omega_{12}$. Assuming $(\omega_{12} - kc)^2 \gg \omega_{0*}^2$, we obtain approximately

$$\Omega^2(\mathbf{k}) - \omega_{12}^2 = k^2c^2 - \omega_{12}^2; \quad \frac{\omega_{0_*}^2(\nu_2 - \nu_1)\,\omega_{12}^2}{k^2c^2 - \omega_{12}^2}. \quad (5.119)$$

The first value of (5.119) corresponds to a refractive index of unity and, as easily seen, is invalid for negative energy; the second value yields

$$\frac{\partial}{\partial\omega}\omega^2\varepsilon^t = -2\,\frac{(k^2c^2 - \omega_{12}^2)^2}{\omega_{0_*}^2(\nu_2 - \nu_1)\,\omega_{12}}. \quad (5.120)$$

Since we have assumed $\Delta\omega \gg \gamma$ we have

$$\Delta k = \left| k - \frac{\omega_{12}}{c} \right| \ll \frac{\omega_{0_*}^2}{\gamma} \text{ and } \Delta k \gg \omega_{0_*}.$$

We now examine the one-dimensional theory of the development of nonlinear instability for the case in which a high-frequency transverse wave has negative energy. Because $\omega_{12} \gg \omega_{0e}$ the terms $N_1^t$ and $N_2^t$ are unimportant, and for positive-energy waves $(\partial/\partial\omega)\omega^2\varepsilon^t \simeq 2\omega_{12}$. The equations now take the form

$$\gamma_1\widetilde{N}_1^t + \frac{\partial\widetilde{N}_1^t}{\partial t} = \widetilde{\beta}^t N^l(\widetilde{N}_1^t + N_2^t); \quad (5.121)$$

$$-\gamma_2 N_2^t + \frac{\partial N_2^t}{\partial t} = \widetilde{\beta}^t N^l(\widetilde{N}_1^t + N_2^t); \quad (5.122)$$

$$\frac{\partial N^l}{\partial t} = \widetilde{\beta}^l N^l(\widetilde{N}_1^t + N_2^t), \quad (5.123)$$

where

$$\widetilde{\beta}^t = \frac{e^2\omega_{0e}^3\,\omega_{0_*}^2(\nu_2 - \nu_1)}{8\pi m_e^2(k^2c^2 - \omega_{12}^2)^2}; \quad \widetilde{\beta}^l = \widetilde{\beta}^t\left(\frac{\omega_{12}}{\omega_{0e}}\right)^3;$$

$N^l$, $\widetilde{N}_1^t$, and $\widetilde{N}_2^t$ are the one dimensional distribution functions for the quanta.

Let the initial intensity of the longitudinal waves be large enough that nonlinear effects dominate over linear effects and the initial transverse-wave intensities are small.* Neglecting $\gamma_1$ and $\gamma_2$ on the left-hand sides of (5.121) and (5.122) and making use of the relations $\widetilde{N}_1^t - N_2^t = \text{const} = \widetilde{N}_{10}^t - N_{20}^t$ и $\widetilde{N}_1^t - \frac{\widetilde{\beta}^t}{\widetilde{\beta}^l} N^l = \widetilde{N}_{10}^t - \frac{\widetilde{\beta}^t}{\widetilde{\beta}^l} N_0^l$, we find the solution of the set of equations

$$N^l = \frac{N_0^l\left(-\widetilde{N}_{10}^t - N_{20}^t + 2\frac{\widetilde{\beta}^t}{\widetilde{\beta}^l} N_0^l\right)}{-e^{2\widetilde{\beta}^t N_0^l t}(\widetilde{N}_{10}^t + N_{20}^t) + 2\frac{\widetilde{\beta}^t}{\widetilde{\beta}^l} N_0^l}. \tag{5.124}$$

Here we have set $\frac{2\widetilde{\beta}^t}{\widetilde{\beta}^l} N_0^l \gg N_{10}^t + N_{20}^t$, which shows that in order of magnitude $W_0^t \ll (\omega_{0e}/\omega_{12}) W_0^l$.

This result implies that the number of quanta of all types grows without limit, because the denominator of (5.124) vanishes. Setting the denominator of (5.124) equal to zero, we obtain the characteristic buildup time $t_0$ of the wave intensity:

$$t_0 = \frac{1}{2\widetilde{\beta}^t N_0^l} \ln \frac{2\widetilde{\beta}^t N_0^l}{\widetilde{\beta}^l(\widetilde{N}_{10}^t + N_{20}^t)}, \tag{5.125}$$

which is of order

$$t_0 \simeq \frac{1}{\omega_e} \frac{\delta\Omega\,(k^2c^2 - \omega_{12}^2)^2}{\omega_{12}^3 \omega_{0*}^2 (\nu_2 - \nu_1)} \frac{n m_e c^2}{W_0^l} \ln \frac{W^l \omega_{12}}{W^t \omega_{0e}}, \tag{5.126}$$

where $d\Omega$ is the spectral width of the negative-energy waves. As $t$ approaches $t_0$, the number of quanta $N^l$ and $N^t$ vary according to

$$N^l = \frac{N_0^l}{(\widetilde{N}_{10}^t + N_{20}^t)\,\beta^l (t_0 - t)}; \qquad \widetilde{N}_1^t \approx \frac{\widetilde{\beta}^t}{\widetilde{\beta}^l} N^l. \tag{5.127}$$

The growth here is more rapid than exponential (Fig. 22). The growth is given by

* In this connection saturation effects, which tend to equalize the populations and eliminate the frequency interval in which the energy is negative, are of little consequence initially.

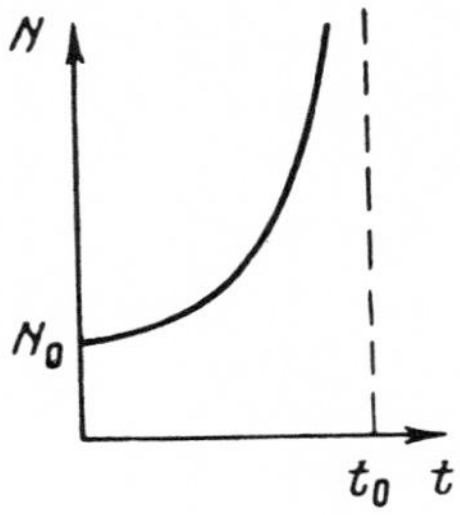


Fig. 22. Variation of the number of wave quanta caused by nonlinear decay instability of the negative-energy waves.

$$\frac{\partial N^l}{\partial t} \sim \frac{1}{t - t_0} N^l,$$

i.e., the characteristic growth rate

$$\gamma \sim \frac{1}{t_0 - t}$$

tends to infinity. The limitation on $N^l$ occurs, for example, from the conditions for applicability of the random-phase approximation or from equalization of the populations as a result of transverse-wave excitation.

It is a simple matter to estimate the maximum energy of the longitudinal and transverse waves from the conditions for applicability of the random-phase approximation. The random-phase condition is

$$\gamma < \left(\frac{\omega_{0e}}{\omega_{12}}\right)^3 \delta\Omega,$$

i.e.,

$$(t - t_0)_{\min} = \left(\frac{\omega_{12}}{\omega_{0e}}\right)^3 \delta\Omega^{-1},$$

which on substitution into (5.127) yields

$$W^l_{\max} \simeq nm_e c^2 \frac{W^l_0}{W^t_0} \frac{(\delta\Omega)^2}{\omega^2_{12}} \frac{\omega_{0e}(kc - \omega_{12})^2}{\omega^2_{0*}(v_2 - v_1)\omega_{12}}. \tag{5.128}$$

Since the maximum value of $|kc - \omega_{12}|$ and of $\delta\Omega$ is $\omega_{0*}$, we have to order of magnitude

$$W^l_{\max} \simeq \frac{nm_e c^2 W^l_0 \omega_{0e} \omega^2_{0*}}{W^t_0 \omega^3_{12}}.$$

The maximum transverse-wave energy has an order of magnitude

$$W^t_{\max} \simeq \frac{\omega_{12}}{\omega_{0e}} W^l_{\max}.$$

Thus, using the results of Chap. IV, we can estimate the change in the energy level populations related to the excitation of transverse waves due to nonlinear instability.

## 16. ASTROPHYSICAL APPLICATIONS OF DECAY INTERACTIONS

The study of decay interactions has a direct bearing on the analysis of astrophysical problems. We merely indicate some applications.

**1.** Some time ago in studying solar flares, in which plasma streams "break through" the surface layers of the sun and generate plasma oscillations, plasma emission was observed at frequencies near $2\omega_{0e}$ [27]. A logical explanation for this effect would be the process described above, $l + l \to t$. We stress that the coalescence of thermal-level Langmuir waves (combination scattering by thermal Langmuir waves) is a small effect compared with that of two strong Langmuir waves.

**2.** The presence of strong Langmuir waves in plasma can result in considerable broadening of their spectrum. If $\delta\Omega \gg \omega_{0e}$, this broadening is described by a diffusion relation and becomes larger the greater the depth of plasma penetrated by the radiation.

**3.** The scattering of electromagnetic waves by plasma oscillations must be a very widespread phenomenon. The scattering of 21-cm waves transmitted near the sun has recently been observed [28].*

**4.** The wavelength migration of electromagnetic waves in a region containing random plasma oscillations through the pro-

* The scattering of radio waves in a turbulent plasma must clearly be taken into account in the interpretation of radar observations of the sun [34]. The possible scattering effects that might be observed for quasars and the remnant sheaths of supernovae and the sun are discussed in [35, 36]. Scattering in a turbulent plasma can appreciably reduce the optical depth for plasma and other (synchrotron) radiation mechanisms.

cesses $t \rightleftarrows t + l$ and $t \rightleftarrows t + s$ may lead to a progressive increase in the average t-wave energy, which may occur in conditions of large densities and intense oscillations, as, for example, in supernova stars [29, 37].*

5. The effects of transformation of longitudinal into transverse waves are of considerable importance, because they provide a potentially useful explanation of the brightness of astronomical sources (e.g., supernovae, solar flares, etc.).

6. Finally, the effects of nonlinear generation of transverse waves in a partially ionized plasma in the presence of strong longitudinal waves could have applications for the interpretation of the anomalous radio emission of molecular lines [31].

## Literature Cited

1. V. N. Tsytovich and A. B. Shvartsburg, "Theory of nonlinear wave interaction in a magnetoactive anisotropic plasma," Zh. Éksp. Teor. Fiz., 49:796 (1965).
2. A. A. Galeev and V. I. Karpman, "Turbulent theory of a slightly nonequilibrium rarefied plasma and the structure of shock waves," Zh. Éksp. Teor. Fiz., 44:592 (1962).
3. V. A. Liperovskii and V. N. Tsytovich, "Decay of longitudinal Langmuir modes of a plasma into ion-acoustic modes," Zh. Prikl. Mekh. i Tekh. Fiz., No. 5 (1965); FIAN Preprint A-54 (1964).
4. G. I. Suramlishvili, "Wave kinetics in plasma," Dokl. Akad. Nauk SSSR, 153:317 (1963).
5. A. A. Vedenov, "Introduction to the theory of weakly turbulent plasma," Problems in Plasma Theory, Vol. 3. Atomizdat (1963), p. 203.
6. I. A. Akhiezer, N. L. Daneliya, and N. N. Tsintsadze, "Theory of the transformation and scattering of electromagnetic waves in nonequilibrium plasma," Zh. Éksp. Teor. Fiz., 46:300 (1964).
7. I. A. Akhiezer, "Scattering and transformation of electromagnetic waves in turbulent plasma," Zh. Éksp. Teor. Fiz., 48:1159 (1965).
8. V. A. Liperovskii and V. N. Tsytovich, "Nonlinear interaction of plasma waves in the presence of strong transverse waves," Izv. Vuzov, Radiofizika, 9:469 (1966).
9. B. B. Kadomtsev, Plasma turbulence, Academic Press, New York (1965).
10. V. N. Tsytovich, "Nonlinear generation of plasma waves by a beam of transverse waves," Zh. Tekh. Fiz., 34:773 (1965).
11. V. D. Fedorchenko, B. N. Rutkevich, V. N. Muratov, and B. M. Chernyi, "Low-frequency plasma oscillations in a magnetic field," Zh. Tekh. Fiz., 32:958 (1962).

* Under conditions such that the average increase in the energy of the quanta is small the radiation migration effect can lead to broadening of the spectral lines.

12. V. D. Fedorchenko, B. N. Rutkevich, V. N. Muratov, "Low-frequency plasma oscillations in a magnetic field," Plasma Physics and Problems of Controlled Thermonuclear Fusion, No. 2. Izd. Akad. Nauk Ukr. SSR (1963), p. 133.
13. V. D. Fedorchenko, V. N. Muratov, and B. N. Rutkevich, "Exchange of energy between high- and low-frequency oscillations in a plasma," Yadernyi Sintez, 4:300 (1964).
14. M. N. Rosenbluth and N. Rostocker, "Scattering of electromagnetic waves by a nonequilibrium plasma," Phys. Fluids, 5:776 (1962).
15. A. I. Akhiezer, I. A. Akhiezer, and A. G. Sitenko, "On the theory of fluctuations in plasma," Zh. Éksp. Teor. Fiz., 41:644 (1961).
16. E. E. Salpeter, "Electron density fluctuations in a plasma," Phys. Rev., 120:1528 (1960); J. Geophys. Res., 66:928 (1961); 69:869 (1964).
17. A. I. Akhiezer, I. G. Prokhoda, and A. G. Sitenko, "Scattering of electromagnetic waves in plasma," Zh. Éksp. Teor. Fiz., 33:730 (1957).
18. I. S. Danilkin, L. M. Kovrizhnykh, M. D. Raizer, and V. N. Tsytovich, "Nonlinear effects of the interaction of high-intensity electromagnetic radiation with plasma," Trudy FIAN, 32:112 (1966).
19. L. M. Kovrizhnykh, "Interaction of transverse waves with turbulent plasma," Zh. Éksp. Teor. Fiz., 49:1332 (1965).
20. V. N. Tsytovich, "Possibilities for the detection of high-frequency turbulence of a cosmic plasma," Astron. Zh., 95:992 (1964).
21. B. A. Demidov and S. D. Fanchenko, "Detection of combination scattering of electromagnetic waves in the shf range in a turbulent plasma," Pis'ma ZhÉTF, 2(12):533 (1965).
22. R. W. Waniek, D. G. Swanson, and R. T. Grannan, "Intense microwave radiation from collective effects in a plasma," Phys. Rev. Lett., 15:444 (1965).
23. J. P. Dougherty and D. T. Farley, "A theory of incoherent scattering of radio waves by a plasma," Proc. Roy. Soc., A259:79 (1960); 263:238 (1961).
24. Y. G. Chen, R. F. Leheny, and T. Marshall, "Combination scattering of microwaves from space energy waves," Phys. Rev. Lett., 15:184 (1965).
25. R. A. Stern and N. Tzoar, "Incoherent microwave scattering from resonant plasma oscillations," Phys. Rev. Lett., 15:485 (1965).
26. A. D. Ryutov, "Radiation of electromagnetic waves in the nonlinear interaction of surface oscillations in a plane plasma layer," Dokl. Akad. Nauk SSSR, 164(6):1273 (1965).
27. V. V. Zheleznyakov, Radioemission of the Sun and Planets. Pergamon, New York.
28. V. V. Vitkevich, "New data on the solar supercorona," Astron. Zh., 35:52 (1958).
29. V. N. Tsytovich, "Statistical acceleration of photons in a turbulent plasma," Izv. Vuzov, Radiofizika, 7:622 (1965).
30. V. M. Dikasov, L. I. Rudakov, and D. D. Ryutov, "Interaction of negative-energy waves in a weakly turbulent plasma," Zh. Éksp. Teor. Fiz., 48:913 (1965).
31. V. N. Tsytovich, "Nonlinear instability of optical frequencies in a partially ionized plasma," Zh. Éksp. Teor. Fiz., 51:1385 (1966).
32. A. M. Messiaen and P. E. Vandenplas, "Nonlinear resonance effect at high power in a cylindrical plasma," Phys. Lett., 25A:339 (1967).

33. J. Olivain and M. Perulli, "Linear propagation and nonlinear processes in a cylindrical plasma column in a strong magnetic field," Eighth International Conference on Phenomena in Ionized Gases, Vienna (1967), p. 398.
34. J. James, "Radar studies of the sun at 38 Mc/sec," Astrophys. J., 146:356 (1966).
35. I. M. Gordon, "Possible detection by the state parameter of a plasma in the sheaths of supernova stars," Astron. Zh., 44:702 (1967).
36. S. A. Kaplan and V. N. Tsytovich, "Interpretation of the trough in the millimeter radio-emission spectrum of the sun," Astron. Zh., 44:1036 (1967).

*Chapter VI*

# Induced Emission of Waves by Plasma Particles

## RESONANT PARTICLES

In the preceding chapters we have been concerned with the induced emission of waves by waves. Effects of this type lead to a redistribution of energy between the various wave modes. The plasma particles in this case can acquire or lose energy only insofar as they participate in the waves.*

The situation can arise, however, when the motion of the particles in a wave field has a resonance character. This happens when the phase velocity of the wave coincides with the particle velocity (more precisely, with the projection of the particle velocity in the direction of wave propagation). The wave then exerts a steady force on the particle, so that the particle continuously gains or loses energy. Clearly, either the wave can lose energy, thereby increasing the energy of the resonant particles, or, conversely, it can gain energy from the resonant particles. The kinetic energy of the resonant particles is not related to the field of the propagating waves in the same way as the nonresonant particles. How-

* The propagation of any wave in a plasma is accompanied by polarization or displacement in the plasma, reflecting the reaction of the plasma to the applied field In addition to the main thermal motion, the plasma particles execute small oscillations in the wave field. The amplitude of these oscillations is completely determined by the strength of the wave field.

ever, the number of resonant particles must in general be small in comparison with the number of nonresonant particles taking part in the wave. The nonlinear interactions discussed earlier referred to the case when there were no resonant particles.

Moreover, in plasma there are a great many effects in which resonant particles play a very important role. This is particularly true of beam instability. If a beam of particles is present in a plasma, with a concentration much smaller than that of the plasma, the propagation of waves can be determined by the plasma particles, whereas the beam particles can be resonant with the plasma waves.

Let us write the resonance condition for the case in which there are no external magnetic fields in the plasma. The component of the particle velocity in the direction of wave propagation is

$$v\cos\theta = v\frac{(\mathbf{kv})}{kv} = \frac{(\mathbf{kv})}{k}, \tag{6.1}$$

and the phase velocity of the wave is $\omega/k$. The resonance condition is

$$\omega = (\mathbf{kv}). \tag{6.2}$$

With an external magnetic field present a new type of resonance appears, namely, gyroresonance. It occurs when the wave frequency is near the frequency of gyration of the particles in the external magnetic field. The wave field in this case does not follow the rectilinear motion of the particle [as was the case for Eq. (6.2)], but its gyration. Resonance can also occur when the frequency of the field is a multiple of the particle rotation frequency. The resonance condition for particles not moving in the direction of the magnetic field may be written in the form

$$\omega = \nu\omega_{H\alpha}; \quad \omega_{H\alpha} = \frac{eH}{m_\alpha c}; \quad \nu = 1,2,3,\ldots \tag{6.3}$$

If the particle moves with constant velocity in the direction of the magnetic field, the frequency that it sees differs from that in the coordinate system in which the wave is observed (Doppler effect). The Doppler effect is easily included if we note that the resonance condition (6.2) is tantamount to the requirement that the wave frequency in the reference system in which the particle is at rest be zero:

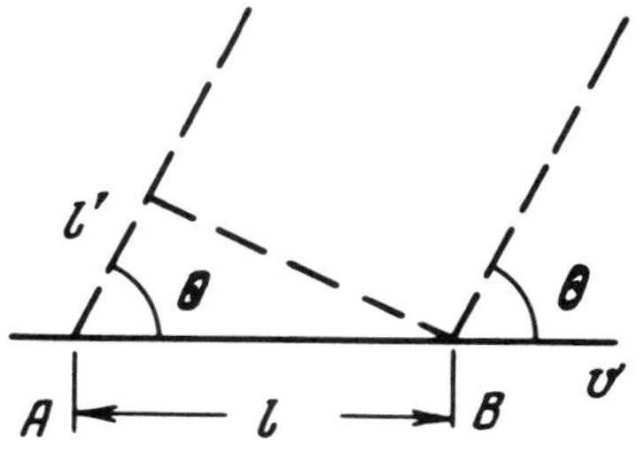


Fig. 23. Cerenkov radiation of waves by charged particles.

$$\omega' = \omega - \mathbf{kv} = 0.$$

For gyroresonance $\omega' = \nu\omega_{H\alpha}$, so that

$$\omega - k_z v_z = \nu\omega_{H\alpha}. \tag{6.4}$$

Here the subscript z indicates the components of the vectors **k** and **v** in the direction of the external magnetic field. The resonance condition (6.2) corresponds to the condition for Cerenkov radiation, and (6.4) to that for cyclotron radiation.

Now follow some straightforward arguments from Tamm and Frank [1] on the mechanism of Cerenkov radiation. Consider a uniformly moving charge whose velocity is **v**. Choose two points on its trajectory, A and B (Fig. 23). At the time it occupies these points (as at all other points along the trajectory) the charge creates an electromagnetic field around itself. Every field can be expanded into plane waves. Let us consider the waves propagating from A and B at an angle $\theta$ to the charge trajectory. The fields created by the charge at different points of the trajectory interfere with one another. If the velocity of the charge is small (less than the phase velocity of the waves), this interference causes the fields far from the charge trajectory to cancel, until only the field moving in unison with the charge remains. But if the velocity of the charge is greater than the wave phase velocity, the fields do not cancel one another at distant points. In fact, the phase difference of waves emanating from A and B can be zero. During a time t the charge moves a distance $l = vt$, while the wave moves a distance $l' = v_\varphi t = \omega t/k$. It is clear from Fig. 23 that the wave originating from B will be in phase with the wave from A if $l' = l \cos\theta$ or $\omega/k = v\cos\theta$; $\omega = kv\cos\theta$. This is the same as condition (6.2). Since the point B is arbitrary, the charge at any point of the trajectory will create waves in phase with those emanating from all other points of the trajectory. In other words, the particle continuously generates waves.

This type of emission is called Cerenkov (Vavilov–Cerenkov in USSR) radiation. Its theory has been expounded by Tamm and Frank (see the survey paper [2]).

We now show how simple it is to derive the Cerenkov condition (6.2) from the laws of energy and momentum conservation in the emission of waves by particles [3]. Suppose that prior to emission the particle has momentum **p** and energy $\varepsilon_p$ (for a nonrelativistic particle $\varepsilon_p = p^2/2m$, for a relativistic particle $\varepsilon_p = c\sqrt{p^2 + m^2c^2}$). The momentum and energy of the emitted wave are $\hbar\mathbf{k}$ and $\hbar\omega$. After emission the energy and momentum of the particle is $\mathbf{p}'$ and $\varepsilon_{p'}$:

$$\mathbf{p}' = \mathbf{p} + \hbar\mathbf{k}; \tag{6.5}$$

$$\varepsilon_{p'} = \varepsilon_p + \hbar\omega. \tag{6.6}$$

Substituting (6.5) into (6.6), we have

$$\varepsilon_{p+\hbar k} = \varepsilon_p + \hbar\omega. \tag{6.7}$$

Usually the momentum acquired by the wave, $\hbar\mathbf{k}$, is small compared with the particle momentum **p**. Expanding in $\hbar\mathbf{k}$, we have

$$\varepsilon_{p+\hbar k} = \varepsilon_p + \hbar\mathbf{k}\,\frac{\partial\varepsilon_p}{\partial\mathbf{p}}. \tag{6.8}$$

In the general case $\partial\varepsilon_p/\partial\mathbf{p} = \mathbf{v}$ the particle velocity (for a nonrelativistic particle, for example, $\partial(p^2/2m)/\partial\mathbf{p} = \mathbf{p}/m = \mathbf{v}$). Substituting (6.7) into (6.8), we obtain

$$\hbar\,(\mathbf{k}\mathbf{v}) = \hbar\omega, \tag{6.9}$$

which agrees with (6.2).

Note that the conservation laws (6.5) and (6.6) are similar to those for decay processes. Whereas decay processes describe the emission of waves by other waves, Cerenkov radiation processes describe the emission of waves by particles. As in the case of decay processes, we speak of the induced emission of waves by particles. In other words, we arrive at the possibility of induced Cerenkov emission and absorption of waves and induced cyclotron emission and absorption. The exchange of energy between waves and particles is indeed associated with these effects.

We stressed above the analogy between the induced emission of waves by waves with a negative-temperature system. Exactly the same analogy exists for induced Cerenkov or cyclotron emission and absorption of waves by particles. In other words, a system containing resonant particles can generate electromagnetic waves. Generation of this type corresponds, for example, to beam instability.

## 2. THE QUASILINEAR APPROXIMATION

An equation describing the effects of induced emission and absorption of waves by particles is called quasilinear. This name stems from the fact that the equation does not account for nonlinear wave interactions themselves (of which decay interactions are an example), but describes the influence of oscillations generated or initially present on the distribution of the plasma particles. The quasilinear equation was first investigated by Vedenov, Velikhov, and Sagdeev [4], Drummond and Pines [5], and others [6-9].

As we saw in the example of a two-level system, the induced effects of waves generated by an inverted population tend to equalize the population and cut off generation. The quasilinear equation describes saturation effects essentially analogous to these. The linear approximation corresponds to predetermined fixed population distributions. The distinction between quasilinear effects and the saturation effects of a two-level system lies in the emission mechanism. This in the two-level system is a spontaneous transition between levels, whereas in a plasma it is the spontaneous emission of Cerenkov and cyclotron radiation. Any change in the population, on the other hand, corresponds to a change in the energy and momentum distribution for the particles.

Thus, our problem is to find out how the distribution function for the plasma particles changes under the influence of oscillations and how the plasma oscillations are excited or damped for a given particle distribution. The derivation is the same in principle as that for the corresponding two-level system. For simplicity we as-

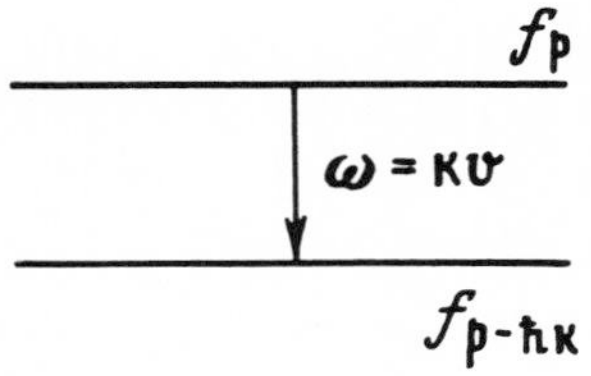


Fig. 24. Change of state of particles in Cerenkov radiation.

sume that there are no magnetic fields present. We introduce the probability $w_{\mathbf{p}}^{\sigma}(\mathbf{k})$ of the spontaneous Cerenkov emission of a wave $\sigma$ by a particle with momentum $\mathbf{p}$. The increase in the number of waves $N_{\mathbf{k}}^{\sigma}$ by emission is

$$\int (N_{\mathbf{k}}^{\sigma} + 1)\, w_{\mathbf{p}}^{\sigma}(\mathbf{k})\, f_{\mathbf{p}} \frac{d\mathbf{p}}{(2\pi)^3}, \tag{6.10}$$

and the reduction due to absorption (Fig. 24) is

$$\int N_{\mathbf{k}}^{\sigma} w_{\mathbf{p}}^{\sigma}(\mathbf{k})\, f_{\mathbf{p}-\hbar\mathbf{k}} \frac{d\mathbf{p}}{(2\pi)^3}. \tag{6.11}$$

Since $\hbar\mathbf{k} \ll \mathbf{p}$, we have

$$f_{\mathbf{p}-\hbar\mathbf{k}} \simeq f_{\mathbf{p}} - \hbar\mathbf{k}\frac{\partial f_{\mathbf{p}}}{\partial \mathbf{p}}. \tag{6.12}$$

Hence

$$\frac{dN_{\mathbf{k}}^{\sigma}}{dt} = \frac{\partial N_{\mathbf{k}}^{\sigma}}{\partial t} + \mathbf{v}_{\mathrm{rp}}^{\sigma} \frac{\partial N_{\mathbf{k}}^{\sigma}}{\partial \mathbf{r}} =$$

$$= N_{\mathbf{k}}^{\sigma} \int w_{\mathbf{p}}^{\sigma}(\mathbf{k})\, \hbar\mathbf{k} \frac{\partial f_{\mathbf{p}}}{\partial \mathbf{p}} \frac{d\mathbf{p}}{(2\pi)^3} + \int w_{\mathbf{p}}^{\sigma}(\mathbf{k})\, f_{\mathbf{p}} \frac{d\mathbf{p}}{(2\pi)^3}. \tag{6.13}$$

The left-hand side of (6.13) includes the translational variation of the number of quanta, and the right-hand side describes the variation due to emission and absorption. The first term in (6.13) corresponds to the second term of (4.11) for a two-level system, and the last term corresponds to the first term of (4.11). In the present case it describes the spontaneous Cerenkov emission.

Notice that whereas the second term of (4.11) contained the population difference between the two levels, in (6.13), owing to the closeness of the "separation between levels," we encounter the derivative of the particle distribution function with respect to their momenta. Consequently, the derivative $\partial f_{\mathbf{p}}/\partial \mathbf{p}$ describes the character of the population levels. Population inversion is roughly equivalent to $\partial f_{\mathbf{p}}/\partial \mathbf{p} > 0$, i.e., to the case when there are more particles with larger momentum than with smaller momentum. With population inversion, however, the plasma is not always unstable. The instability criterion involves the growth rate $\gamma_k$. If the intensity of the oscillations is high enough, so that the effect of spontaneous emission can be neglected, then

$$\frac{dN_{\mathbf{k}}^{\sigma}}{dt} = N_{\mathbf{k}}^{\sigma}\gamma_{\mathbf{k}}^{\sigma}; \quad \gamma_{\mathbf{k}}^{\sigma} = \int w_{\mathbf{p}}^{\sigma}(\mathbf{k})\, \hbar\mathbf{k}\, \frac{\partial f_{\mathbf{p}}^{\alpha}}{\partial \mathbf{p}} \frac{d\mathbf{p}}{(2\pi)^3}. \tag{6.14}$$

The instability condition is $\gamma_k^{\sigma} > 0$. If $\partial f_{\mathbf{p}}/\partial \mathbf{p}_i > 0$, the number of particles in a given momentum interval increases with $\mathbf{p}$. However, since f $\to 0$ as $\mathbf{p} \to \infty$, there must be an interval in which $\partial f_{\mathbf{p}}/\partial p_i < 0$. The growth rate (6.14) describes the total effect originating both from particles for which $\partial f_{\mathbf{p}}/\partial p_i > 0$ and from those for which $\partial f_{\mathbf{p}}/\partial p_i < 0$. In the presence of population inversion, therefore, both the buildup and damping of oscillations are possible. We now see that the growth rate expression (6.14) is analogous to the result (5.12) for the diffusion approximation in the emission of waves by waves. We can use the diffusion approximation for the study of particle–wave interaction, because $\hbar k \ll p$.

It is also important to note that although the growth rate (6.14) contains $\hbar$ the result is still classical, because the probability is inversely proportional to $\hbar$. This is quickly substantiated when we recall that the intensity of spontaneous Cerenkov radiation is given by a classical expression. On the other hand, we obtain the following for the intensity of spontaneous emission from (6.13):

$$\frac{\partial W^{\sigma}}{\partial t} = \int \frac{\hbar\Omega^{\sigma}}{(2\pi)^3} \frac{\partial N^{\sigma}}{\partial t}\, d\mathbf{k} = \int \frac{\hbar\Omega^{\sigma} w_{\mathbf{p}}^{\sigma}(\mathbf{k})}{(2\pi)^6} f_{\mathbf{p}}\, d\mathbf{p}\, d\mathbf{k}. \tag{6.15}$$

We now consider the manner in which the particle distribution function varies. The corresponding equation is analogous to Eq. (4.15) for the two-level system and to Eq. (5.16) for the interaction of waves with waves. For $N_{\mathbf{k}}^{\sigma} \gg 1$

$$\frac{\partial f_{\mathbf{p}}}{\partial t} = \int [w_{\mathbf{p}}(\mathbf{k})(f_{\mathbf{p}+\hbar\mathbf{k}} - f_{\mathbf{p}}) - w_{\mathbf{p}}(\mathbf{k})(f_{\mathbf{p}} - f_{\mathbf{p}-\hbar\mathbf{k}})]\, N_{\mathbf{k}}^{\sigma} \frac{d\mathbf{k}}{(2\pi)^3}. \tag{6.16}$$

The expansion in $\hbar\mathbf{k}$ does not differ in any respect from the expansion in the derivation of (5.16). We obtain the diffusion relation

$$\frac{\partial f_{\mathbf{p}}}{\partial t} = \frac{\partial}{\partial p_i} D_{ij} \frac{\partial f_{\mathbf{p}}}{\partial p_j} + \frac{\partial}{\partial p_i} A_i f_{\mathbf{p}}; \tag{6.17}$$

$$D_{ij} = \int \hbar^2 k_i k_j N_{\mathbf{k}}^{\sigma} w_{\mathbf{p}}^{\sigma} (\mathbf{k}) \frac{d\mathbf{k}}{(2\pi)^3} \; ; \quad A_i = \int k_i w_{\mathbf{p}} (\mathbf{k}) \frac{d\mathbf{k}}{(2\pi)^3} . \tag{6.18}$$

It is reasonable to say that the effects of population equalization reduce to a diffusion of the particle distribution in the field of the oscillations and decrease the derivative $\partial f_{\mathbf{p}}/\partial \mathbf{p}$ , i.e., the growth rate $\gamma_{\mathbf{k}}^{\sigma}$. Consequently, during the generation of oscillations their inverse action on the particle distribution function causes a decrease of the growth rate. This effect, as in the case of wave–wave interaction, can result in stabilization of the instability. The coefficients $A_i$ in (6.18) describe the influence of spontaneous emission on the particle distribution. This effect is normally small, being of the same order as that due to collisions. Usually only those equations which include induced effects are considered as quasilinear.

## 3. CONSERVATION LAWS

Consider the time variation of the total particle energy $E^{\alpha}$

$$E^{\alpha} = \int \varepsilon_{\mathbf{p}} f_{\mathbf{p}}^{\alpha} \frac{d\mathbf{p}}{(2\pi)^3} . \tag{6.19}$$

Here $f_{\mathbf{p}}$ is normalized as follows: $\int f_{\mathbf{p}} \frac{d\mathbf{p}}{(2\pi)^3} = n$ , where n is the particle density.

We recall that the wave energy is

$$W^{\sigma} = \int \hbar \Omega^{\sigma} N_{\mathbf{k}}^{\sigma} \frac{d\mathbf{k}}{(2\pi)^3} . \tag{6.20}$$

Integrating by parts the equation

$$\frac{\partial f_{\mathbf{p}}}{\partial t} = \frac{\partial}{\partial p_i} D_{ij} \frac{\partial f_{\mathbf{p}}}{\partial p_j} ,$$

we obtain

$$\frac{\partial}{\partial t} E^{\alpha} = \int \varepsilon_{\mathbf{p}} \frac{\partial}{\partial p_i} D_{ij} \frac{\partial f_{\mathbf{p}}^{\alpha}}{\partial p_j} \frac{d\mathbf{p}}{(2\pi)^3} = - \int D_{ij} \frac{\partial f_{\mathbf{p}}^{\alpha}}{\partial p_j} \frac{\partial \varepsilon_{\mathbf{p}}}{\partial p_i} \frac{d\mathbf{p}}{(2\pi)^3} =$$

$$= - \int v_i D_{ij} \frac{\partial f_{\mathbf{p}}^{\alpha}}{\partial p_j} \frac{d\mathbf{p}}{(2\pi)^3} . \tag{6.21}$$

Inserting the expression for the diffusion coefficient (6.18) into (6.21), we obtain

$$\frac{\partial}{\partial t} E^\alpha = -\int \hbar^2 (\mathbf{kv})\, \omega_\mathbf{p}(\mathbf{k})\, N_\mathbf{k}^\sigma \left(\mathbf{k} \frac{\partial f_\mathbf{p}^\alpha}{\partial \mathbf{p}}\right) \frac{d\mathbf{k} d\mathbf{p}}{(2\pi)^6}. \tag{6.22}$$

It is also a simple matter to find the variation of the wave energy from (6.14):

$$\frac{\partial W^\sigma}{\partial t} = \int \hbar \Omega^\sigma \frac{\partial N_\mathbf{k}^\sigma}{\partial t} \frac{d\mathbf{k}}{(2\pi)^3} = \int \hbar^2 \Omega^\sigma \omega_\mathbf{p}(\mathbf{k})\, N_\mathbf{k}^\sigma \left(\mathbf{k} \frac{\partial f_\mathbf{p}^\alpha}{\partial \mathbf{p}}\right) \frac{d\mathbf{p}\, d\mathbf{k}}{(2\pi)^6} \tag{6.23}$$

Comparing (6.22) and (6.23), we verify that the right-hand sides differ only in sign, by virtue of the Cerenkov condition $\omega^\sigma = (\mathbf{kv})$, so that

$$\frac{\partial}{\partial t}(E^\alpha + W^\sigma) = 0. \tag{6.24}$$

This conservation law shows that the processes of induced emission and absorption of waves by particles proceeds in such a fashion that the particles and waves continuously exchange energy.

It is also easy to find the rate of change of the particle momentum:

$$\frac{\partial}{\partial t} P_i^\alpha = -\int D_{ij} \frac{\partial f_\mathbf{p}^\alpha}{\partial p_j} \frac{d\mathbf{p}}{(2\pi)^3} = -\hbar^2 \int k_i \left(\mathbf{k} \frac{\partial f_\mathbf{p}^\alpha}{\partial \mathbf{p}}\right) N_\mathbf{k}^\sigma \omega_\mathbf{p}(\mathbf{k}) \frac{d\mathbf{k}\, d\mathbf{p}}{(2\pi)^6};$$

$$\frac{\partial}{\partial t} P_i^\sigma = \int \hbar^2 k_i \left(\mathbf{k} \frac{\partial f_\mathbf{p}^\alpha}{\partial \mathbf{p}}\right) N_\mathbf{k}^\sigma \omega_\mathbf{p}(\mathbf{k}) \frac{d\mathbf{k}\, d\mathbf{p}}{(2\pi)^6} = -\frac{\partial P_i^\alpha}{\partial t}.$$

Thus the momentum conservation law is satisfied:

$$\frac{\partial}{\partial t}(\mathbf{P}^\alpha + \mathbf{P}^\sigma) = 0. \tag{6.25}$$

Finally, it is easily confirmed that the total number of particles is also conserved:

$$\frac{\partial n}{\partial t} = 0. \tag{6.26}$$

These conservation laws are analogous with those involved in decay processes.

## 4. PROBABILITIES FOR THE EMISSION OF WAVES BY PLASMA PARTICLES

It is sufficient to determine the probability of Cerenkov emission to consider the spontaneous processes. We use (6.15), assuming that the number of quanta is small, $N_{\mathbf{k}}^{\sigma} \rightarrow 0$. As indicated by (6.15), the radiation intensity is the sum of the radiation intensities from the individual particles:

$$\frac{\partial W^{\sigma}}{\partial t} = \int W_{\mathbf{p}}^{\sigma} f_{\mathbf{p}} \frac{d\mathbf{p}}{(2\pi)^3}, \tag{6.27}$$

where $W_{\mathbf{p}}^{\sigma}$ is the radiation intensity from an individual charge with momentum $\mathbf{p}$:

$$W_{\mathbf{p}}^{\sigma} = \int \hbar \Omega^{\sigma} w_{\mathbf{p}}^{\sigma}(\mathbf{k}) \frac{d\mathbf{k}}{(2\pi)^3}. \tag{6.28}$$

The radiation from an individual charge can be found if the work done on the charge by the field it creates is known [10] (cf. the analysis of decay probabilities). The expression for the current produced by a uniformly moving charge is

$$\mathbf{j} = \rho \mathbf{v} = e\mathbf{v}\delta(\mathbf{r} - \mathbf{v}t). \tag{6.29}$$

As the current is known, the calculation of the radiation generated by the current (6.29) presents no special difficulties [10]. By comparison with (6.28) we find an expression for the probability of emission (Appendix 3):

$$w_{\mathbf{p}}^{\sigma}(\mathbf{k}) = \frac{e^2}{\pi\hbar} \frac{|\mathbf{v}\mathbf{e}_{\mathbf{k}}^{\sigma}|^2 (2\pi)^3}{\left.\frac{\partial}{\partial\omega}\omega^2\varepsilon^{\sigma}\right|_{\omega=\Omega^{\sigma}(\mathbf{k})}} \delta(\Omega^{\sigma}(\mathbf{k}) - \mathbf{k}\mathbf{v}). \tag{6.30}$$

Let us consider, for example, an isotropic plasma in which only transverse and longitudinal waves can occur. In a fully ionized plasma the transverse waves have phase velocities greater than the velocity of light, and since the particle velocities are less than the velocity of light, Cerenkov emission of transverse waves is not allowed. In a partially ionized plasma, however, it is possible. The Tamm–Frank formula for this case is easily derived. Since $\mathbf{e}_{\mathbf{k}}^{t}$, the unit polarization vector, is perpendicular to $\mathbf{k}$, we have

$$(\mathbf{e}_{\mathbf{k}}^{t}\mathbf{v})^2 = v^2 - \frac{(\mathbf{k}\mathbf{v})^2}{k^2} = v^2 - \frac{(\Omega^t)^2}{k^2} = v^2 - v_{\Phi}^2;$$

$$W_{\mathbf{p}}^{t} = \frac{e^2}{\pi}\int \frac{\Omega^t \delta(\Omega^t - \mathbf{k}\mathbf{v})(v^2 - v_{\varphi}^2)}{\frac{\partial}{\partial\omega}\,\omega^2\varepsilon^t\Big|_{\omega=\Omega^t(\mathbf{k})}}\,d\mathbf{k}. \qquad (6.31)$$

In (6.31) the integration over direction is easily carried out:

$$W_{\mathbf{p}}^{t} = e^2 v \int \frac{\Omega^t\left(1 - \frac{v_{\varphi}^2}{v^2}\right)}{\frac{\partial}{\partial\omega}\,\omega^2\varepsilon^t\Big|_{\omega=\Omega^t(\mathbf{k})}}\,dk^2. \qquad (6.32)$$

Recognizing that $dk^2 = \left(\frac{d}{d\omega}\,\omega^2\varepsilon^t\right)d\omega$, we obtain the result of Tamm and Frank [1]

$$W_{\mathbf{p}}^{t} = e^2 v \int \omega\, d\omega \left(1 - \frac{v_{\varphi}^2}{v^2}\right). \qquad (6.33)$$

Longitudinal waves in an isotropic plasma can have phase velocities much smaller than the velocity of light. For Langmuir waves, for example,

$$v_{Te} < v_{\varphi}^{l} < \infty, \qquad (6.34)$$

and for ion-acoustic waves

$$v_{Ti} < v_{\varphi}^{s} < \sqrt{\frac{m_e}{m_i}}\,v_{Te}, \qquad (6.35)$$

where $v_{Ti}$ and $v_{Te}$ are the mean thermal velocities of the plasma ions and electrons. Consequently, the Cerenkov effect for plasma and ion-acoustic waves can occur for a wide range of particle velocities. This is why beam instabilities of a plasma are so important. The longitudinal-wave polarization vector is $\mathbf{e}_k = \mathbf{k}/k$. Hence

$$\mathbf{e}_k\mathbf{v} = \frac{(\mathbf{k}\mathbf{v})}{k} = \frac{\omega}{k}.$$

For Langmuir waves having frequencies close to $\omega_{0e}$ and for $\varepsilon \simeq 1 - \omega_{0e}^2/\omega^2$ we obtain

$$w_{\mathbf{p}}^{l}(\mathbf{k}) = \frac{(2\pi)^2 e^2\omega_{0e}}{\hbar k^2}\,\delta(\omega_{0e} - \mathbf{k}\mathbf{v}), \qquad (6.36)$$

and for ion-acoustic waves with $\varepsilon \simeq 1 - \frac{\omega_{0i}^2}{\omega^2} + \frac{1}{k^2\lambda_{De}^2}$

$$w_{\mathbf{p}}^{s}(\mathbf{k}) = \frac{(2\pi)^2 e^2 \Omega_s^3}{\hbar k^2 \omega_{0i}^2} \delta(\Omega^s - \mathbf{kv}); \quad \Omega^s = \frac{kv_s}{\sqrt{1 + k^2\lambda_{De}^2}}. \tag{6.37}$$

Using (6.36), we readily find the intensity of spontaneous Cerenkov emission of plasma waves by a charge:

$$W_{\mathbf{p}}^{l} = \int_0^\infty \frac{e^2\omega_{0e}^2}{2\pi k^2} 2\pi k^2\, dk \int_{-1}^{1} dx\, \delta(\omega_{0e} - kvx) =$$

$$= \int_{\frac{\omega_{0e}}{v}}^{k_{\max} \sim \frac{\omega_{0e}}{v_{Te}}} \frac{e^2\omega_{0e}^2\, dk}{vk} = \frac{e^2\omega_{0e}^2}{v} \ln \frac{v}{v_{Te}}. \tag{6.38}$$

As a rule, the energy losses by the charge in the emission of Langmuir waves are of the same order as those caused by collisions. In order for the charge to be able to emit plasma waves, its velocity must be greater than $v_{Te}$.

The spontaneous emission of ion-acoustic waves differs somewhat for $v < v_s$ and $v > v_s$. In the former instance

$$W_{\mathbf{p}}^{s} = \frac{e^2\omega_{0i}^2}{v} \ln \frac{v}{v_{Ti}}, \tag{6.39}$$

and in the latter

$$W_{\mathbf{p}}^{s} = \frac{e^2\omega_{0i}^2}{2v} \left( \ln \left( \frac{T_e}{T_i} \right) + \frac{T_i}{T_e} \right). \tag{6.40}$$

## 5. LANDAU WAVE DAMPING

The damping of Langmuir waves, first investigated by Landau [11], has a very simple interpretation. It is related to the induced Cerenkov absorption and emission of Langmuir waves by plasma particles. As we know, Langmuir waves are slightly damped and can exist in a plasma only if their phase velocity is larger than the mean thermal velocity of the electrons. This is because in this case the relative number of resonant particles having velocities on the order of the wave, given a Maxwellian velocity distribution, is small. Consequently, for $v_\varphi \gg v_{Te}$ the bulk of the nonresonant particles participates in wave propagation, while only a small fraction of particles in the tail of the Maxwellian distribution are resonant in nature and participate in wave absorption. In the case of Langmuir waves this absorption occurs on the plasma electrons.

Substituting into (6.14) the probability (6.36) and the Maxwellian distribution

$$f_{\mathbf{p}}^{\alpha} = (2\pi)^{3/2} n \frac{\exp(-v^2/2v_{Te}^2)}{m_e^3 v_{Te}^3},$$

we obtain

$$\gamma_{\mathbf{k}}^{l} = -\int \frac{e^2\omega_{0e}^2 n\delta(\omega_{0e} - \mathbf{kv})}{k^2 m_e v_{Te}^5} \sqrt{2\pi} \exp(-v^2/2v_{Te}^2)\, d\mathbf{v}. \qquad (6.41)$$

An elementary integration of (6.41) yields the familiar expression for the Landau damping:

$$\gamma_{\mathbf{k}}^{l} = -\sqrt{\frac{\pi}{2}}\, \omega_{0e} \frac{\omega_{0e}^3}{k^3 v_{Te}^3} \exp\left(-\frac{\omega_{0e}^2}{2k^2 v_{Te}^2} - \frac{3}{2}\right). \qquad (6.42)$$

The damping of ion-acoustic waves is calculated in the same way. As their phase velocity is greater than $v_{Ti}$, they are damped by ions in a manner similar to that described by Eq. (6.42). Considering the fact that the phase velocity of ion-acoustic waves is much smaller than the mean thermal velocity of the electrons, it might appear that every electron is able to participate in a resonant interaction with a wave. But then the damping is not weak. Actually the number of resonant electrons for this type of wave is small compared with the total number of electrons.

Let us consider, for example, electrons having a velocity close to the phase velocity of the wave. The distribution of these electrons is independent of the velocity by virtue of $v_\varphi \leq v \ll v_{Te}$. Consequently, the relative proportion of such particles in the distribution corresponds to the ratio of the phase volumes $\sim v_\phi^3/v^3 \ll 1$. But if we consider electrons whose velocities are much larger than the phase velocities of the wave and are, in particular, near the thermal velocity, it turns out that the only particles capable of resonant interaction with a given wave are those moving almost perpendicularly to the wave direction. In fact, the Cerenkov condition $\omega = kv\cos\theta$ now becomes

$$\cos\theta = \frac{v_\varphi}{v} \sim \frac{v_\varphi}{v_{Te}} \ll 1.$$

The number of such electrons as a fraction of the total number is of order $v_\varphi/v_{Te} \ll 1$. We therefore arrive at the conclusion that the number of resonant electrons is a small fraction of the total. Consequently, the Landau damping for ion-acoustic waves can be found by analogy with (6.41). From (6.37) and (6.14) we obtain

$$\gamma_{\mathbf{k}}^s = -\sqrt{2\pi}\int \frac{e^2\Omega_s^4\delta(\Omega^s - \mathbf{kv})}{k^2\omega_{0i}^2 m_e v_{Te}^5} \exp\left(-\frac{v^2}{2v_{Te}^2}\right) d\mathbf{v}. \tag{6.43}$$

An elementary integration of this expression yields

$$\gamma_{\mathbf{k}}^s = -\sqrt{\frac{\pi}{2}}\,\Omega^s \sqrt{\frac{m_e}{m_i}}\,\frac{1}{(1 + k^2\lambda_{De}^2)^{3/2}}\,. \tag{6.44}$$

This makes it clear that indeed $\gamma \ll \Omega^s$.

Let us now explore in more detail why plasma waves are damped. The change in the number of quanta, as we have already seen, arises from a balance between the induced Cerenkov absorption and emission. The resulting damping arises because the number of absorbing particles is greater than the number of emitting particles. In absorption the particle acquires energy, in emission it loses it, i.e., for damping to occur the number of particles of lower energy must be greater than those of higher energy. However, if the wave intensity is high enough, the domination of absorption over emission tends to increase the particle energy and equalize the populations. In this case the derivative of the Maxwellian distribution decreases, hence so also does the absorption [4, 12]. This saturation effect for Landau absorption is exactly analogous to the saturation effect for the two-level system or the saturation effect discussed earlier in connection with the mutual interaction of waves. The effect of flattening the distribution is described by the diffusion equation (6.17).

## 6. THE ONE-DIMENSIONAL QUASILINEAR EQUATIONS

We now wish to examine a number of effects associated with the one-dimensional treatment of the induced emission and absorption of waves. Let us assume that the particle distribution can be described by a distribution function $f_p$, where p is the momentum of the particles in a given direction. We propose to investigate the emission and absorption of waves moving only in this direction. This type of situation is realized in a strong magnetic field, which impedes the motion of plasma particles across the field, or in the case of large growth (damping) in the direction of the field. The diffusion equation in the one-dimensional case has the form

$$\frac{\partial f_p}{\partial t} = \frac{\partial}{\partial p} D^\sigma \frac{\partial f_p}{\partial p}; \; f_p = \int f_{\mathbf{p}} \frac{d\mathbf{p}_\perp}{(2\pi)^2}, \tag{6.45}$$

where

$$D^\sigma = \int \hbar^2 k^2 N_k^\sigma \omega_p(k) \frac{dk}{2\pi}; \; N_k^\sigma = \int N_{\mathbf{k}}^\sigma \frac{d\mathbf{k}_\perp}{(2\pi)^2}. \tag{6.46}$$

For Langmuir waves

$$D^l = 2\pi e^2 \hbar \frac{N^l(v)}{v} \omega_{0e}, \tag{6.47}$$

where

$$N_k^l = N^l\left(\frac{\omega_{0e}}{k}\right) = N^l(v_\varphi^l). \tag{6.48}$$

We also write the one-dimensional equation for the change in the number of quanta:

$$\frac{\partial N^l(v)}{\partial t} = \gamma^l(v) N^l(v);$$

$$\gamma^l\left(\frac{\omega_{0e}}{k}\right) = \int \frac{\omega_p(k)\hbar k}{2\pi} \frac{\partial f_p}{\partial p} dp = \frac{2\pi e^2 m_e \omega_{0e}}{k^2} \frac{\partial f_p}{\partial p}\bigg|_{v=\omega_{0e}/k};$$

$$\gamma^l(v) = \frac{2\pi e^2 m_e}{\omega_{0e}} v^2 \frac{\partial f_p}{\partial p}. \tag{6.49}$$

Notice that the diffusion of the distribution function as described by (6.45) leads to a reduction of the derivatives $\partial f_p/\partial p$, i.e., to a flattening of the distribution function. If $\partial f_p/\partial p < 0$ (Maxwellian particle distribution), the decrease of this derivative through the influence of oscillations on the particles leads to a reduction of the absorption. Suppose, however, there are regions where $\partial f_p/\partial p > 0$, e.g., if there are beams of particles in the plasma. Then $\gamma^l(v) > 0$, and instability sets in. This instability occurs as the result of the inverted population of the particles with respect to their energy levels. It is analogous to the instability of the two-level system,

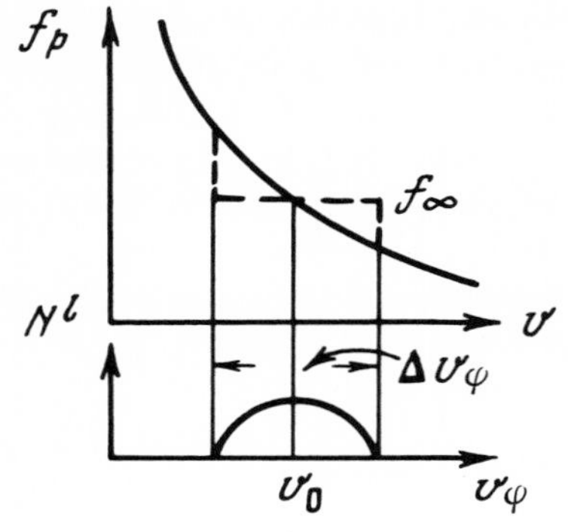


Fig. 25. Change in the plasma particle distribution function such as to produce saturation of Landau damping; $f_\infty$ is the final state. The lower graph shows the oscillation intensity as a function of the phase velocity.

the only difference being that the mechanism of spontaneous emission is not a transition between the two levels, but Cerenkov emission, and that a longitudinal, rather than a transverse, wave is emitted. The inverse action of the oscillations generated in beam instability causes a decrease in $\partial f_p/\partial p$, i.e., a decrease in the growth rate (6.49). A detailed analysis of beam instability is presented later.

## 7. ONE-DIMENSIONAL THEORY OF SATURATION EFFECTS FOR LANDAU WAVE ABSORPTION

As remarked above, a sufficiently intense Langmuir wave can alter the velocity distribution of the particles, resulting in a decrease of the Landau absorption. Using the example of the one-dimensional problem we investigate how the particle distribution changes and what the wave energy must be in order for this change to have an appreciable effect on the wave absorption. Let us assume that the waves have wave numbers lying in some finite interval $\Delta k$. Then the phase velocities of the waves lie in the interval $\Delta v_\varphi$ (Fig. 25) in which at $t = 0$ the particle distribution function is decreasing, $\partial f/\partial p < 0$ (for a Maxwellian distribution $\partial f/\partial p = -vf/m_e v_{Te}^2$). The diffusion of the particle velocities causes a reduction of $\partial f/\partial p$ in the interval $\Delta v_\varphi$, as a result of which a "plateau" $\partial f/\partial v = 0$ is created, i.e., population equalization and saturation of absorption take place.

We assume that $\Delta v_\varphi \ll v_0$ and say that the wave distribution can be approximated in this interval by a parabola (Fig. 25):

$$N^l(v, t) = \frac{\frac{(\Delta v_\varphi)^2}{4} - (v - v_0)^2}{2} N_0^l(t); \quad D^l = \frac{2\pi e^2 \hbar}{v_0} N^l(v, t)\omega_{0e}. \tag{6.50}$$

The number $N_0^l$ can be expressed in terms of the oscillation energy:

$$W^l = \int \frac{\hbar\omega_{0e} N^l(v)\, dk}{2\pi} = \frac{\hbar\omega_{0e}^2 N_0^l}{2v_0^2 \cdot 2\pi} \int\limits_{v_0 - \frac{\Delta v_\varphi}{2}}^{v_0 + \frac{\Delta v_\varphi}{2}} dv \left(\frac{(\Delta v_\varphi)^2}{4} - (v - v_0)^2\right) =$$

$$= \frac{\omega_{0e}^2 N_0^l(t)\, \hbar}{6v_0^2 \cdot 4\pi} (\Delta v_\varphi)^3. \tag{6.51}$$

We also approximate $f_p$ in the interval $\Delta v_\varphi$ by a linear function

$$f_p = f(v_0) - (v - v_0)\, \varphi(t), \tag{6.52}$$

where $\varphi(t)$ is the value of the derivative (with the minus sign) of the distribution function for $v = v_0$. Substitution of (6.52) and (6.50) into Eqs. (6.45) and (6.49) yields [7]

$$\frac{\partial \psi}{\partial t} = -\alpha \psi w; \tag{6.53}$$

$$\frac{\partial w}{\partial t} = -\beta \psi w, \tag{6.54}$$

where

$$\alpha = 12\pi \frac{v_0}{\Delta v_\varphi} \omega_{0e}; \tag{6.55}$$

$$w = \frac{W^l}{n m_e (\Delta v_\varphi)^2}; \quad \beta = \pi\omega_{0e} \left(\frac{v_0}{\Delta v_\varphi}\right)^2; \quad \psi = \frac{\varphi (\Delta v_\varphi)^2 m_e}{n 2\pi}; \tag{6.56}$$

$\psi$ and w are, respectively, the normalized derivative of the distribution function and the normalized energy of the oscillations.

The solution of the system (6.53) and (6.54) is easy if we recognize the fact that $\psi/\alpha - \omega/\beta = \text{const}$:

$$w(t) = w(0) \frac{\left(\psi(0) - \frac{\alpha}{\beta} w(0)\right) \exp\left\{-\beta t \left[\psi(0) - \frac{\alpha}{\beta} w(0)\right]\right\}}{\psi(0) - \frac{\alpha}{\beta} w(0) \exp\left\{-\beta t \left[\psi(0) - \frac{\alpha}{\beta} w(0)\right]\right\}}. \tag{6.57}$$

It follows from (6.57) that for a small initial energy

$$w(0) \ll \frac{\beta}{\alpha}\psi(0)$$

the damping is linear:

$$w(t) = w(0)\, e^{-\beta t \psi(0)} \tag{6.58}$$

and the saturation effect is unimportant. Conversely, for a large initial energy

$$w(0) \gg \frac{\beta}{\alpha}\psi(0) \tag{6.59}$$

the absorption becomes small:

$$w(t) \simeq w(0) - \frac{\beta}{\alpha}\psi(0)\left(1 - e^{-\alpha w(0) t}\right). \tag{6.60}$$

The last result is easily understood when we realize that the wave energy in this process is transferred into particle energy. The total energy that the particles can acquire corresponds to the difference between the energies of the initial state with nonzero derivative and the final "plateau" state. The energy of the final state is greater than that of the initial state, because some of the low-energy particles have increased their energy. However, this difference is finite and cannot be greater than the energy of the resonant particles in the final state. Therefore, if the wave energy is greater than the energy that can be acquired by the particles, the "plateau" occurs very quickly, and further energy transfer is halted.

Notice that the time during which energy is absorbed in the case (6.59) is now determined by the initial oscillation energy $w(0)$, and no longer by the damping rate associated with the particle distribution $\beta\psi(0)$. The time during which energy is absorbed is $1/\gamma$ where

$$\gamma = \alpha w(0) = 12\pi\omega_{0e}\frac{v_0}{\Delta v_\varphi}\frac{W^l}{nm_e(\Delta v_\varphi)^2}. \tag{6.61}$$

Comparing (6.61) with the decay time (5.56), we see at once that conditions are possible such that (6.61) is much larger than the reciprocal time for the decay of Langmuir waves into sound waves. It is important to note that (6.61) describes the time in which ab-

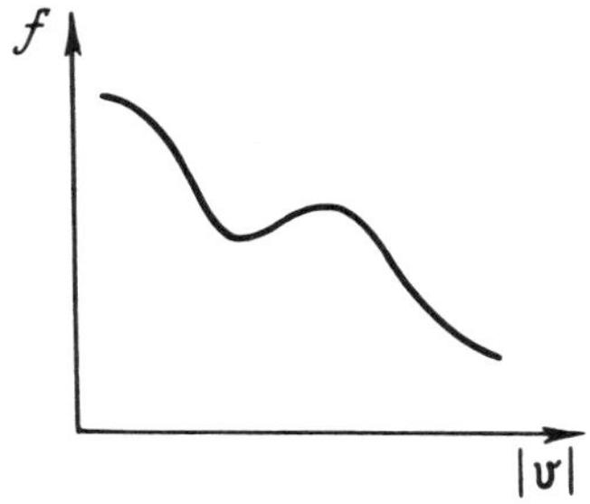


Fig. 26. "Humped" distribution function for isotropically distributed particles.

sorption vanishes. For a thermal distribution of the particles the role of the resonant particles becomes exponentially small if the phase velocity of the waves is much larger than the mean thermal velocity of the particles. For waves of this type the Landau absorption is negligible in comparison with other effects, particularly with decay effects.

It should also be pointed out that the one-dimensional model considered above is clearly invalid in the absence of a magnetic field. As a matter of fact, in the one-dimensional case the only resonant particles are those whose velocities coincide with the phase velocities of the waves, whereas in the three-dimensional case all particles whose velocities are greater than the phase velocities of the waves, but whose velocity projections on the direction of wave propagation coincide with the phase velocities, are also resonant.

## 8. STABILITY OF AN ISOTROPIC PARTICLE DISTRIBUTION

As mentioned, the one-dimensional approximation is possible only in the presence of a strong magnetic field. In all other situations the variation of the direction of particle motion is most important. We pose the following question. Will the system of particles be unstable if their distribution is completely isotropic? It would seem that the answer should depend on the form of the particle distribution function; in particular, the presence of a "hump" on the distribution curve for $f(\varepsilon)$,* as in Fig. 26 (where $\varepsilon$ is the

* In the isotropic case f depends only on the absolute value of the velocity, i.e., only on the particle energy $\varepsilon$.

particle energy), could in principle cause instability. However, it is easily demonstrated that any isotropic particle distribution is stable with respect to the generation of Langmuir and ion-acoustic waves.

Thus, because

$$\frac{\partial f}{\partial \mathbf{p}} = \frac{\partial f}{\partial \varepsilon}\frac{\partial \varepsilon}{\partial \mathbf{p}} = \mathbf{v}\frac{\partial f}{\partial \varepsilon}$$

and $\mathbf{kv} = \omega$, the growth rate $\gamma_k$ may be written for any longitudinal waves in the form [see (6.14) and (6.30)]

$$\gamma_{\mathbf{k}} = \int \frac{e^2(\Omega^\sigma)^3 c^{-4}}{\pi k^2 \frac{\partial}{\partial \omega}\,\omega^2\varepsilon^\sigma\Big|_{\omega=\Omega^\sigma}}\,\delta(\Omega^\sigma - kv\cos\theta)\,\frac{\partial f}{\partial \varepsilon}\,2\pi d\cos\theta v\varepsilon^2\,d\varepsilon,$$

which, after integration over direction, yields

$$\gamma_{\mathbf{k}} = \frac{2e^2(\Omega^\sigma)^3}{c^4\pi k^3 \frac{\partial}{\partial \omega}\,\omega^2\varepsilon^\sigma\Big|_{\omega=\Omega^\sigma}}\int_{mc^2}^{\infty}\varepsilon^2\frac{\partial f}{\partial \varepsilon}\,d\varepsilon. \tag{6.62}$$

Equation (6.62) is applicable for any particles, relativistic particles included. Integration of (6.62) by parts putting $f(\infty) = 0$ gives

$$\gamma_{\mathbf{k}} = -\frac{2e^2(\Omega^\sigma)^3}{\pi k^3 \frac{\partial}{\partial \omega}\,\omega^2\varepsilon^\sigma\Big|_{\omega=\Omega^\sigma}}\left[m^2 f(mc^2) + \frac{2}{c^4}\int_{mc^2}^{\infty}\varepsilon f(\varepsilon)\,d\varepsilon\right]. \tag{6.63}$$

The stability of an isotropic distribution implied by (6.63) is a result of the fact that absorption always prevails over emission. It is impossible to formulate an $f(\varepsilon)$ with $\partial f/\partial\varepsilon > 0$ and without regions where $\partial f/\partial\varepsilon < 0$ if $f(\infty) = 0$. According to (6.62) those particles for which $\partial f/\partial\varepsilon > 0$ give growth, those for which $\partial f/\partial\varepsilon < 0$ provide damping. We have merely proved that the damping by particles with $\partial f/\partial\varepsilon < 0$ always prevails over growth by particles with $\partial f/\partial\varepsilon > 0$.

The conclusion that stability occurs even for relativistic particles is important for problems relating to the origin of cosmic radiation. It is well known that charged particles in cosmic rays very precisely fit an isotropic distribution. This distribution is described for $T = \varepsilon - m_p c^2 \gg m_p c^2$ (where $m_p$ is the proton mass) by a decreasing power spectrum $f(T) = k/T^\gamma$. The spectrum has a maximum at $T \sim m_p c^2$, and for $T < m_p c^2$ the function $f(T)$ again decreases. The cause of the maximum in the distribution of cosmic rays is not sufficiently clear at the present time. It is not known whether the decrease in the number of low-energy particles is related to certain phenomena in the vicinity of the sun, or whether this maximum belongs to the spectrum of galactic cosmic rays. The latter should not be excluded simply because of the stability of an isotropic distribution of particles in a plasma.

The stability of an arbitrary isotropic plasma particle distribution has been demonstrated by Ginzburg [13].

## 9. ACCELERATION OF PARTICLES INTERACTING WITH PLASMA WAVES

The stability of an isotropic particle distribution has an important consequence regarding the absorption of oscillations by arbitrarily distributed isotropic particles. By the conservation of energy this absorption must produce an increase in the energy of those particles with which the oscillations interact. In the present section we investigate in more detail the problem of finding when the energy of particles interacting with waves increases. As before, we assume that the density of resonant particles is small in comparison with the density of nonresonant particles, and that the energy density of the resonant particles is small compared with the mean thermal energy of the plasma $n(T_e + T_i)$. The implication is that we are interested in the acceleration of a small group of plasma particles. This formulation is of special interest not only for laboratory situations, in which the acceleration of a small fraction of the plasma particles is observed, but also for the origin of cosmic radiation.

Let us assume, consistent with the statement of the problem, that there are fairly intense oscillations in the plasma, their spectrum and intensity being allowed to vary arbitrarily with time. The only constraint is that these variations are to be independent of the variables characterizing the accelerated particles. In the above example of the one-dimensional problem we examined the interaction of strong longitudinal waves with plasma particles and showed that such interaction yields a "plateau" in the particle distribution, i.e., an increase in the number of particles having higher energies. We demonstrated that the formation of the "plateau" depends on the oscillation energy if the energy density of the oscillations greatly exceeds the energy density of the resonant particles. It can be shown in general that this requirement is sufficient for us to regard the energy density of the oscillations as independent of the distribution of the accelerated particles. Accordingly, we regard the energy density of the oscillations as fully defined by a time-dependent function that is independent of the accelerated-particle distribution. Its time-dependence is determined by the external sources and nonlinear effects.

As shown above, nonlinear effects tend to make the wave distribution isotropic. For example, Langmuir waves can become isotropic due to their decay instability. We shall see later that the oscillations also become isotropic when involved in nonlinear (not quasilinear) interaction with plasma particles. In this connection it is useful to examine the acceleration of particles by isotropically distributed oscillations.

Following these preliminaries we pose the question: what implications for the stated problem of particle acceleration can be deduced from the conservation of energy and the stability of an isotropic particle distribution demonstrated above? It might be inferred, for example, that with an arbitrary isotropic distribution of particles their total energy and, hence, the mean energy of each particle, will increase. There is no justification, however, for assuming that the particle distribution will not change with time. On the contrary, we know that such a variation is caused by acceleration. The change in the particle distribution for intense oscillations is regulated by the distribution and variation of the oscillations, and there is no reason to believe that if the particle distribution were initially isotropic, it would remain isotropic thereafter. All we can say is that if the particles are initially isotropically distributed, their mean energy will increase. We can show that if intense plasma oscillations are distributed isotropically, then for an arbitrary distribution of the accelerated particles their mean energy increases [14].

As we know, the variation of the distribution of resonant particles is governed by the diffusion relation

$$\frac{\partial f_{\mathbf{p}}}{\partial t} = \frac{\partial}{\partial p_i} D_{i,j} \frac{\partial f_{\mathbf{p}}}{\partial p_j} . \tag{6.64}$$

The diffusion coefficient $D_{i,j}$ depends on the distribution of the oscillations and is a function only of the momenta $p_i$. If the oscillations are isotropically distributed, there will be no preferred direction other than that of the momentum **p**. The only second-rank tensor that can be formed from the vector $p_i$ is $p_i p_j$. For isotropic oscillations, therefore, $D_{i,j}$ can be composed only of the tensor $p_i p_j/p^2$ and the unit tensor $\delta_{i,j}$. Combining them, we write

$$D_{i,j} = D^l(p^2)\frac{p_i p_j}{p^2} + D^t(p^2)\left(\delta_{i,j} - \frac{p_i p_j}{p^2}\right). \tag{6.65}$$

Equation (6.65) is a definition of $D^l$ and $D^t$. The physical meaning of these quantities is as follows: the coefficient $D^l$, as we shall verify presently, describes the variation of the particle energy in the oscillation field, and $D^t$ describes the scattering of the particles, i.e., the change in the direction of the momentum. Scattering plays a significant part only if the particle distribution is not isotropic.

Consider the variation of the particle energy. According to (6.64)

$$\frac{\partial}{\partial t}E = \frac{\partial}{\partial t}\int \frac{\varepsilon_{\mathbf{p}} f_{\mathbf{p}} d\mathbf{p}}{(2\pi)^3} = \int \frac{\varepsilon_{\mathbf{p}}\frac{\partial f_{\mathbf{p}}}{\partial t}d\mathbf{p}}{(2\pi)^3} = \int \frac{\varepsilon_{\mathbf{p}}}{(2\pi)^3}\frac{\partial}{\partial p_i}D_{i,j}\frac{\partial f_{\mathbf{p}}}{\partial p_j}d\mathbf{p}. \tag{6.66}$$

Integrating (6.66) by parts and recognizing that $\partial\varepsilon_{\mathbf{p}}/\partial p_i = v_i$, we obtain

$$\frac{\partial}{\partial t}E = -\int v_i D_{i,j}\frac{\partial f_{\mathbf{p}}}{\partial p_j}\frac{d\mathbf{p}}{(2\pi)^3} = \int\left(\frac{\partial}{\partial p_j}v_i D_{i,j}\right)\frac{f_{\mathbf{p}}\,d\mathbf{p}}{(2\pi)^3}. \tag{6.67}$$

Introducing the mean value of a quantity L

$$\langle L\rangle = \frac{\int L f_{\mathbf{p}}\frac{d\mathbf{p}}{(2\pi)^3}}{\int f_{\mathbf{p}}\frac{d\mathbf{p}}{(2\pi)^3}}, \tag{6.68}$$

we write (6.67) in the form of (6.65):

$$\frac{\partial}{\partial t}\langle\varepsilon\rangle = \left\langle\frac{\partial}{\partial p_j}v_i D_{i,j}\right\rangle. \tag{6.69}$$

We now consider the fact that by virtue of (6.65)

$$v_i D_{i,j} = \frac{p_i c^2}{\varepsilon_{\mathbf{p}}} D_{i,j} = \frac{p_j c^2}{\varepsilon_{\mathbf{p}}} D^l = v_j D^l;$$

$$\frac{\partial}{\partial p_j} v_j D^l = \frac{\partial}{\partial p_j} p_j \frac{D^l c^2}{\varepsilon_{\mathbf{p}}} = \frac{3D^l c^2}{\varepsilon_{\mathbf{p}}} + p \frac{\partial}{\partial p} \frac{D^l c^2}{\varepsilon_{\mathbf{p}}} = \frac{1}{p^2} \frac{\partial}{\partial p} p^2 v D^l, \qquad (6.70)$$

i.e.,

$$\frac{\partial}{\partial t} \langle \varepsilon \rangle = \left\langle \frac{1}{p^2} \frac{\partial}{\partial p} p^2 v D^l \right\rangle.$$

We see that of the two diffusion coefficients introduced above, $D^l$ and $D^t$, only $D^l$ characterizes the variation of the particle energy. Equations (6.65), (6.30), and (6.18) lead to a more general expression for the diffusion coefficient $D^l$ for longitudinal waves:

$$D^l = \frac{p_i p_j}{p^2} D_{i,j} = \frac{1}{(2\pi)^3} \int \hbar^2 \frac{(\mathbf{kv})^2}{v^2} \hbar^2 N_{\mathbf{k}}^{\sigma} w_{\mathbf{p}}(\mathbf{k})\, d\mathbf{k} =$$

$$= \int \frac{\hbar (\Omega^{\sigma})^2}{v^2} N_{\mathbf{k}}^{\sigma}\, d\mathbf{k} \frac{e^2}{\pi} \frac{(\Omega^{\sigma})^2}{k^2 \frac{\partial}{\partial \omega} \omega^2 \varepsilon^{\sigma} \big|_{\omega=\Omega^{\sigma}(\mathbf{k})}} \delta(\Omega^{\sigma} - \mathbf{kv}).$$

Because of the assumed isotropy of the oscillations, we can integrate (6.70) for all directions:

$$D^l = \frac{2}{v^3} \int\limits_{k > \frac{\Omega^{\sigma}(\mathbf{k})}{v}} e^2 \frac{dk}{k} N_{\mathbf{k}}^{\sigma} \frac{(\Omega^{\sigma}(\mathbf{k}))^4 \hbar}{\frac{\partial}{\partial \omega} \omega^2 \varepsilon^{\sigma} \big|_{\omega=\Omega^{\sigma}(\mathbf{k})}}. \qquad (6.71)$$

From (6.71) we can easily check the assertion of a systematic increase in the mean energy of particles interacting with isotropically distributed oscillations. In fact, according to (6.70) it suffices for this to show that $\rho = (\partial/\partial p)\, p^2 v D^l > 0$. By (6.71)

$$\rho = \frac{\partial}{\partial p} (p^2 + m^2 c^2) \int\limits_{k > \frac{\Omega^{\sigma}(\mathbf{k})}{v}} 2e^2 \frac{dk}{k} \hbar N_{\mathbf{k}}^{\sigma} \frac{(\Omega^{\sigma}(\mathbf{k}))^4}{\frac{\partial}{\partial \omega} \omega^2 \varepsilon^{\sigma} \big|_{\omega=\Omega^{\sigma}(\mathbf{k})}} =$$

$$= 4p \int\limits_{k_{\min}} e^2 \frac{dk}{k} \hbar N_{\mathbf{k}}^{\sigma} \frac{(\Omega^{\sigma}(\mathbf{k}))^4}{\frac{\partial}{\partial \omega} \omega^2 \varepsilon^{\sigma} \big|_{\omega=\Omega^{\sigma}(\mathbf{k})}} +$$

$$+ \frac{m^2}{pv} \frac{\Omega^{\sigma}(k_{\min})}{1 - \frac{v_{\mathrm{gr}}^{\sigma}}{v}} \frac{e^2}{k_{\min}} \hbar N_{k_{\min}}^{\sigma} \frac{(\Omega^{\sigma}(k_{\min}))^4 c^2}{\frac{\partial}{\partial \omega} \omega^2 \varepsilon^{\sigma} \big|_{\omega=\Omega^{\sigma}(k_{\min})}}. \qquad (6.72)$$

Here $k_{\min}$ is the solution of $k_{\min} = \Omega^{\sigma}(k_{\min})/v$. The positiveness of (6.72) ($v_{\mathrm{gr}}^{\sigma} < v_{\varphi}^{\sigma} < v$) proves the stated proposition.

A very simple expression is obtained for the systematic acceleration of ultrarelativistic particles (cosmic rays, for example), when $v \to c$ and (6.71) is independent of p

$$\frac{\partial}{\partial t}\langle\varepsilon\rangle=\left\langle\frac{4e^2}{\varepsilon}\int\limits_{k>\Omega^\sigma/v}\hbar\frac{dk}{k}N_{\mathbf{k}}^{\sigma}\frac{(\Omega^\sigma(\mathbf{k}))^4}{\frac{\partial}{\partial\omega}\omega^2\varepsilon^\sigma\,|_{\omega=\Omega^\sigma}}\right\rangle. \tag{6.73}$$

From Eq. (6.71) it is an easy task to find general expressions for the acceleration of particles with arbitrary velocities by Langmuir oscillations [14, 15]:

$$\frac{\partial}{\partial t}\langle\varepsilon\rangle=\frac{e^2\omega_{0e}^3}{p}\left\{\frac{1-v^2/c^2}{v^2}N^l(v)\hbar+\frac{2\hbar}{c^2}\int\limits_0^v\frac{N^l(v_\varphi)\,dv_\varphi}{v_\varphi}\right\} \tag{6.74}$$

and by ion-acoustic oscillations [15]:

$$\frac{\partial}{\partial t}\langle\varepsilon\rangle=\left\langle\frac{e^2\omega_{0e}^3}{p}\left\{2\int\limits_0^{v}\frac{dv_\varphi^s}{v_\varphi^s}\hbar N^s(v_\varphi^s)\left(\frac{v_s^2-v_\varphi^{s2}}{v_{Te}^2}\right)^{3/2}\frac{1}{c^2}+\right.\right.$$
$$\left.\left.+\frac{1-v^2}{v^2}\left(\frac{v_s^2-v^2}{v_{Te}^2}\right)^{3/2}\frac{1}{2}\left(1+\frac{v_s-v}{|v_s-v|}\right)\hbar N^s(v)\right\}\right\rangle;\quad \tilde{v}=\min(v,\ v_s). \tag{6.75}$$

The acceleration of high-energy particles by ion-acoustic oscillations is less efficient than that by high-frequency oscillations. The Fermi mechanism is an analogy of acceleration by low-frequency magnetohydrodynamic oscillations.

The acceleration of high-energy particles by ion-acoustic oscillations is less efficient than that by high-frequency oscillations. The Fermi mechanism is an analogy of acceleration by low-frequency magnetohydrodynamic oscillations. Despite the lower efficiency of the acceleration, low-frequency oscillations can play an important part as an injection mechanism, i.e., for the preliminary acceleration of particles to energies such that they can be accelerated by high-frequency oscillations. For example, plasma oscillations can accelerate only particles whose velocities are greater than $v_{Te}$. For ions, the energy of such particles $m_i v_{Te}^2 = m_i T_e/m_e$ must be $m_i/m_e$ times the mean thermal energy of the electrons. Consequently, the acceleration of ions by low-frequency oscillations is an important mechanism for enhancing the mean energy of the ions. Notice in this connection the important part played in acceleration problems by various nonlinear effects that result in the generation of low-frequency oscillations from the high-frequency oscillations discussed previously.

It must be emphasized that in many cases a systematic increase in the particle energy can be inefficient, but then the increased energy spread of the particles that always accompanies an increase in their mean energy can become important. Carrying out an analysis similar to that in the determination of $\partial\langle e\rangle/\partial t$, we easily verify that in the case of small $\langle \partial\varepsilon/\partial t\rangle$

$$\frac{\partial}{\partial t}\langle\varepsilon^2\rangle = \langle 2D^l v^2\rangle. \tag{6.76}$$

An increase in the spread also produces fast particles. This kind of acceleration may be called fluctuational acceleration. If, for example, the distribution of Langmuir waves does not contain waves whose phase velocity is equal to the particle velocity, the systematic increase in the mean energy (6.74) for nonrelativistic particles is smaller than in the case $v_\varphi = v$ by a factor $v^2/c^2$ for comparable oscillation energy densities. However, the fluctuational acceleration estimated from (6.76) turns out to be of the same order as that for $v_\varphi = v$. In precisely the same way, systematic acceleration by ion-acoustic oscillations falls off rapidly for $v > v_s$, but the fluctuational acceleration in this case does not change significantly. For the acceleration of relativistic particles by Langmuir waves the second term of (6.74) is the decisive one.

Of utmost importance is the fact that the efficiency of acceleration of relativistic particles, as of particles in general whose velocities are much greater than the oscillation phase velocity, is determined not only by the energy of the oscillations, but also by their phase velocity distribution. The significance of the latter is evident from the following example. As shown earlier, nonlinear interactions of Langmuir and ion-acoustic waves can result in an increase of the Langmuir phase velocities while only slightly altering their energy. This can have an appreciable effect, however, on the acceleration of particles. Let us estimate the coefficient $D^l$ for relativistic particles in the case when the phase velocity interval $\Delta v_\varphi$ in which oscillations occur is of the same order as the characteristic mean phase velocity $\bar{v}_\varphi$ of the oscillations:

$$D^l \simeq e^2\omega_{0e}^3 \int \frac{N^l(v_\varphi)\,dv_\varphi}{v_\varphi c^3} \sim e^2\omega_{0e}^3 N^l\,(\bar{v}_\varphi)\,\hbar/c^3. \tag{6.77}$$

The energy of the oscillations can be estimated as

$$W^l = \int \frac{\hbar\omega\;{}_eN_{\mathbf{k}}4\pi k^2 dk}{(2\pi)^3} = \frac{1}{2\pi^2}\int \frac{\hbar\omega_{0e}^4 N\,(v_\varphi)\,dv_\varphi}{v_\varphi^4} \simeq \frac{\hbar\omega_{0e}^4}{2\pi^2}\,N^l\,(\bar{v}_\varphi)\,\frac{1}{(\bar{v}_\varphi)^3}, \tag{6.78}$$

i.e.,

$$D^l \sim \frac{e^2 2\pi^2 W^l (\bar{v}_\varphi)^3}{\omega_{0e} c^3}. \tag{6.79}$$

As apparent from (6.79), for a given oscillation energy $W^l$ the diffusion coefficient varies very strongly with $\bar{v}_\varphi$, approximately as $\bar{v}_\varphi^3$. Consequently, the nonlinear transfer of oscillations to large phase velocities can affect the acceleration significantly. Below we gain some insight into other nonlinear interactions (for example, scattering) which are also capable of causing this nonlinear transfer.

## 10. ISOTROPIZATION OF PLASMA PARTICLES

Closely linked with the acceleration of particles is their isotropization. If the particle velocity greatly exceeds the wave phase velocity, then, as we show later on, the principal result of its interaction with the waves is to change the direction of its momentum, but not its magnitude. If nonisotropically distributed particles are present in plasma containing isotropically distributed oscillations, then, as explained earlier, they are accelerated. The same acceleration will also occur for an isotropic distribution. However, the isotropization of anisotropically distributed particles will in general proceed more rapidly than the increase of their energy. The isotropization is determined by the coefficient $D^t$ in (6.65). In fact, this coefficient does not reflect any change in the mean particle energy, i.e., it describes those changes in the momentum which do not affect its magnitude. It is easily shown that for $v \gg v_\varphi$ the relation $D^t \gg D^l$ holds. Thus, from (6.65) we obtain

$$D^t = \frac{D_{i,i} - D^l}{2} = \int \frac{\hbar^2}{2}\left(k^2 - \frac{\omega^2}{v^2}\right) N_{\mathbf{k}}^\sigma w_{\mathbf{p}}(\mathbf{k}) \frac{d\mathbf{k}}{(2\pi)^3}. \tag{6.80}$$

For $\omega/k \ll v$ we have $k^2 \gg \omega^2/v^2$, and since the coefficient $D^l$ contains only $\omega^2/v^2$, we have in order of magnitude

$$D^t \approx \frac{v^2}{v_\varphi^2} D^l. \tag{6.81}$$

This shows that the acceleration of particles must necessarily be accompanied by their rapid isotropization. This could account for the high degree of isotropy of cosmic rays recorded on the earth [16].

## 11. ASTROPHYSICAL APPLICATIONS OF ACCELERATION AND ISOTROPIZATION OF PARTICLES BY PLASMA WAVES

Here we describe briefly some astrophysical applications of the acceleration effects investigated above.*

The acceleration mechanisms associated with the generation of fast particles under cosmic conditions are not only widespread, but may also be of significance in cosmological evolution [17]. They appear in the dynamics of flares from supernova stars and radio-galaxies and play an important part in the nuclei of galaxies, quasars, etc. We note that acceleration by high-frequency oscillations can be very significant [20] for the acceleration of electrons and ions in the radiation belts around the earth [21, 22], as well as for the interpretation of the radiation from the Crab nebula [20, 23], quasars [24], and the radio source in Cygnus A [25]. Acceleration by high-frequency oscillations suggests the possibility of a selective accleration of electrons, thus removing many of the usual power-input problems [23]. Ginzburg and Syrovatskii [26] have recently shown that galactic electrons cannot be secondary electrons. This is borne out by the latest experimental data on the electronic component of cosmic radiation recorded on earth.

The acceleration mechanism in question could play some part in the generation of cosmic rays on the sun, specifically in chromospheric flares. Here again we can mention the problem of interstellar acceleration, by Langmuir waves in particular.†

As explained earlier, isotropization and acceleration represent different aspects of the same process of interaction between accelerated particles and plasma oscillations. Estimates show that the characteristic time for isotropization by Langmuir oscilla-

* See [17-19] regarding other acceleration effects.

† This problem has been discussed by the author in collaboration with V. L. Ginzburg. Estimates show that an energy source for Langmuir oscillations, besides those listed above, might be the influx of intergalactic gas into our galaxy [15].

tion is of order $\frac{1}{\tau^l} \sim \omega_{0e} \frac{W^l}{nm_e c^3} v_\varphi^l$ and by ion-acoustic oscillations of order $\frac{1}{\tau^s} \approx \omega_{0e} \left(\frac{m_e}{m_i}\right)^2 \frac{W^s}{nm_e} \frac{v_{Te}}{c^3}$. For $v_{Te} \sim 10^{-3}c$, $n \sim 1\ cm^{-3}$, and $W^s \sim nT/10$ the bulk of cosmic rays $\varepsilon \sim m_i c^2$ becomes isotropic in a period $\tau^s \sim 100$ yr.

## 12. ON THE ACCELERATION OF ELECTRONS AND IONS UNDER LABORATORY CONDITIONS

The acceleration mechanisms described can be used to interprete a number of results from laboratory experiments. Here we can mention experiments in which the acceleration of a group of particles has been observed. They can be attributed to the action of plasma oscillations. However, owing to the complexity of the experimental conditions, this is not the only interpretation possible at present. We therefore direct our attention mainly to the numerous experiments in which accelerated electrons and ion with a broad energy spectrum have been observed.

1. It has been established that the x-ray and neutron emissions observed from impulsive discharges in many experiments arise from accelerated electrons and ions [27, 28]. Much experimental data (such as the equal mean energies of electrons and ions, etc.) might be explained by the hypothesis of acceleration by Langmuir oscillations [29].*

2. Accelerated electrons have been observed during the growth of beam instability [30, 31]. They can be explained in terms of acceleration by Langmuir waves, which are excited by the beam and are then rendered isotropic by nonlinear effects. Under these conditions the increase in the phase velocities of the oscillations is also important, as it leads to diffusion of the electron distribution toward higher particle velocities.

3. The accleration of ions in the development of beam instability has been observed in [32, 33]. Often the yield of accelerated

---

* We point out that the acceleration of particles by plasma oscillations is possible when they have an anisotropic distribution. If, for example, the oscillation intensity is highest in some preferred direction, particles whose velocities are almost perpendicular to that direction will be predominantly accelerated.

ions is correlated with the generation of low-frequency waves, which are capable of accelerating ions efficiently.

**4.** The acceleration of ions when a plasma is acted upon by intense high-frequency radiation has been observed in experiments on the radiative acceleration of a plasma [34]. It could be attributed to the generation of strong Langmuir and low-frequency oscillations when the plasma is acted upon by a strong high-frequency oscillation. Another possible mechanism for the generation of oscillations is the beam instability which occurs when the leading edge of an accelerated bunch is reflected from a region containing a high-frequency field.

## 13. EFFECTIVE COLLISION FREQUENCY OF PARTICLES AND WAVES

Since when $v \gg v_\varphi$ particles are mainly scattered by waves without much altering their energy, their interactions with waves in this case are very similar to the quasi-elastic collisions of particles. A plasma particle might clearly be visualized as colliding with a wave and thereby altering the direction and very slightly the magnitude of its momentum. It is very useful for practical estimates, therefore, to introduce an effective collision frequency of particles with waves. For comparison we give without derivation a formula describing the elastic collisions of electrons with infinitely heavy ions (this formula is easily obtained from the so-called Landau collision integral):

$$\frac{\partial f_{\mathbf{p}}}{\partial t} = \frac{\partial}{\partial p_i}\left(\frac{2\pi e^2 e_i^2\left(\delta_{ij} - \frac{v_i v_j}{v^2}\right)}{v} L n_i\right)\frac{\partial f_{\mathbf{p}}}{\partial p_j}. \tag{6.82}$$

Here $L = \ln(r_{max}/r_{min})$ is the Coulomb logarithm, and $n_i$ the ion concentration. Equation (6.82) can also be written in a form analogous to the formula obtained for the interaction of particles and waves:

$$\frac{\partial f_{\mathbf{p}}}{\partial t} = \frac{\partial}{\partial p_i} D^t \left(\delta_{ij} - \frac{p_i p_j}{p^2}\right)\frac{\partial f_{\mathbf{p}}}{\partial p_j}, \tag{6.83}$$

where $D^t = m_e^2 v^2 \nu_{eff}^{Coul}$, and $\nu_{eff}^{Coul}$ is expressed by the familiar relation for the effective number of Coulomb collisions:

$$\nu_{\text{eff}}^{\text{Coul}} = \frac{e^4 n}{m_e^2 v^3} 2\pi \ln \frac{r_{\max}}{r_{\min}} \sim \frac{\omega_{0e}}{8\pi} \frac{L}{n\lambda_{De}^3} \left(\frac{v_{Te}}{v}\right)^3. \tag{6.84}$$

Proceeding from this analogy, we introduce the concept of the effective number of particle–wave collisions:

$$\nu_{\text{eff}} = D^t / p^2. \tag{6.85}$$

For $v \gg v_\varphi$ we have for longitudinal waves

$$\nu_{\text{eff}} \simeq \frac{e^2}{\pi p^2} \int \frac{(\Omega^\sigma)^2 N_{\mathbf{k}}^\sigma}{\left.\frac{\partial}{\partial \omega} \omega^2 \varepsilon^\sigma \right|_{\omega = \Omega^\sigma(k)}} \delta(\Omega^\sigma - \mathbf{kv}) \frac{\hbar}{2} d\mathbf{k}, \tag{6.86}$$

or, after angular integration,

$$\nu_{\text{eff}} \simeq \frac{e^2}{p^2 v} \int \frac{(\Omega^\sigma)^2 N_k^\sigma k\,dk}{\left.\frac{\partial}{\partial \omega} \omega^2 \varepsilon^\sigma \right|_{\omega = \Omega^\sigma(k)}} \hbar. \tag{6.87}$$

We now estimate the effective collision frequency of particles and Langmuir waves and compare it with the Coulomb collision frequency (6.84). Assuming $\Delta k$ is of the same order as $k$, we obtain

$$\nu_{\text{eff}}^l \simeq \frac{e^2 \omega_{0e}}{p^2 v} k^2 N_k^l \hbar; \quad W^l \sim \frac{\omega_{0e} k^3 N_k^l \hbar}{2\pi^2},$$

in other words

$$\nu_{\text{eff}}^l \simeq \frac{\pi}{4} \frac{m_e \omega_{0e}^2 W^l}{p^2 v k n},$$

or for nonrelativistic electrons

$$\nu_{\text{eff}}^l \simeq \frac{\pi}{4} \omega_{0e} \frac{v_\Phi^l}{v} \frac{W^l}{n m_e v^2}. \tag{6.88}$$

For comparison we write $\nu_{\text{eff}}^{\text{Coul}}$ in the form

$$\nu_{\text{eff}}^{\text{Coul}} \simeq \frac{\pi}{4} L \omega_{0e} \frac{v_{Te}}{v} \frac{W_{\text{T}}^l}{n m_e v^2}, \tag{6.89}$$

where

$$W_{\text{T}} \simeq \frac{T k^3}{2\pi^2} \sim \frac{T}{\lambda_{De}^3 2\pi^2}$$

is the energy density of the plasma waves corresponding to thermal equilibrium at a temperature T.

Notice that if the energy of the plasma waves is much greater than their thermal level, $W^l \gg W_T^l$, i.e., if the level of the oscillations is higher than the thermal level, the collision of particles with waves is more frequent than the mutual collisions of particles.

A similar estimate of the effective collision frequency for ion-acoustic oscillations yields

$$\nu_{\text{eff}} \simeq \frac{\pi}{4} \omega^s \frac{m_i}{m} \frac{v_\varphi^s}{v} \frac{W^s}{nmv^2}. \tag{6.90}$$

Frequently a magnetohydrodynamical equation is used for the description of the slow motions of a plasma in magnetic fields. The plasma is assumed infinitely conducting, i.e., collisional effects are neglected. The high frequency of particle–wave collisions can produce more rapid penetration of the magnetic field into the plasma, as well as rapid dissipation of the magnetic fields. It is possible, for example, for the so-called freezing-in of the force lines of a magnetic field in a plasma to be upset. The description of a plasma by means of the magnetohydrodynamical equations might prove impossible if there are strong high-frequency oscillations present in the plasma.

These results should have important astrophysical applications. We note, for example, that in chromospheric flares a strong dissipation of the magnetic field is observed simultaneously with particle acceleration. The correlation of these two effects has a logical explanation if we postulate that both effects arise from the excitation of oscillations by instabilities created in the development of the chromospheric flares.

## 14. QUASILINEAR THEORY OF BEAM INSTABILITY IN A PLASMA

The instability of beams in a plasma is a frequently encountered mechanism for the generation of plasma oscillations [35, 36]. The physical interpretation of the instability described by the quasilinear approximation arises from a population inversion (the beam particles have a higher energy than the plasma particles) and

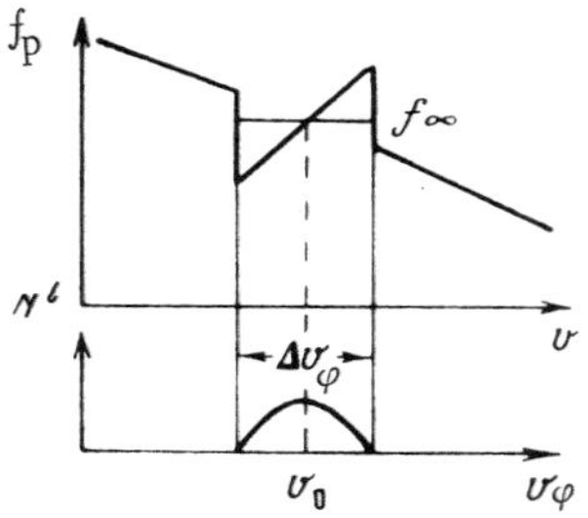


Fig. 27. Change in the plasma particle distribution function in the presence of a slow beam.

the possibility of induced Cerenkov emission, resulting in the avalanche multiplication of the quanta.* In the one-dimensional case instability occurs for $\partial f/\partial p > 0$. An example of such a distribution is given in Fig. 27. This distribution differs from the one in Fig. 25 in the sign of the derivative in the interval $\Delta v_\varphi$. The development of instability leads to the generation of oscillations. If the oscillation intensity becomes large, an appreciable diffusion of the particles occurs, taking the particles from higher to lower energies. The value of the derivative decreases, and the instability vanishes. Consequently, in this case the saturation effect eliminates the instability.

The dynamics of the development of beam instability can be described by (6.53) and (6.54) with $\psi(0)$ replaced by $-\psi(0)$, where $\psi(0) > 0$. If $\psi(0) \gg (\alpha/\beta)\, w(0)$, we obtain in place of (6.57)

$$w(t) = w(0)\,\frac{\left(\psi(0) + \frac{\alpha}{\beta}\,w(0)\right) e^{\beta t \psi(0)}}{\psi(0) + \frac{\alpha}{\beta}\,w(0)\,e^{\beta t \psi(0)}}\,. \tag{6.91}$$

The energy generated by the beam as $t \to \infty$ amounts to $w(\infty) = (\beta/\alpha)\,\psi(0)$:

---

* Nonlinear effects, especially the decay interactions and induced scattering effects analyzed below, can play a major role in the development of beam instability. They can terminate beam instability [37]. This is extremely important from the experimental point of view, because the analysis of nonlinear effects makes it possible to predetermine the interaction efficiency of beams with a plasma and to estimate the fraction of the energy lost by the beam. The successive alternation of oscillation buildup and cutoff is often observed experimentally, an effect that can be attributed to nonlinear effects. It is also important to note that, given the same beam parameters, the efficiency of its interaction with the plasma differs owing to nonlinear effects, depending on whether or not oscillations are present.

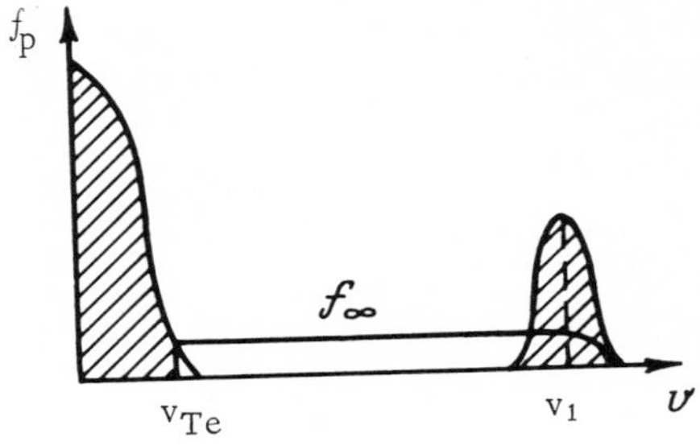


Fig. 28. Interaction of a fast particle beam with a plasma.

$$W^l = \frac{v_0}{12\Delta v_\varphi} nm_e (\Delta v_\varphi)^2 \psi(0). \tag{6.92}$$

It follows from (6.92) that $W^l$ is equal to the change in the energy of the beam particles. This shows that effective interaction of beams with plasma is possible and that a considerable fraction of the beam energy is transferred to the plasma oscillations.

Consider a beam whose particle velocity is much greater than the thermal velocities of the plasma electrons [38, 39] (Fig. 28). In this case the distribution function will necessarily become flatter and flatter, until the particles diffuse to thermal velocities. Suppose that the initial spread $\Delta v$ in the beam particle velocity is considerably smaller than their mean velocity $v_1$. We wish to determine the energy lost by the beam and the distribution of the oscillations generated by it. We denote the beam particle distribution function corresponding to the final plateau state by $f_\infty$. By the conservation of the number of particles

$$\int_{v_{min}}^{v_1} f_\infty \frac{dp}{2\pi} = n_1 = f_\infty \frac{m_e}{2\pi}(v_1 - v_{min}) \simeq f_\infty \frac{m_e v_1}{2\pi}. \tag{6.93}$$

The energy of the beam particles in the final state is

$$E = \int_{v_{min}}^{v_1} \frac{m_e v^2}{2} f_\infty \frac{m_e\, dv}{2\pi} = \frac{n_1 m_e v_1^2}{6}, \tag{6.94}$$

which is one third of the initial beam energy. By the law of conservation the energy of the excited oscillations is two thirds of the initial beam energy:

$$W^l = \frac{n_1 m v_1^2}{3}. \qquad (6.95)$$

The distribution of these oscillations with wave number can be found from the so-called quasilinear integral. Substituting (6.49) into (6.45), we have

$$\frac{\partial f_p}{\partial t} = \frac{\partial}{\partial v} \frac{\hbar}{v^3} \omega_{0e}^2 \frac{\partial N^l}{\partial t} \frac{1}{m_e^2}. \qquad (6.96)$$

Integrating (6.96) with respect to time, we obtain

$$f_p(\infty) - f_p(0) = \frac{\partial}{\partial v} \frac{\hbar\omega_{0e}^2}{v^3 m_e^2} (N^l(v, \infty) - N^l(v, 0)). \qquad (6.97)$$

Assuming $N^l(v, 0) \ll N^l(v, \infty)$, we integrate over v from $v_{min}$ to $v < v_1$, i.e., over the interval in which $f_p(0) = 0$; we have

$$\int\limits_{v_{min}}^{v} f_\infty \, dv = \frac{\hbar\omega_{0e}^2}{v^3 m_e^2} N_\infty^l(v) = \frac{n_1 2\pi v}{m_e v_1}. \qquad (6.98)$$

Hence

$$W^l(v) = \frac{\hbar\omega_{0e}^2 N_\infty^l}{2\pi v^2} = \frac{n_1 v^2 m_e}{v_1}.$$

so that

$$W^l = \int W^l(v)\, dv = \frac{n_1 v_1^2 m_e}{3}. \qquad (6.99)$$

This result corresponds exactly to (6.95).

Notice that half the energy in the plasma oscillations is in the electric field of the waves, the other half being in the oscillations of the plasma particles:

$$W^l = W_E + W_{pl}; \; W_E \simeq \frac{n_1 m_e v_1^2}{6} \simeq W_{pl}. \qquad (6.100)$$

The characteristic time for energy loss by the beam can be estimated from the growth rate

$$\gamma \simeq \frac{2\pi e^2}{\omega_{0e}} v^2 \frac{\partial f}{\partial v} \simeq \frac{2\pi e^2 v_1^2}{\omega_{0e}} \frac{f}{\Delta v};$$

$$\int f m_e \, dv \sim 2\pi n_1 \sim f m_e \Delta v,$$

i.e.,

$$\gamma \sim \frac{e^2 v_1^2 (2\pi)^2 n_1}{\omega_{0e} m_e (\Delta v)^2};$$

$$\gamma \sim \pi \frac{n_1}{n_0} \omega_{0e} \left( \frac{v_1}{\Delta v} \right)^2. \tag{6.101}$$

For the particles to lose energy of the same order as they had originally, we require that $\Delta v \sim v$, i.e., that the characteristic time be determined by the slowest final stage:*

$$\gamma \sim \frac{n_1}{n_0} \omega_{0e}. \tag{6.102}$$

The same estimate can be obtained from the diffusion relation, assuming $W \sim n_1 m v_1 \Delta v$. Thus,

$$\gamma \sim \frac{D}{m_e^2 (\Delta v)^2} \simeq \frac{\omega_{0e} e^2 \hbar N 2\pi}{m_e^2 (\Delta v)^2 v_1}, \tag{6.103}$$

or

$$W \sim \int \frac{N\hbar\omega_{0e}^2 \, dv}{v^2 2\pi} \sim \frac{N\hbar\omega_{0e}^2 \Delta v}{v_1^2 2\pi} \text{ and } \gamma \sim \frac{\pi\omega_{0e} v_1 W}{n m_e (\Delta v)^3} \sim \pi\omega_{0e} \frac{n_1 v_1^2}{n (\Delta v)^2}.$$

For $\Delta v \sim v$ we arrive at (6.102).

There are various conditions under which certain factors will stabilize beam instability. Typical of these factors are nonlinear wave interaction effects. We can show, in particular, that decay interactions are capable of stabilizing beam instability. The mechanism of this process is analogous to that discussed in the stabilization of transverse-wave instability; specifically it involves the transfer of oscillatory energy from the generation interval into the interval of weak interaction with the beam. We have seen that the decay of a Langmuir wave into a sound wave can reverse the direction of propagation of the Langmuir wave. The decay-induced Langmuir waves immediately become nonresonant with the beam particles. The characteristic time for this one-

* A more thorough analysis [40] reveals that the development of quasilinear relaxation can occur with the formation of a nonlinear "shock" wave in velocity space.

dimensional decay can be estimated from Eq. (5.56), which may be written in the form ($\Delta k/k = -\Delta v/v$)

$$\gamma^{\text{nlr}} = \frac{\pi}{4}\frac{v_1}{\Delta v}\omega_{0e}\frac{W^l}{nm_e v_{Te}^2}\cdot\frac{v_1}{3v_{Te}}\sqrt{\frac{m_e}{m_i}}. \qquad (6.104)$$

Substituting $W^l \sim n_1 m v_1 \Delta v$, we have

$$\gamma^{\text{nlr}} = \frac{\pi}{4}\omega_{0e}\frac{v_1^2}{v_{Te}^2}\frac{n_1}{n_0}\cdot\frac{v_1}{3v_{Te}}\sqrt{\frac{m_e}{m_i}}.$$

Comparing this expression with (6.101), we see that for $(\Delta v)^2 \gtrsim 12v_{Te}^2 \times \frac{v_{Te}}{v_1}\sqrt{\frac{m_i}{m_e}}$ decay interaction begins to prevail, i.e., the oscillations are transported more rapidly from the interval $\Delta k$ by decay interaction than they are generated therein by the beam instability. In this situation the instability is stabilized. In order to use (6.104) we must have

$$\Delta k = \frac{\omega_{0e}}{v_1^2}\Delta v < k_0 = \frac{\omega_{0e}}{3v_{Te}}\sqrt{\frac{m_e}{m_i}},$$

i.e.,

$$\left(\frac{m_i}{m_e}\right)^{1/4}\left(\frac{v_{Te}}{v_1}\right)^{1/2}\sqrt{12}v_{Te} < \Delta v < \frac{v_1^2}{3v_{Te}}\sqrt{\frac{m_e}{m_i}},$$

which is possible only if $v_1 > 3v_{Te}\,(m_i/m_e)^{3/5}$. Moreover, for decay to be possible, we must have $v_1 < 3v_{Te}\,(m_i/m_e)^{1/2}$.

The above estimates illustrate the conceptual possibility that the instability becomes stabilized, thereby resulting in a less efficient interaction between the beam and the plasma.

## 15. EFFECTS OF OSCILLATION PILEUP IN THE DEVELOPMENT OF BEAM INSTABILITY

We have demonstrated that beam instability in a plasma is a simple consequence of population inversion and that its stabilization is possible by nonlinear wave interaction. Let us now examine in more detail the effect of oscillation pileup [41], which can promote the stabilization of beam instability to a powerful degree. It was postulated above that the particle beam has initially the same velocity distribution at any point of an unconfined plasma. In real situations, however, the beams are often nonuniform, particularly axially, if they are externally injected into the plasma.

Let us see how beam instability behaves when the velocity distribution of the particles is nonuniform along the beam. The oscillations generated by the beam move very slowly through the plasma. The rate of energy transfer of the oscillations is determined by the wave group velocity. For Langmuir waves

$$v_{gr}^{l} = \frac{d\omega}{dk} \simeq \frac{3v_{Te}^2}{v_{\varphi}} \sim \frac{3v_{Te}^2}{v_1} \ll v_{Te} \tag{6.106}$$

for $v_1 \gg v_{Te}$. This implies that the beam particles, as they move with velocity $v_1$ from one point in space to another, are acted upon not only by the waves that they themselves have generated, but also by those generated earlier by other beam particles. This is because the rate of energy transfer of the oscillations is small compared to the beam velocity. Consequently, the beam particles arriving at a given point in space are not in equilibrium with those leaving it. In fact, the newly arrived particles begin to generate waves from the level generated by the preceding particles. The increase in the number of quanta, on the other hand, is proportional to their initial number ($dN \sim \gamma N dt$), so that the new arrivals can generate a larger number of quanta. This promotes an increase in the number of quanta at the specified point. This pileup effect of the oscillations generated by different particles of the beam can cause a markedly nonuniform spatial distribution of the oscillations. A small nonuniformity in the spatial distribution of the beam particles is adequate for this process to develop into an avalanche.

The simplest example of the pileup of oscillation energy is in the injection of a beam into a semi-infinite plasma (Fig. 29). The first beam particles to enter the plasma create oscillations, the ensuing particles then enter the oscillating plasma and amplify those oscillations. As a result a narrow layer with intense oscillations must be formed at the plasma boundary, its thickness rapidly decreasing. Through the avalanche behavior of the oscillation pileup the thickness of the layer must decrease exponentially. This does not mean that the oscillation intensity outside the layer fails

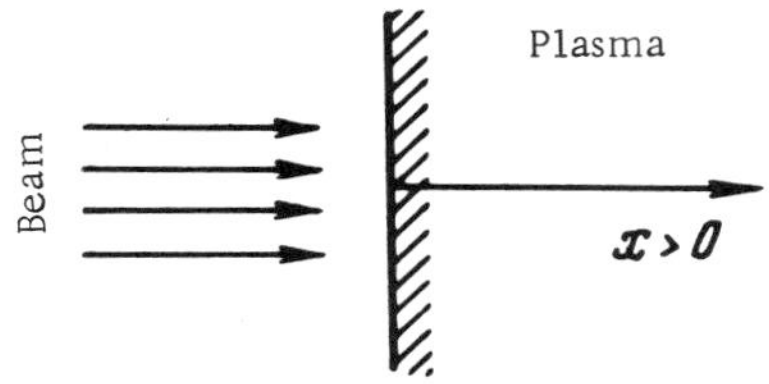


Fig. 29. Injection of a beam into a semi-infinite plasma.

to increase; the effective width of the intense-oscillation layer as defined, for example, in terms of the distance from the plasma boundary such that the oscillation intensity decreases by one half, simply decreases. With continuous injection of the beam this width must decrease exponentially with time [41]:

$$x = x_0 e^{-\gamma t}, \tag{6.107}$$

where it is clear from physical considerations that $\gamma$ is of the same order as the linear growth rate.

We now show that this result can be obtained from an elementary calculation [42]. For this we use the simple model described in Secs. 7 and 14. We make use of Eqs. (6.53) and (6.54), taking account of the nonuniform axial distribution of the beam and oscillations:

$$\frac{\partial \psi}{\partial t} + v_1 \frac{\partial \psi}{\partial x} = -\alpha \psi w: \tag{6.108}$$

$$\frac{\partial w}{\partial t} + v_{\text{gr}}^{l} \frac{\partial w}{\partial x} = \beta \psi w. \tag{6.109}$$

Let us first neglect the effect of oscillation transfer, assuming $v_{\text{gr}}^{l} = 0$. Then the solution of Eq. (6.109) may be written in the form

$$w = w_0 e^z; \quad z = \beta \int_0^t \psi(x, t)\, dt. \tag{6.110}$$

We have assumed that at t = 0 the noise intensity is constant and equal to $w_0$ over the length of the plasma. Substitution of (6.110) into (6.108) yields a single nonlinear equation in $\psi$, which is best written in terms of z, recognizing that $\psi = \frac{1}{\beta}\frac{dz}{dt}$,

$$\frac{\partial}{\partial t}\left(\frac{\partial z}{\partial t} + v_1 \frac{\partial z}{\partial x}\right) = -\alpha \frac{\partial z}{\partial t} e^z w_0 = -\alpha \frac{\partial}{\partial t} e^z w_0. \tag{6.111}$$

It follows from (6.111) that

$$\frac{\partial z}{\partial t} + v_1 \frac{\partial z}{\partial x} + \alpha w_0 e^z = \text{const.} \tag{6.112}$$

The constant (6.112) depends on x and can be found from the initial condition. For $t = 0$, $z = 0$, and therefore, $\partial z/\partial x = 0$. We have also $\partial z/\partial t|_{t=0} = \beta\psi(x, 0)$. Let us suppose that $t = 0$ is the time at which the beam first enters the plasma, i.e., for $t = 0$ and $x > 0$ in the plasma $\psi(x, 0) = 0$. Then

$$\frac{\partial z}{\partial t} + v_1 \frac{\partial z}{\partial x} + \alpha w_0 (e^z - 1) = 0. \tag{6.113}$$

We seek z as a function of $\xi = x - v_1 t$ and t, viz., $z = \tilde{z}(\xi, t)$. Then (6.113) becomes

$$\frac{\partial \tilde{z}}{\partial t} = -\alpha w_0 (e^{\tilde{z}} - 1). \tag{6.114}$$

The integration of this equation is trivial after the separation of variables:

$$\int \frac{d\tilde{z}}{e^{\tilde{z}} - 1} = \ln(1 - e^{-\tilde{z}}) = -\alpha w_0 t + c(\xi). \tag{6.115}$$

The "constant" $c(\xi)$ is determined from the boundary conditions with the assumption that the beam is continuously injected into the plasma. For $x = 0$

$$z = \beta \int_0^t \psi(0, t)\, dt = \beta\psi_0 t.$$

We have

$$c(-v_1 t) = \alpha w_0 t + \ln(1 - e^{-\beta\psi_0 t}),$$

i.e.,

$$c(\xi) = -\alpha w_0 \frac{\xi}{v_1} + \ln\left(1 - e^{\frac{\beta}{v_1}\psi_0 \xi}\right). \tag{6.116}$$

Substituting (6.116) into (6.115), we readily find the desired solution:

$$w = w_0 e^z = \frac{w_0}{1 - e^{-\alpha w_0 \frac{x}{v_1}} + e^{-\alpha w_0 \frac{x}{v_1} + \frac{\beta}{v_1}\psi_0 (x - v_1 t)}}. \tag{6.117}$$

This solution applies for $\xi < 0$, i.e., for $x < v_1 t$, as well as for $x > 0$ and $t > 0$. It is in this interval that pileup occurs, hence it is the interval in which we are now most interested.

The expression (6.117) describes the space–time distribution of the oscillations. How does the result (6.117) differ qualitatively from that deduced earlier, when we investigated the spatially unbounded problem? The difference is that previously the growth of the oscillations was limited by saturation, with the straightforward physical implication that the oscillation energy could not exceed that of the beam particles, whereas according to (6.117) the oscillation energy can grow without limit. We shall discover that in reality there are limitations and the unlimited growth of (6.117) is a result of the factors originally neglected, although the energy density of the oscillations in this instance far exceeds that of an unbounded plasma. This has a simple physical interpretation, as the oscillations at any point are generated by different beam particles, i.e., pileup occurs.

We now show that near the plasma boundary the energy density of the oscillations does in fact increase continuously. For this we consider the case $x = 0$:

$$w = w_0 e^{\beta \psi_0 t}. \tag{6.118}$$

This equation shows that the oscillation energy at the actual plasma boundary increases by a law corresponding to a linear growth rate. Let us estimate the distance $x_{eff}$ over which this rapid growth is sustained. We assume $\alpha w_0(x/v_1) \ll 1$ and $x \ll v_1 t$. Then

$$w = \frac{w_0}{\alpha \omega_0 \frac{x}{v_1} + e^{-\beta \psi_0 t}}. \tag{6.119}$$

It is obvious that

$$x_{\text{eff}} \lesssim \frac{v_1}{\alpha w_0} e^{-\beta \psi_0 t}, \tag{6.120}$$

i.e., the layer narrows exponentially with time. It must be stressed that the increase in the energy of the oscillations in a narrow region causes a rapid increase in $dw/dx$, which is of order $w/x_{eff}$. The removal of the oscillations from this layer is determined by $v_{gr}\, dw/dx$, and even though $v_{gr}$ is small, for large gradients this becomes decisive. After a sufficient time, i.e., for prolonged injection, a stationary situation must ultimately develop, where all variables depend only on x. In order to obtain this stationary distribution, which is described by

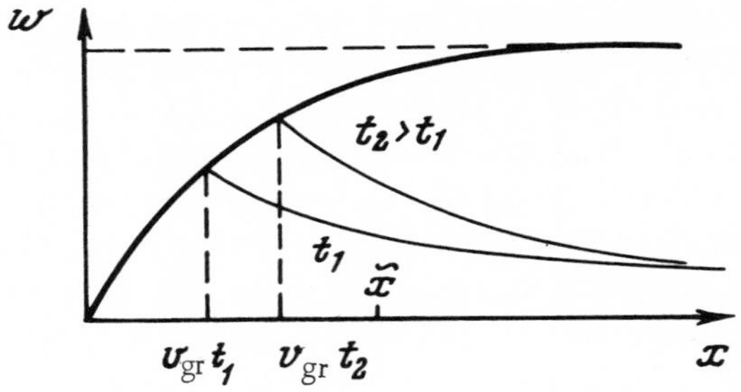


Fig. 30. Oscillation energy as a function of the distance from the plasma boundary after different time intervals.

$$\frac{\partial \psi}{\partial x} = -\frac{\alpha}{v_1} \psi w; \quad \frac{\partial w}{\partial x} = \frac{\beta}{v_{gr}^l} \psi w, \tag{6.121}$$

it suffices to replace t by x, $\alpha$ by $\alpha' = \alpha/v_1$, and $\beta$ by $\beta' = \beta/v_{gr}^l$ in the previously found time-dependent solutions. For $w(0) \ll \psi(0)\beta'/\alpha'$ we have

$$w(x) \simeq w(0) \frac{\left(\psi(0) + \frac{\alpha}{\beta} \frac{v_{gr}^l}{v_1} w(0)\right) \exp\left(\beta \psi(0) \frac{x}{v_{gr}^l}\right)}{\psi(0) + \frac{\alpha}{\beta} \frac{v_{gr}^l}{v_1} w(0) \exp\left(\beta \psi(0) \frac{x}{v_{gr}^l}\right)}. \tag{6.122}$$

A typical representation of w(x) is shown in Fig. 30. "Saturation" of the oscillation energy, $w_\infty$, is reached at a distance $\tilde{x}$ from the plasma boundary:

$$\tilde{x} \simeq \frac{v_{gr}^l}{\beta \psi(0)} = \frac{v_{gr}^l}{\gamma}, \tag{6.123}$$

where $\gamma$ is the linear growth rate, and

$$w_\infty = \frac{v_1}{v_{gr}^l} \psi(0) \frac{\beta}{\alpha}; \tag{6.124}$$

$w_\infty$ is $v_1/v_{gr}^l$ times the saturation energy developed in the uniform case. The factor $v_1/v_{gr}^l$ is of the order $v_1^2/v_{Te}^2 \gg 1$. The quantity $w_\infty$ in (6.124) is precisely the upper limit on the growth of the oscillations near the plasma boundary.

The overall pattern of the development of instability in the injection of a beam into a plasma may be portrayed as follows. The energy of the oscillations rapidly increases in a narrow region near the plasma boundary through the pileup effect, reaching its limiting value (6.124). The further growth of the oscillations is attended by a gradual movement of the saturation "front" away from the boundary and by the formation of a stationary distribution. Figure 30 also shows the successive distributions of the oscillations as the saturation "front" propagates away from the plasma boundary.

Therefore, as a result of the pileup effect, the oscillations can acquire very large energies. This important fact suggests that the nonlinear interactions of the generated oscillations with each other and with other oscillations can be substantial. We have not taken these interactions into consideration. They can lead to a situation in which the maximum values of $W^l$ are determined, not by (6.124), but by nonlinear effects.

An especially intriguing aspect of pileup is the fact that many of the known plasma instabilities have a beam character. It is evident that effects similar to those described above are possible in an unconfined plasma for a nonuniform initial axial distribution of the oscillations or for a nonuniform initial distribution of the beam particles in the plasma, etc.

## 16. LIMITS OF APPLICABILITY OF THE QUASILINEAR APPROXIMATION; HYDRODYNAMIC BEAM INSTABILITIES OF LONGITUDINAL WAVES

When is it possible in the description of beam instability to use the concept that wave–particle interaction originates from induced wave emission and absorption? This problem is closely akin to that associated with a beam of waves. The decay interactions of waves are analogous to Cerenkov emission and absorption of waves by particles. In passing, we note the analogy in either case with negative-temperature systems. We have seen that for a sufficiently high wave intensity the decay instabilities are modified, becoming nonlinear hydrodynamic instabilities, for which the decay conditions are only approximately fulfilled.

A similar situation is observed for particle beams. For a sufficiently high concentration of beam particles or in the case of a monoenergetic beam the whole concept of interaction as the induced emission and absorption of waves loses its validity.

A necessary criterion for the applicability of the quasilinear approach is related to the admissibility of a statistical description, i.e., with the admissibility of the concept of phase randomness on the part of the waves interacting with the particles. This requires that the reciprocal of the characteristic time for the process be much smaller than the characteristic width $\delta\Omega$ of the spectrum of the interacting waves. For example, for beam instability

$$\gamma \sim \pi \frac{n_1}{n_0} \omega_{0e} \left( \frac{v_1}{\Delta v} \right)^2. \tag{6.125}$$

The width of the oscillation spectrum in a reference system associated with the plasma is

$$\delta\Omega = \frac{3}{2} \frac{(k_1^2 - k_2^2)\, v_{Te}^2}{\omega_{0e}} \sim \frac{3k_{Te}^2 k^2}{\omega_{0e}} \frac{\Delta k}{k} = 3\omega_{0e} \frac{v_{Te}^2}{v_1^2} \frac{\Delta v}{v_1}.$$

We are concerned with the width of the spectrum in the reference system associated with the beam, because only if that is sufficiently wide can we talk about a particle being situated in a random-wave field. Since by the Doppler effect, $\omega_1 = \omega - kv_0$,

$$\delta\Omega_1 = \delta\Omega - \frac{\Delta k}{k} \frac{k}{\omega_0} v_1 \omega_0 \simeq 3\omega_{0e} \frac{v_{Te}^2}{v_1^2} \frac{\Delta v}{v_1} + \omega_{0e} \frac{\Delta v}{v_1} \sim \omega_{0e} \frac{\Delta v}{v_1}.$$

It follows from $\gamma \ll \delta\Omega_1$ that

$$\frac{\Delta v}{v_1} > \left( \frac{n_1}{n_0} \right)^{1/3}. \tag{6.126}$$

This condition implies that the beam particles can be in resonance with several waves.

Let us see now how the beam parameters change if condition (6.126) is violated. We first examine the hydrodynamic instabilities of longitudinal waves. In the range of applicability of (6.126) the growth rate can be found neglecting thermal motion both in the beam and in the plasma (hydrodynamic approximation). In the presence of the beam the longitudinal waves are determined by:

$$\varepsilon = 1 - \frac{\omega_{0e}^2}{\omega^2} - \frac{\omega_{0e}^2 \frac{n_1}{n_0}}{(\omega - \mathbf{kv}_1)^2} = 0. \tag{6.127}$$

This takes account of the Doppler effect ($\omega_1 = \omega - \mathbf{kv}_1$) for the beam. The solution of (6.127) is quickly found for the most unstable modes by setting $\omega \simeq \mathbf{kv}_1 + \delta\omega = \omega_{0e} + \delta\omega$:

$$\frac{2\delta\omega}{\omega_{0e}} - \frac{\omega_{0e}^2 \frac{n_1}{n_0}}{(\delta\omega)^2} = 0; \quad \delta\omega = \left(\frac{1}{2} \frac{n_1}{n_0} \omega_{0e}^3\right)^{1/3}.$$

Hence, since

$$\mathrm{Im}\,(\delta\omega) = \frac{\sqrt{3}}{2^{4/3}} \left(\frac{n_1}{n_0}\right)^{1/3} \omega_{0e}; \quad \Omega_{\mathbf{k}} - \mathbf{kv}_1 = \frac{-\omega_{0e}}{2^{4/3}} \left(\frac{n_1}{n_0}\right)^{1/3}. \tag{6.128}$$

If we insert the limiting value of $\Delta v$ determined from (6.126) into the kinetic instability growth rate (6.125), we obtain the hydrodynamic growth rate (6.128), correct to a coefficient of order unity. This shows that the hydrodynamic instability (6.128) and the kinetic instability (6.125) pass continuously into one another.

We now find out what happens when the beam has a small velocity spread $\Delta v/v_1 \ll (n_1/n_0)^{1/3}$ [43]. This type of beam is reasonably termed "monoenergetic." In the quasilinear stage, when the opposite inequality holds, the beam is affected by the oscillations it generates to the extent, as we know, that it increases the velocity spread $\Delta v$. Since the kinetic and hydrodynamic instabilities of the beams continuously interchange, it is logical also to expect an increase in the velocity spread in the hydrodynamic stage. If this is true, however, the hydrodynamic stage must rapidly go over to a quasilinear stage. Indeed the hydrodynamic growth rates are very large, and therefore the velocity spread increases very rapidly. As soon as (6.126) is satisfied, the quasilinear stage begins. The energy lost by the beam in the hydrodynamic stage in this case is small compared with its initial energy. Every particle of the beam changes its velocity at most by $\Delta v = v_1(n_1/n_0)^{1/3}$. The energy change for an individual particle amounts to $mv_1\Delta v = mv_1^2 \cdot (n_1/n_0)^{1/3}$, and for all particles in the beam it is $\Delta W = n_1 m v_1^2 (n_1/n_0)^{1/3}$. The ratio of $\Delta W$ to the initial beam energy

$$\frac{\Delta W}{W} \simeq \left(\frac{n_1}{n_0}\right)^{1/3} \ll 1 \tag{6.129}$$

by virtue of the assumption $n_1/n_0 \ll 1$. Since the energy lost by the beam is converted into plasma oscillations, the energy of the oscillations generated by the beam must not exceed $n_1 m v_1^2 (n_1/n_0)^{1/3}$ in the hydrodynamic stage:

$$W^l < n_1 m v_1^2 (n_1/n_0)^{1/3}. \tag{6.130}$$

We now show that the velocity spread of the beam particles does in fact grow in the hydrodynamic stage. Note that the characteristic time for the velocity spread on the amplitudes of the fields, since the beam particles acquire additional velocity only under the influence of the electric fields developed as a result of their instability. The essential thing is that, unlike during the quasilinear stage, the rate does not depend on the detailed velocity distribution of the particles nor even on the width $\Delta v$ of that distribution, provided only that $\Delta v \ll v_1(n_1/n_0)^{1/3}$. In the "broadening" of the beam, therefore, the rate cannot change appreciably (for example, if the velocity spread in the beam greatly exceeds the initial velocity distribution).

We now consider the case in which the relationship between the current and field strength may be described with good accuracy by the linear approximation. This does not mean (and this point has already been made clear) that the electric field only weakly perturbs the distribution of the beam particles. The current **j** is an average characteristic of the particle distribution and in the given instance is practically independent of the width of that distribution, although the width can depend on the electric fields.

The kinetic equation describing the variation of the particle distribution function in the one-dimensional case for longitudinal waves has the form

$$\frac{\partial f_{\mathbf{p}}}{\partial t} + v\frac{\partial f_{\mathbf{p}}}{\partial x} + \frac{e}{m}E\frac{\partial f_{\mathbf{p}}}{\partial v} = 0 \tag{6.131}$$

where $f_{\mathbf{p}}$ is the distribution function, $\int f_{\mathbf{p}} \frac{d\mathbf{p}}{(2\pi)^3} = n.$

Integrating (6.131) with respect to velocity, we obtain

$$\frac{\partial n}{\partial t} + \frac{\partial}{\partial x} n\bar{v} = 0, \tag{6.132}$$

where

$$\bar{v} = \frac{\int v f_{\mathbf{p}}\, d\mathbf{p}}{\int f_{\mathbf{p}}\, d\mathbf{p}} \tag{6.133}$$

is the average particle velocity. Equation (6.132) is the equation of continuity. Multiplying (6.131) by v and by $v^2$ and integrating with respect to velocity, we obtain

$$\frac{\partial}{\partial t} n\bar{v} + \frac{\partial}{\partial x} n\overline{v^2} = \frac{e}{m} En; \tag{6.134}$$

$$\frac{\partial}{\partial t} n\overline{v^2} + \frac{\partial}{\partial x} n\overline{v^3} = 2\frac{e}{m} En\bar{v}. \tag{6.135}$$

By assumption the initial distribution of the beam and plasma particles is uniform. Consequently, the change in the average quantities as a function of x results from the action of the oscillations, hence $\partial/\partial x$ is of order $1/\lambda$, where $\lambda$ is the characteristic wavelength of the oscillations. We average the resulting equations over lengths L greatly exceeding $\lambda$. We denote this averaging procedure by the angular brackets < >. From (6.132) we obtain

$$\frac{\partial}{\partial t}\langle n\rangle = 0, \tag{6.136}$$

and from (6.134) and (6.135)

$$\frac{\partial}{\partial t}\langle n\bar{v}\rangle = \frac{e}{m}\langle En\rangle; \tag{6.137}$$

$$\frac{\partial}{\partial t}\langle n\overline{v^2}\rangle = 2\frac{e}{m}\langle En\bar{v}\rangle. \tag{6.138}$$

Equations (6.136)-(6.138) are useful when we are concerned not with the detailed picture of the spatial distribution of the unknown quantities, but merely with finding some spatially averaged beam characteristics. The conservation laws can be derived from (6.136)-(6.138). These equations can be written both for the plasma particles and for the beam particles. Adding them, we obtain

$$\frac{\partial}{\partial t}(\langle n_1\bar{v}_1\rangle + \langle n_0\bar{v}_0\rangle) = \frac{1}{m}\langle E\rho\rangle; \tag{6.139}$$

$$\frac{\partial}{\partial t}\left(\frac{\langle n_1\overline{v_1^2}\rangle}{2} + \frac{\langle n_0\overline{v_0^2}\rangle}{2}\right) = \frac{1}{m}\langle Ej\rangle, \tag{6.140}$$

where $\rho = e_1 n_1 + e_0 n_0$; $j = e_1 n_1 \bar{v}_1 + e_0 n_0 \bar{v}_0$. The subscript 1 refers to the beam, 0 to the plasma.

The total current $\mathbf{j}$ is easily expressed in terms of $\mathbf{E}$ by means of Maxwell's equations. Taking account of the electrostatic nature of the longitudinal modes, rot $\mathbf{E} = 0$, we have

$$\mathbf{j} = -\frac{1}{4\pi}\frac{\partial \mathbf{E}}{\partial t};$$

from (6.140) we find

$$\frac{\partial}{\partial t}\left(\frac{\langle n_1 m v_1^2\rangle}{2} + \frac{\langle n_0 m v_0^2\rangle}{2} + \frac{\langle E^2\rangle}{8\pi}\right) = 0, \tag{6.141}$$

i.e., (6.141) in fact expresses the conservation of energy.

It is easy to show that (6.139) describes the conservation of momentum and that (6.136) expresses the conservation of the number of particles. The right-hand sides of (6.139) and (6.140) include the total charge and total current, which (like the charge and current of the plasma and beam separately) may be expressed, by our assumption, in terms of the field through linear relations. Consequently, it is possible to determine the variation of the average characteristics of the beam and plasma. We observe that the only requirement is the validity of the linear approximation for the current and charge in the plasma. We need not assume that the distribution function can be separated into slowly and rapidly varying parts.

Let us investigate, for example, the variation of the beam particle momentum. Equation (6.137) in this case has the form

$$\frac{\partial}{\partial t}\langle n_1 v_1\rangle = \frac{1}{m_e}\langle E\rho_1\rangle. \tag{6.142}$$

We write $\rho_1$ as an expansion in plane waves within an interval of length L:

$$\rho_1 = \sum_k \rho_{1k} e^{-i\omega_k t + ikx} = \sum_k \frac{j_{1k} k}{\omega_k} e^{-i\omega_k t + ikx}. \tag{6.143}$$

Allowance is made here, based on the equation of continuity $\partial\rho_1/\partial t + \partial j_1/\partial x$, for the fact that $\rho_{1k} = j_{1k} k/\omega_k$. We use a linear relation between $j_k$ and $E_k$: $j_{1k} = \sigma_{1k} E_k$, where $\sigma_{1k}$ is the part of the electrical conductivity due to the beam:

$$\sigma_{1k} = \frac{\omega_k(\varepsilon_1 - 1)}{4\pi i} = \frac{in_{10}e^2\omega_k}{m_e(\omega_k - kv_1)^2}. \quad (6.144)$$

Consequently, Eq. (6.143) becomes

$$\frac{\partial}{\partial t}\langle n_1 v_1\rangle = \frac{n_{10}}{m_e^2}\left\langle \sum_{kk'} \frac{ie^2 k E_k E_{k'}}{(\omega_k - kv_1)^2} e^{-i(\omega_k + \omega_{k'})t + i(k+k')x}\right\rangle. \quad (6.145)$$

In averaging over the coordinates $\langle e^{i(k+k')x}\rangle$ becomes zero for all k and k' with the exception of $k = -k'$. As $\omega_{-k} = -\omega_k^*$ we have $\omega_k + \omega_{-k} = 2i\gamma_k$, and, as the field is real, $E_{-k} = E_k^*$. Thus

$$\frac{\partial}{\partial t}\langle n_1 v_1\rangle = \frac{n_{10}}{m_e^2}\sum_k \frac{ie^2 k E_k e^{2\gamma_k t} E_k^*}{(\Omega_k - kv_1 + i\gamma_k)^2}, \quad (6.146)$$

where $\omega_k = \Omega_k + i\gamma_k$. Similarly we see that spatial averaging of the electrical energy density yields

$$\langle E^2\rangle = \sum_k |E_k|^2 e^{2\gamma_k t}. \quad (6.147)$$

Turning to integration over k in (6.147) and (6.146) and denoting the energy density of the electric field by $W_k$, i.e., $\langle E^2\rangle = \int W_k dk$, we obtain

$$\frac{\partial}{\partial t}\langle n_1 v_1\rangle = \frac{n_{10}}{m_e^2}\operatorname{Re}\int \frac{ie^2 k W_k dk}{(\Omega_k - kv_1 + i\gamma_k)^2} =$$

$$= \frac{2n_{10}e^2}{m_e^2}\int \frac{\gamma_k(\Omega_k - kv_1)W_k k dk}{[(\Omega_k - kv_1)^2 + \gamma_k^2]^2}. \quad (6.148)$$

Allowance is made here for the fact that only the real part of the right-hand side contributes to the change in the beam particle momentum. The resulting equation enables us to determined the variation of the beam momentum under the influence of the oscillations generated by the beam. Using the value of the growth rate and frequency for the most unstable wave number according to (6.128), we have

$$\frac{\Omega_k - kv_1}{[(\Omega_k - kv_1)^2 + \gamma^2 k]^2} \approx -\frac{1}{\omega_{0e}^3}\frac{n_0}{n_1}; \; k \simeq \frac{\omega_{0e}}{v_1}.$$

Thus

$$\frac{\partial}{\partial t}\langle n_1 v_1\rangle = -\frac{1}{m_e v_1}\int \frac{2\gamma_k W_k dk}{4\pi} = -\frac{\partial}{\partial t}\frac{1}{m_e v_1}\int \frac{W_k dk}{4\pi}. \tag{6.149}$$

It is extremely useful for the ensuing arguments to introduce the concept of particle velocities averaged both with respect to the distribution function $f_p$ according to (6.133) and with respect to space, i.e., to define the average by the relation

$$\bar{u} = \frac{\left\langle \int v f_{\mathbf{p}} d\mathbf{p} \right\rangle}{\left\langle \int f_{\mathbf{p}} d\mathbf{p} \right\rangle}. \tag{6.150}$$

It is seen at once that by (6.150) and (6.133)

$$\bar{u} = \frac{\langle n\bar{v}\rangle}{\langle n\rangle}. \tag{6.151}$$

As $\langle n\rangle$ is conserved, we have $\langle n\rangle = n\,|_{t=0}$. Equation (6.149) can be written in the form

$$n_{10}\frac{\partial}{\partial t}\bar{u}_1 = -\frac{\partial}{\partial t}\frac{1}{m_e v_1}\int \frac{W_k dk}{4\pi}. \tag{6.152}$$

Assuming that the oscillation intensity at time t is much greater than at t = 0, we obtain the following for the change of the directed velocity of the beam particles:

$$\delta u = \bar{u}(t) - \bar{u}(t=0);$$

$$m_e n_{10} v_1 \delta u = -\int \frac{W_k dk}{4\pi}. \tag{6.153}$$

The relation (6.153) is valid only as long as $k\delta u \ll \gamma$, i.e., $\delta u / v_1 \ll (n_1/n_0)^{1/3}$, or

$$W^l \leqslant n_1 m_e v_1^2 \left(\frac{n_1}{n_0}\right)^{1/3}. \tag{6.154}$$

This last inequality corresponds to (6.130). It is reasonable to expect that when the directed velocity of the beam changes by $\delta u$, the velocities should also acquire a spread on the same order or larger.

In order to answer the question posed, we analyze Eq. (6.138) for the beam particles:

$$\frac{\partial}{\partial t}\langle n_1 \overline{v_1^2}\rangle = 2\,\frac{1}{m_e}\langle E j_1\rangle. \tag{6.155}$$

Substituting the current $j_1$ from the linear approximation, we obtain after averaging in the manner of (6.148)

$$\frac{\partial}{\partial t}\langle n_1 \overline{v_1^2}\rangle = 2\,\frac{n_{10}}{m_e}\,\mathrm{Re}\int \frac{ie^2\omega_k W_k dk}{(\omega_k - kv_1)^2}. \tag{6.156}$$

We now find the magnitude of the mean-square spread of the beam particles under the influence of the oscillations generated by them, $\overline{u^2} - (\bar{u})^2 = \overline{\Delta u^2}$, where $\bar{u}$ is determined by (6.151) and according to (6.151) $\overline{u^2}$ is

$$\overline{u_1^2} = \frac{\langle n_1 \overline{v_1^2}\rangle}{\langle n_1\rangle} = \frac{\left\langle \int f_p v_1^2 dp \right\rangle}{\langle f_p dp\rangle}. \tag{6.157}$$

By the conservation of $\langle n_1\rangle$ we have $\langle n_1\rangle = n_{10}$, and (6.156) becomes

$$\frac{\partial}{\partial t}\overline{u_1^2} = \frac{2}{m_e}\,\mathrm{Re}\int \frac{ie^2\omega_k W_k dk}{(\omega_k - kv_1)^2}. \tag{6.158}$$

Further

$$\frac{\partial}{\partial t}(\bar{u})^2 = 2\bar{u}\,\frac{\partial \bar{u}}{\partial t}$$

and from

$$\frac{\partial}{\partial t}\bar{u}_1 = \frac{1}{m_e}\,\mathrm{Re}\int \frac{ie^2 k W_k dk}{(\omega_k - kv_1)^2} \tag{6.159}$$

we have

$$\frac{\partial}{\partial t}\overline{\Delta u^2} = \frac{\partial}{\partial t}[\overline{u_1^2} - (\bar{u}_1)^2] = \frac{2}{m_e^2}\,\mathrm{Re}\int \frac{ie^2(\omega_k - \bar{u}_1 k) W_k dk}{(\omega_k - kv_1)^2}. \tag{6.160}$$

When the inequality (6.154) is satisfied, the change in the average directed velocity of the beam is small since $k\delta u \ll \gamma$, hence it is permissible in (6.160) to set $\bar{u}_1 \simeq v_1$. Hence,

$$\frac{\partial}{\partial t}[\overline{u_1^2} - (\bar{u}_1)^2] = \frac{2e^2}{m_e^2}\int \frac{\gamma_k W_k dk}{(\Omega_k - kv_1)^2 + \gamma_k^2}. \tag{6.161}$$

Substituting the values of $\Omega_k$ and $\gamma_k$ for the most unstable wave into the denominator of (6.161), we obtain ( $\partial/\partial t\, W_k = 2\gamma_k W_k$ )

$$\frac{\partial}{\partial t}\overline{\Delta u^2}=\frac{\partial}{\partial t}[\overline{u_1^2}-(\overline{u}_1)^2]=\frac{2}{n_1 m_e}\left(\frac{n_1}{2n_0}\right)^{1/3}\frac{\partial}{\partial t}\int\frac{W_k dk}{4\pi}. \tag{6.162}$$

Assuming that $\overline{\Delta u^2}$ at time t greatly exceeds $\overline{\Delta u^2}$ at time zero and that the initial oscillation energy is small, we have

$$\overline{\Delta u^2}=\frac{2}{n_1 m_e}\left(\frac{n_1}{2n_0}\right)^{1/3}\int\frac{W_k dk}{4\pi}. \tag{6.163}$$

Comparing (6.163) and (6.153), we find the ratio of the velocity spread to the change in the average directed velocity of the beam particles:

$$\frac{\overline{\Delta u^2}}{\delta u^2}=\frac{\left(\frac{n_1}{2n_0}\right)^{1/3} n_1 m_e v_1^2}{\int\frac{W_k dk}{8\pi}}\simeq\frac{\left(\frac{n_1}{n_0}\right)^{1/3} n_1 m_e v_1^2}{W^l}. \tag{6.164}$$

When (6.154) holds, $\overline{\Delta u^2}\gg\delta u^2$, i.e., the increase in the velocity spread greatly exceeds the change in the directed velocity of the beam particles. We note that (6.154) can also be obtained from (6.163) as a condition for the validity of the hydrodynamic description, given the requirement $k^2\overline{\Delta u^2}\ll\gamma^2$.

Thus, in the hydrodynamic stage the appearance of a particle velocity spread is the most important effect, greater than the change in the directed beam velocity. The two effects are comparable at the limit of applicability of the hydrodynamic stage and result in the loss of only a small fraction of beam energy in the hydrodynamic stage. It is also quickly verified that the energy acquired by the plasma in the hydrodynamic stage is small in comparison with the beam energy. The hydrodynamic stage therefore represents a rapidly evolving stage in the development of beam instability, one in which the beams lose but a slight fraction of their energy.

This stage rapidly goes over to the kinetic stage, when the instability may be represented as the result of a balance between the effects of induced wave emission and absorption.

The above discussion of the theory of hydrodynamic instability is similar to the quasilinear treatment in the sense that the fields are assumed in both cases to be linear. However, consideration is given to the inverse action of the oscillations on the beam, which causes a considerable change in its distribution (the final velocity spread greatly exceeds the initial spread). We also stress that we have determined only the average characteristics of the distribution, whereas the initial equation (6.133) permits a more detailed determination of the variation of the beam distribution as long as E is governed by the linear approximation and is independent of the detailed distribution. Finally, we have confined our attention above to the case of a low-density beam, $n_1/n_0 \ll 1$, which is the case of greatest practical significance.

## 17. HYDRODYNAMIC INSTABILITIES OF BEAMS CAUSED BY ANISOTROPY OF THE DISTRIBUTION FUNCTION

In a plasma in which beams are present there can be hydrodynamic instabilities of waves other than longitudinal waves. Such instabilities result in the generation of magnetic as well as electric fields. Also, it turns out that the magnetic fields in this case can greatly exceed the electric fields.*

These instabilities occur when the presence of the beam renders the total particle distribution function (beam plus plasma) anisotropic [44, 45].

As a result of the anisotropy of the total particle distribution function the electric field parallel to the beam creates a component of space-charge current perpendicular to the beam (as in any anisotropic medium), and the field perpendicular to the beam generates a current component along the beam. A straightforward calculation analogous to that in Chap. I for a plasma without beams present gives for the present situation†

$$j_{k,x} = -\sum_\alpha \frac{e^2 n_\alpha}{m i\omega} E_{k,x} + \sum_\alpha \frac{e^2 n_\alpha}{i\omega m} \frac{k_x v_\alpha}{\omega - k_z v_\alpha} E_{k,z}; \tag{6.165}$$

* The comparison here is between the electric and magnetic fields of the particular type of instability. The electric fields due to hydrodynamic instability of longtudinal waves can greatly exceed their magnetic fields.

† The field $E_k$ is a wave with $k_y = 0$. This can always be achieved by a suitable choice of the coordinate axes.

$$j_{k,z} = \sum_{\alpha} \frac{e^2 n_\alpha}{i\omega m} \cdot \frac{k_x v_\alpha}{(\omega - k_z v_\alpha)} E_{k,x} -$$

$$- \sum_{\alpha} \frac{e^2 n_\alpha}{mi\omega} \left( \frac{\omega^2}{(\omega - k_z v_\alpha)^2} + \frac{k_x^2 v_\alpha^2}{(\omega - k_z v_\alpha)^2} \right) E_{k,z}; \qquad (6.166)$$

$$j_{k,y} = \sum_{\alpha} \frac{e^2 n_\alpha}{mi\omega} E_{k,y}. \qquad (6.167)$$

Here z is the direction of relative motion of the beam and plasma. The specific instability associated with the anisotropy of the distribution function is particularly striking in the analysis of waves propagating transverse to the beam, i.e., for $k_z = 0$. It is convenient to choose a coordinate system in which the plasma is not at rest, but has a small velocity $v_0 = -v_1 n_1/n_0$. In this reference system the total momentum of the plasma and beam is zero. Then for $k_z = 0$ the off-diagonal elements vanish: $\sum_\alpha v_\alpha n_\alpha = 0$.

Maxwell's equation

$$c^2 \operatorname{rot}\operatorname{rot} \mathbf{E} = -4\pi \frac{\partial \mathbf{j}}{\partial t} - \frac{\partial^2 \mathbf{E}}{\partial t^2}$$

in projection on the z axis for plane waves becomes

$$(k_x^2 - \omega^2) E_{k,z} = 4\pi i \omega j_{k,z},$$

or, taking account of the explicit value of the current, $j_{k,z}$,

$$c^2 k_x^2 - \omega^2 + \omega_{0e}^2 \left(1 + \frac{n_1}{n_0}\right) + \frac{\omega_{0e}^2}{\omega^2} k_x^2 \left(v_1^2 \frac{n_1}{n_0} + v_0^2\right) = 0. \qquad (6.168)$$

Since $v_0 = -v_1 n_1/n_0$ we have $v_0^2 \ll v_1^2 n_1/n_0$ for $n_1/n_0 \ll 1$, and the solution of the equation has the form

$$\omega^2 \approx \frac{1}{2}(k_x^2 c^2 + \omega_{0e}^2) \pm \frac{1}{2} \sqrt{(k_x^2 c^2 + \omega_{0e}^2)^2 + 4 k_x^2 \omega_{0e}^2 v_1^2 \frac{n_1}{n_0}}. \qquad (6.169)$$

One of the solutions (plus sign) corresponds to a transverse wave $\omega^2 \approx k_x^2 c^2 + \omega_{0e}^2$, while the second describes the hydrodynamic instabilities we are looking for:

$$\omega^2 = -\frac{k_x^2 v_1^2 \omega_{1e}^2 \frac{n_1}{n_0}}{k_x^2 c^2 + \omega_{0e}^2}.$$

Thus, the growth rate associated with the anisotropy of the distribution function has the form

$$\gamma_a = \omega_{0e}\left(\frac{n_1}{n_0}\right)^{1/2}\frac{v_1}{c}\frac{k}{\sqrt{k^2+\omega_{0e}^2/c^2}}. \qquad (6.170)$$

Because the last factor of (6.170) is always smaller than unity, the maximum growth rate is $\omega_{0e}(n_1/n_0)^{1/2}v_1/c$, which is considerably smaller than that of the hydrodynamic beam instability $\omega_{0e}(n_1/n_0)^{1/3}$. For high-velocity beams ($v_1 \sim c$), however, the increment (6.170) can approach $\omega_{0e}(n_1/n_0)^{1/2}$, which is larger than the quasilinear growth rate.

Let us investigate how the beam parameters are affected by the development of this instability. We note that the equations derived referred to the case for which $v_1 \gg v_{Te}$ and the velocity spread of the particles in the beam is negligible. This implies that $|\omega| \gg k_z\Delta v_z$ and $k_x\Delta v_\perp$. For waves propagating transverse to the beam $k_z = 0$, and the condition for applicability of (6.170) is

$$v_\varphi \gg v_{Te}; \quad \gamma_a \gg k\Delta v_\perp. \qquad (6.171)$$

If the spread $\Delta v_1$ in the beam is greater than in the plasma, we then have from (6.171) and (6.169)

$$\frac{\Delta v_{\perp 1}}{v_1} \ll \left(\frac{n_1}{n_0}\right)^{1/2}\frac{\omega_{0e}}{\sqrt{\omega_{0e}^2+k^2c^2}}. \qquad (6.172)$$

This last condition often cannot be met. As we have already seen, the effective isotropization of cosmic rays can occur by quasilinear effects of particle scattering by plasma oscillations. We point out that (6.172) is not fulfilled for cosmic rays ($\Delta v_\perp \sim v_1 \sim c$, and $n_1 \ll n_0$), and isotropization by the instability under consideration here is impossible.

Suppose that the velocity spread $\Delta v_{\perp 1}$ is so small that (6.172) is satisfied. It is logical to expect the development of instability to cause an increase in the spread $\Delta v_{\perp 1}$ and violation of (6.172). It is important to emphasize that an increase in $\Delta v_{\perp 1}$ may be attributed both to these instabilities and to longitudinal-wave instabilities (see Sec. 16), which can develop far more rapidly. In fact, the longitudinal-wave hydrodynamic instability of the beam produces a spread $\Delta v_\parallel$ in the velocities exclusively along the beam in the one-dimensional case, which can be realized only in strong magnetic fields. Analysis shows that the anisotropic instability is suppressed in strong magnetic fields. But if the magnetic fields are not strong, then oblique longitudinal waves develop in hydrodynamic instability, resulting in both $\Delta v_{\parallel}$ and $\Delta v_{\perp}$ in the beam. If $\Delta v_{\perp}$ is of the same order as $\Delta v_{\parallel}$ and the latter reaches $v_0(n_1/n_0)^{1/2}$, then (6.172) is violated.

The expression (6.172) can also be violated as a result of the development of the instability (6.170), for example, if the initial beam has a large longitudinal velocity spread, so that $\Delta v_{\parallel}/v_{\parallel} \gg (n_1/n_0)^{1/3}$ and the spread $\Delta v_{\perp}$ is small. If $\gamma_a > \gamma_k$, where $\gamma_k$ is the quasilinear growth rate for longitudinal waves, it may be assumed that the velocity spread $\Delta v_{\perp}$ due to the instability (6.170) shows up more rapidly than that due to the quasilinear instability of longitudinal waves.

The cause of the spread $\Delta v_{\perp}$ is easily understood from a simple argument. As implied by (6.168), this instability corresponds to the excitation of transverse waves. It is clear that excitation is possible only when the electric field excited during the instability does work on the beam. As the wave propagation vector is perpendicular to the beam, E is parallel to the beam, and H is perpendicular to it. The magnetic field excited during the instability "uncoils" the beam particles, which acquire components $v_{\perp}$ perpendicular to the original direction of motion.

The equations describing the mean spread are obtained in a similar way to the above for longitudinal waves taking into account the Lorentz force. If the beam particle distribution were axisymmetric, no mean velocities perpendicular to the original direction of the beam could occur during the development of the instability. Therefore, the average spread $\overline{(\Delta v_{\perp})^2}$ is determined by $\overline{v_{\perp}^2}$. It is easy to derive an equation, to take the place of (6.138) for this case, for the mean square of the beam particle velocity $v_x$ in the direction of the wave:

$$\frac{\partial}{\partial t}\overline{u_x^2} = \frac{\partial}{\partial t}\frac{\langle nv_x^2\rangle}{\langle n\rangle} = \frac{2e}{m\langle n\rangle}\left\langle\left(E_x + \left[\frac{v}{c}H\right]_x\right)\right\rangle. \qquad (6.173)$$

Since the field E is parallel to the beam (z axis), $E_x = 0$, i.e., it plays the role of nothing more than the Lorentz force. Allowing for $H_z = 0$ and assuming $v_z \approx v_1$, we obtain

$$\frac{\partial}{\partial t}\overline{u_x^2} = -\frac{2v_1}{mn_1}\langle j_x H_y\rangle;$$

$$j_{x,k,1} = \frac{\varepsilon_{x,z,1}\omega}{4\pi i}E_{z,k};\ E_{z,k} = -\frac{\omega}{k}H_{y,k};\ \varepsilon_{x,z,1} = -\frac{4\pi e^2 n_1}{\omega^3 m_e}kv_1. \tag{6.174}$$

Thus,

$$\frac{\partial}{\partial t}\overline{u_x^2} = \frac{\omega_{0e}^2 v_1^2}{m_e n_0}\int\frac{2|H_k|^2 dk}{4\pi\gamma}. \tag{6.175}$$

The maximum growth rate $\gamma$ according to (6.170) occurs for $k \gg \omega_{0e}/c$. For this case $\gamma_{max}$ = const; $2\gamma|H_k|^2 = \frac{\partial}{\partial t}|H_k|^2$, and therefore

$$\frac{\overline{u_x^2}}{2} = \frac{\omega_{0e}^2 v_1^2}{n_0 m\gamma_{max}^2}W_H;\quad W_H = \int\frac{|H_k|^2 dk}{8\pi}, \tag{6.176}$$

or

$$n_1 m_e\frac{\overline{u_x^2}}{2} = W_H, \tag{6.177}$$

i.e., the energy in the transverse motion of the particles is of the same order as the magnetic field energy developed in the instability. Setting $\overline{u_x^2} = \Delta v_\perp^2$, we obtain from (6.177)

$$\frac{2W_H}{n_1 m_e v_1^2} < \frac{n_1}{n_0}\frac{\omega_{0e}^2}{\omega_{0e}^2 + k^2c^2}. \tag{6.178}$$

As the energy $\Delta W$ lost by the beam cannot be greater than $W_H$, it follows that $\Delta W/n_1 m_e v_1^2 \ll 1$, i.e., the beam cannot lose a significant fraction of its energy. If (6.171) is not met, the beam instabilities for transverse waves do not vanish. However, they may be described by quasilinear equations [46].

According to [46], during the quasilinear stage, as opposed to the hydrodynamic stage, the change in the magnetic energy is small compared with the kinetic energy of the particles perpendicular to the original direction of the beam. The change in the transverse energy is also small. Quasilinear diffusion reduces $\gamma_k$ almost to zero.

## Literature Cited

1. I. E. Tamm and I. M. Frank, "Coherent radiation from a fast electron in a medium," Dokl. Akad. Nauk SSSR, 14:107 (1937).
2. B. M. Bolotovskii, "Theory of the Vavilov–Cerenkov effect," Uspekhi Fiz. Nauk, 62:201 (1957).
3. V. L. Ginzburg, "Quantum theory of the radiation from an electron moving uniformly in a medium at a velocity faster than light," Zh. Éksp. Teor. Fiz., 10:589 (1940).
4. A. A. Vedenov, E. P. Velikhov, and R. Z. Sagdeev, "Quasilinear theory of plasma oscillations," Yadernyi Sintez, Suppl. Part 2, p. 465 (1962).
5. W. E. Drummond and D. Pines, "Nonlinear stability of plasma oscillations," Nuclear Fusion, Suppl. Part 3, p. 1049 (1962).
6. Yu. A. Romanov and G. F. Filippov, "Interaction of fluxes of fast electrons with longitudinal plasma waves," Zh. Éksp. Teor. Fiz., 40:123 (1961).
7. A. A. Vedenov, "Quasilinear plasma theory (theory of a weakly turbulent plasma)," Atomnaya Énergiya, 13:5 (1962).
8. Yu. L. Klimantovich, "Kinetic description of quasi-equilibrium turbulent processes in a plasma," Dokl. Akad. Nauk SSSR, 144:1022 (1962).
9. E. Frieman, S. Bodner, and P. Rutherford, "Some new results on the quasilinear theory of plasma instabilities," Phys. Fluids, 6:1298 (1963).
10. L. D. Landau and E. M. Lifshits, Electrodynamics of Continuous Media. Addison-Wesley, Reading, Mass. (1960).
11. L. D. Landau, "Oscillations of an electronic plasma," Zh. Éksp. Teor. Fiz., 16:949 (1955).
12. G. Knorr, "Numerische Integration der nichtlinearen Vlasov Gleichung [Numerical integration of the nonlinear Vlasov equation]," Z. Naturforsch., 16a:1320 (1961); 18a:1304 (1963).
13. V. L. Ginzburg, "Aspects of the theory of radiation at speeds faster than light," Uspekh. Fiz. Nauk, 69:537 (1959).
14. V. N. Tsytovich, "Acceleration of charged particles by plasma waves," Izv. Vuzov Radiofizika, 6:1641 (1963).
15. V. N. Tsytovich, "Problems in the statistical acceleration of particles in a turbulent plasma," Paper at the Conference on Ionization Phenomena in a Plasma, Belgrad (1965). FIAN Preprint (1965).
16. V. N. Tsytovich, "Isotropization of cosmic rays," Astron. Zh., 43:528 (1966).
17. V. L. Ginzburg and S. I. Syrovatskii, "Origin of cosmic rays," Uspekhi Fiz. Nauk, 74:521 (1961).
18. S. A. Kaplan, "Theory of the accleration of charged particles by isotropic gas-magnetic turbulent fields," Zh. Éksp. Teor. Fiz., 29:406 (1955).
19. S. B. Pikel'ner and S. A. Kaplan, The Interstellar Medium. Moscow (1963).
20. V. N. Tsytovich, "Acceleration of radiation and problems in the generation of fast particles under cosmic conditions," Astron. Zh., 40:612 (1963); 41:7 (1963).
21. V. N. Tsytovich, "Acceleration of electrons in the radiation belts of earth," Geomagnetizm i Aéronomiya, 4:616 (1963).

22. S.N. Vernov, A.E. Chudakov, P.V. Vavilov, and Yu.N. Logachev, "Study of earth corpuscular radiation and cosmic rays during the flight of a space rocket," Dokl. Akad. Nauk SSSR, 125:304 (1959).
23. I.S. Shklovskii, Cosmic Radio Waves. Harvard (1960).
24. V.N. Tsytovich, "Mechanisms of the growth of plasma turbulence and acceleration of cosmic rays," Astron. Zh., 42:33 (1965).
25. A. Gailitis and V.N. Tsytovich, "Acceleration by radiation and problems in the generation of fast particles under cosmic conditions (III): Time variations of the spectra of radio-emission sources," Astron. Zh., 41:452 (1964).
26. V.L. Ginzburg and S.I. Syrovatskii, "Gamma rays and the x-ray magnetobremsstrahlung of galactic and metagalactic space," Zh. Éksp. Teor. Fiz., 45:35 (1963).
27. S. Yu. Luk'yanov and I. M. Podgornyi, Hard x-rays accompanying discharge in a gas, Atomnaya Énergiya Vol. 3, p. 97 (1956).
28. I.G. Koval'skii, I.M. Podgornyi, and S.Kh. Khvoshchevskii, "The energy of x-rays emitted by a powerful impulsive discharge in hydrogen," Zh. Éksp. Teor. Fiz., 35:940 (1958).
29. M.D. Raizer and V.N. Tsytovich, "Mechanisms of x-ray and neutron radiations from powerful impulsive discharges," Atomnaya Énergiya, 17(9) (1964).
30. Ya.B. Fainberg, A.K. Berezin, G.N. Berezina, and L.I. Bolotin, "Interaction of pulsed high-current beams with a plasma in a magnetic field," Atomnaya Énergiya, 14:249 (1963).
31. I. Alexeff and R. Neidigh, "Observation of ionic sound waves in a plasma," Phys. Rev., 129:516 (1963); I. Alexeff, R. Neidigh, W. Peed, and E. Shipley, "Hot-electron plasma by beam-plasma interaction," Phys. Rev. Lett., 10:273 (1963).
32. M.V. Nezlin and A.M. Solntsev, "Acceleration of ions in plasma beams," Zh. Éksp. Teor. Fiz., 45:840 (1963).
33. E.K. Zavoiskii, et al., "New data on the turbulent heating of a plasma," Zh. Éksp. Teor. Fiz., 46:511 (1964).
34. V.I. Veksler, I.R. Hekker, et al., "Interaction of plasmoids with an electromagnetic wave," Paper at the conference in Dubna, No. IT-6317 (1963); Atomnaya Énergiya, 18:14 (1965).
35. I.A. Akhiezer and Ya.B. Fainberg, "High-frequency oscillations of an electron plasma," Zh. Éksp. Teor. Fiz., 21:1262 (1951).
36. D. Bohm and E.P. Gross, "Theory of plasma oscillations," Phys. Rev., 75:1851 (1949).
37. V.N. Tsytovich and V.D. Shapiro, "Nonlinear stabilization of plasma beam instabilities," Yadernyi Sintez, 5:228 (1965).
38. V.D. Shapiro, Nonlinear Theory of the Interaction of Charged Particle Beams with a Plasma (author's abstract of dissertation). OIYaI LVÉ, Dubna (1962).
39. A.A. Vedenov, Introduction to the Theory of a Weakly Turbulent Plasma, Vol. 3. Atomizdat (1963), p. 203.
40. A.I. Ivanov and L.I. Rudakov, "Dynamics of quasilinear relaxation," Zh. Éksp. Teor. Fiz., 51 (1967).

41. Ya. B. Fainberg and V. D. Shapiro, "Quasilinear theory of the generation of oscillations during the injection of an electron beam into a plasma half-space," Zh. Éksp. Teor. Fiz., 47:1389 (1964).
42. V. N. Tsytovich and V. D. Shapiro, "Theory of the transmission of a charged particle beam through a plasma," Zh. Tekh. Fiz., 35:1925 (1965).
43. V. D. Shapiro, "Nonlinear theory of the transmission of monoenergetic beams through a plasma," Zh. Éksp. Teor. Fiz., 94:613 (1963).
44. S. Neufeld and P. H. Doyle, "Electromagnetic interaction of a beam of charged particles with a plasma," Phys. Rev., 121:654 (1961); 127:846 (1962).
45. A. A. Rukhadze and V. G. Makhan'kov, "Excitation of electromagnetic waves transversely to a beam of particles in a plasma," Yadernyi Sinztez, 2:177 (1962).
46. V. G. Makhan'kov and V. L. Shevchenko, "Quasilinear theory of aperiodic instabilities in beam-plasma interaction," Beams in a Plasma. Izd. Khar'kovskogo Gosudarstvennogo Univ. (1966).

*Chapter VII*

# Induced Scattering of Waves by Plasma Particles

## 1. RESONANCES OF PLASMA PARTICLES WITH SEVERAL WAVES

The induced scattering of waves by plasma particles is one of the most important types of nonlinear wave interactions in plasma [1]. In order to understand fully the physical meaning of such interactions, we consider the simplest example, in which a charged particle is acted upon by two electromagnetic waves $\mathbf{E}_1 = \mathbf{E}_{10}e^{-i\omega_1 t + i\mathbf{k}_1\mathbf{r}}$ and $\mathbf{E}_2 = \mathbf{E}_{20}e^{-i\omega_2 t + i\mathbf{k}_2\mathbf{r}}$. We show that simultaneous resonant interaction is possible between the particle and the two waves. Neglecting the Lorentz force in the interest of simplicity (which is permissible, for example, in the case of longitudinal waves), we write the following equation of motion for the particle in the field of the waves:

$$\frac{m}{e}\frac{d^2\mathbf{r}}{dt^2} = \mathrm{Re}\,\mathbf{E}_{10}e^{-i\omega_1 t + i\mathbf{k}_1\mathbf{r}} + \mathrm{Re}\,\mathbf{E}_{20}e^{-i\omega_2 t + i\mathbf{k}_2\mathbf{r}}. \tag{7.1}$$

This equation is nonlinear in $\mathbf{r}(t)$. If the field weakly perturbs the motion of the particle, Eq. (7.1) can be solved by perturbation methods. In the zeroth approximation the particle moves uniformly in a straight line:

$$\mathbf{r}_0 = \mathbf{v}_0 t. \tag{7.2}$$

Since the fields only weakly perturb the motion, it may be assumed that $\mathbf{r} = \mathbf{r}_0 + \mathbf{r}_\sim$, where $|\mathbf{r}_\sim| \ll |\mathbf{r}_0|$, and $\mathbf{r}_\sim$ describes the oscillations of the particles in the wave field. Expanding in $\mathbf{r}_\sim$, we have

$$\frac{m}{e}\frac{d^2\mathbf{r}}{dt^2} = \mathrm{Re}\,\mathbf{E}_{10}e^{-i(\omega_1-\mathbf{k}_1\mathbf{v}_0)t}(1 + i\mathbf{k}_1\mathbf{r}_\sim) + \mathrm{Re}\,\mathbf{E}_{20}e^{-i(\omega_2-\mathbf{k}_2\mathbf{v}_0)t}(1 + i\mathbf{k}_2\mathbf{r}_\sim). \quad (7.3)$$

If we neglect $\mathbf{r}_\sim$ on the right-hand side of (7.3), we see that the waves exert a periodic driving force on the particle. If this force oscillates rapidly ($\omega - \mathbf{kv}_0 \neq 0$), the displacements of the particle under the influence of the waves are relatively small and also oscillate rapidly with time. If, on the other hand, the Cerenkov resonance condition $\omega = \mathbf{kv}_0$ is met, the particle is subject to prolonged action of the wave field and can therefore change its energy significantly. This type of particle–wave interaction corresponds to the induced wave emission or absorption discussed previously.

Let us now see what happens when $\omega_1 \neq \mathbf{k}_1\mathbf{v}_0$ and $\omega_2 \neq \mathbf{k}_2\mathbf{v}_0$, i.e., when the particle is not resonant with either of the waves affecting it. Is it possible in this case for the particle to be acted upon by a steady, rather than a rapidly oscillating, force? It turns out that such a situation is indeed possible. However, the steady force is of a higher order with respect to the applied field strengths, which were presumed weak. This is not too important, because for $\omega \neq \mathbf{kv}_0$ the force happens to be the first nonvanishing (with respect to the field amplitude) constant force acting on the particle. To find the expression for this force, we use (7.3) to derive an approximate expression for

$$\mathbf{r}_\sim = \frac{-e}{m}\,\mathrm{Re}\,\mathbf{E}_{10}\frac{e^{-i(\omega_1-\mathbf{k}_1\mathbf{v}_0)t}}{(\omega_1-\mathbf{k}_1\mathbf{v}_0)^2} - \frac{e}{m}\,\mathrm{Re}\,\mathbf{E}_{20}\frac{e^{-i(\omega_2-\mathbf{k}_2\mathbf{v}_0)t}}{(\omega_2-\mathbf{k}_2\mathbf{v}_0)^2}) =$$

$$= -\frac{e}{2m}\frac{\mathbf{E}_{10}\left(e^{-i(\omega_1-\mathbf{k}_1\mathbf{v}_0)t} + e^{i(\omega_1-\mathbf{k}_1\mathbf{v}_0)t}\right)}{(\omega_1-\mathbf{k}_1\mathbf{v}_0)^2} - \frac{e}{2m}\frac{\mathbf{E}_{20}\left(e^{-i(\omega_2-\mathbf{k}_2\mathbf{v}_0)t} + e^{i(\omega_2-\mathbf{k}_2\mathbf{v}_0)t}\right)}{(\omega_2-\mathbf{k}_2\mathbf{v}_0)^2}. \quad (7.4)$$

Putting (7.4) into (7.3) and rejecting that part of the additional force proportional to $e^{-2i(\omega-\mathbf{kv}_0)t}$, since it (like the part proportional to $e^{-i(\omega-\mathbf{kv}_0)t}$) is rapidly oscillating, we obtain the following for the relevant part of the additional force:

$$\delta\mathbf{F} = -\frac{e}{2m}\mathrm{Re}\,\frac{i\mathbf{E}_{10}(\mathbf{k}_1\mathbf{E}_{20})}{(\omega_2-\mathbf{k}_2\mathbf{v}_0)^2}e^{-i[\omega_1-\omega_2-(\mathbf{k}_1-\mathbf{k}_2)\mathbf{v}_0]t} -$$

$$-\frac{e}{2m}\mathrm{Re}\,\frac{i\mathbf{E}_{20}(\mathbf{k}_2\mathbf{E}_{10})}{(\omega_1-\mathbf{k}_1\mathbf{v}_0)^2}e^{i[\omega_1-\omega_2-(\mathbf{k}_1-\mathbf{k}_2)\mathbf{v}_0]t}. \quad (7.5)$$

As is apparent from (7.5), the force $\delta\mathbf{F}$ depends on the product of the amplitudes of the two waves and is constant in time if the resonance condition is met, viz.:

$$\omega_1 - \omega_2 = (\mathbf{k}_1 - \mathbf{k}_2)\,\mathbf{v}_0 \tag{7.6}$$

Consequently, a particle out of resonance with both waves can become resonant in the simultaneous presence of both waves. If (7.6) is not satisfied, the particle can be resonant with respect to three waves:

$$\omega_1 - \omega_2 - \omega_3 = (\mathbf{k}_1 - \mathbf{k}_2 - \mathbf{k}_3)\,\mathbf{v}_0\,, \tag{7.7}$$

and so on.

The meaning of the above resonance conditions is very simple. Whereas the first resonance $\omega = \mathbf{kv}$ corresponds to the Cerenkov emission of waves by particles, condition (7.6) corresponds to the scattering of waves by particles, (7.7) to the emission of waves upon scattering, etc. We write, for example, the laws of conservation of energy and momentum of a quantum $\hbar\omega$, $\hbar\mathbf{k}$ and a particle $\varepsilon$ , $\mathbf{p}$ in scattering. The subscript 1 corresponds to the values of the energy and momentum of the quantum before scattering, the subscript 2 to the same variable after scattering; the energy and momentum of the particle after scattering are denoted by primes. We obtain

$$\hbar\mathbf{k}_2 + \mathbf{p}' = \mathbf{p} + \hbar\mathbf{k}_1;$$
$$\hbar\omega_2 + \varepsilon_{\mathbf{p}'} = \varepsilon_{\mathbf{p}} + \hbar\omega_1,$$

or

$$\varepsilon_{\mathbf{p}'} = \varepsilon_{\mathbf{p}+\hbar\mathbf{k}_1-\hbar\mathbf{k}_2} \simeq \varepsilon_{\mathbf{p}} + \hbar\,(\mathbf{k}_1 - \mathbf{k}_2)\,\frac{\partial\varepsilon_{\mathbf{p}}}{\partial\mathbf{p}} =$$
$$= \varepsilon_{\mathbf{p}} + \hbar\,(\mathbf{k}_1 - \mathbf{k}_2)\,\mathbf{v} = \varepsilon_{\mathbf{p}} + \hbar\,(\omega_1 - \omega_2).$$

Equation (7.6) follows directly from this.

The interest in the scattering problem arises since the heavier plasma particles often cannot satisfy the Cerenkov condition $\omega = \mathbf{kv}$ (as explained above), but can meet condition (7.6).

Before proceeding with the analysis of induced scattering, we find a criterion for the validity of this concept. It follows from (7.3) that $\mathbf{k}r_\sim \ll 1$, or

$$r_\sim \ll \lambda, \tag{7.8}$$

i.e., the amplitude of particle oscillation in the wave field must be much smaller than the wavelength. Sometimes this condition may be written in an alternative form. If, for example, $\mathbf{kv}_0 \ll \omega$, then $r_\sim \sim v_\sim/\omega$, i.e.,

$$v_\sim \lll \lambda\omega = v_\varphi, \tag{7.9}$$

where $v_\varphi$ is the phase velocity of the waves.

## 2. THE WAVE KINETIC EQUATION

The derivation of the kinetic equation describing the effects of induced scattering is similar to that for the induced emission [2, 3]. We introduce the probability of scattering of a wave $\sigma$ by a plasma particle $\alpha$ with transformation into a wave $\sigma'$

$$w_{\alpha,\sigma}^{\sigma'}(\mathbf{p},\ \mathbf{k},\ \mathbf{k}'). \tag{7.10}$$

This probability describes the absorption of the wave $\sigma$ and emission of the wave $\sigma'$. We refer to this process in the general case $\sigma' \neq \sigma$ as scattering. The case $\sigma' = \sigma$, when the scattered wave is of the same type as the incident wave, corresponds to scattering without wave transformation. The reduction in the number of waves $\sigma$ due to scattering is

$$-\int N_{\mathbf{k}}^{\sigma}(N_{\mathbf{k}'}^{\sigma'}+1)\, w_{\alpha,\sigma}^{\sigma'}(\mathbf{p},\mathbf{k},\mathbf{k}')\, \frac{f_{\mathbf{p}}^{\alpha}\, d\mathbf{p}\, d\mathbf{k}'}{(2\pi)^6}, \tag{7.11}$$

and the increase due to the inverse process is

$$\int (N_{\mathbf{k}}^{\sigma}+1)\, N_{\mathbf{k}'}^{\sigma'} w_{\alpha,\sigma}^{\sigma'}(\mathbf{p},\mathbf{k},\mathbf{k}')\, f_{\mathbf{p}+\hbar\mathbf{k}-\hbar\mathbf{k}'}^{\alpha}\, \frac{d\mathbf{p}\, d\mathbf{k}'}{(2\pi)^6}. \tag{7.12}$$

Hence

$$\frac{\partial N_{\mathbf{k}}^{\sigma}}{\partial t} = N_{\mathbf{k}}^{\sigma} \sum_{\alpha,\sigma'} \int w_{\alpha,\sigma}^{\sigma'}(\mathbf{p},\mathbf{k},\mathbf{k}')\, \hbar(\mathbf{k}-\mathbf{k}')\, \frac{\partial f_{\mathbf{p}}^{\alpha}}{\partial \mathbf{p}}\, N_{\mathbf{k}'}^{\sigma'}\, \frac{d\mathbf{p}\, d\mathbf{k}'}{(2\pi)^6} +$$

$$+ \sum_{\alpha,\sigma'} \int w_{\alpha,\sigma}^{\sigma'}(\mathbf{p},\mathbf{k},\mathbf{k}')\, N_{\mathbf{k}'}^{\sigma'} f_{\mathbf{p}}^{\alpha}\, \frac{d\mathbf{p}\, d\mathbf{k}'}{(2\pi)^6} - N_{\mathbf{k}}^{\sigma} \sum_{\alpha,\sigma'} \int w_{\alpha,\sigma}^{\sigma'}(\mathbf{p},\mathbf{k},\mathbf{k}')\, f_{\mathbf{p}}^{\alpha}\, \frac{d\mathbf{p}\, d\mathbf{k}'}{(2\pi)^6} \tag{7.13}$$

Only the first term of (7.13) describes the nonlinear interaction corresponding to induced scattering; the last terms correspond to spontaneous scattering effects. In the limit as $N_{\mathbf{k}}^{\sigma} \to 0$ only the spontaneous emission of waves $\sigma$ remains:

$$\left.\frac{\partial N_{\mathbf{k}}^{\sigma}}{\partial t}\right|_{N_{\mathbf{k}}^{\sigma}\to 0} = \sum_{\alpha,\sigma'} \int w_{\alpha,\sigma}^{\sigma'}(\mathbf{p},\mathbf{k},\mathbf{k}')\, N_{\mathbf{k}'}^{\sigma'} f_{\mathbf{p}}^{\alpha} \frac{d\mathbf{p}\, d\mathbf{k}'}{(2\pi)^6}. \tag{7.14}$$

Notice that in (7.13) we have neglected the scattering recoil, assuming that $\hbar \mid \mathbf{k} - \mathbf{k}' \mid \ll \mathbf{p}$. The equation for the change in the particle distribution due to induced scattering processes is similarly simply derived. With only minor modifications this derivation follows that of (5.16) and (6.18). We therefore proceed directly to the results:

$$\frac{\partial f_{\mathbf{p}}^{\alpha}}{\partial t} = \frac{\partial}{\partial p_i} D_{ij}^{\alpha} \frac{\partial f_{\mathbf{p}}^{\alpha}}{\partial p_j}; \tag{7.15}$$

$$D_{ij}^{\alpha} = \sum_{\sigma\sigma'} \int \hbar^2 (k_i - k_i')(k_j - k_j')\, N_{\mathbf{k}}^{\sigma} N_{\mathbf{k}'}^{\sigma'} w_{\alpha,\sigma'}^{\sigma'}(\mathbf{p},\mathbf{k},\mathbf{k}') \frac{d\mathbf{k}\, d\mathbf{k}'}{(2\pi)^6}. \tag{7.16}$$

Equation (7.15) shows that the plasma particles diffuse in momentum space through scattering by the plasma oscillations.

## 3. THE CONSERVATION LAWS

As in emission processes, in scattering there is an exchange of energy between the plasma particles and waves. The law of conservation of the sum of the oscillation and particle energy follows from the laws of conservation in the elementary scattering event:

$$\omega - \omega' = (\mathbf{k} - \mathbf{k}')\mathbf{v}. \tag{7.17}$$

As Eqs. (7.17), (7.13), (7.15), and (7.16) differ formally from their quasilinear counterparts by the substitutions $\omega \to \omega - \omega'$ and $\mathbf{k} \to \mathbf{k} - \mathbf{k}'$, we immediately obtain

$$\frac{\partial}{\partial t}(E^{\alpha} + W^{\sigma} + W^{\sigma'}) = 0 \tag{7.18}$$

and

$$\frac{\partial}{\partial t}(\mathbf{P}^{\alpha} + \mathbf{P}^{\sigma} + \mathbf{P}^{\sigma'}) = 0. \tag{7.19}$$

However, there are other conservation laws besides these "trivial" ones. Let us introduce the total number of quanta $N^{\sigma} = \int N_{\mathbf{k}}^{\sigma}\, d\mathbf{k}/(2\pi)^3$. Writing the equation for induced scattering

$$\frac{\partial N_{\mathbf{k}}^{\sigma}}{\partial t} = N_{\mathbf{k}}^{\sigma} \sum_{\alpha,\sigma'} \int w_{\alpha,\sigma}^{\sigma'}(\mathbf{p},\mathbf{k},\mathbf{k}')\, \hbar(\mathbf{k} - \mathbf{k}') \frac{\partial f_{\mathbf{p}}^{\alpha}}{\partial \mathbf{p}} N_{\mathbf{k}'}^{\sigma'} \frac{d\mathbf{p}\, d\mathbf{k}'}{(2\pi)^6}, \tag{7.20}$$

we readily find

$$\frac{\partial N^{\sigma}}{\partial t}=\sum_{\alpha,\,\sigma'}\int w^{\sigma'}_{\alpha,\,\sigma}(\mathbf{p},\ \mathbf{k},\ \mathbf{k}')\,\hbar\, N^{\sigma}_{\mathbf{k}}N^{\sigma'}_{\mathbf{k}'}(\mathbf{k}-\mathbf{k}')\frac{\partial f^{\alpha}_{\mathbf{p}}}{\partial \mathbf{p}}\frac{d\mathbf{p}d\mathbf{k}d\mathbf{k}'}{(2\pi)^{9}}. \tag{7.21}$$

We are interested in processes in which a wave of type $\sigma$ is scattered by another wave of the same type. The probability $w^{\sigma}_{\alpha,\sigma}$ of this process should not change if we substitute $\mathbf{k}\to\mathbf{k}'$. Indeed this substitution implies that the wave $\mathbf{k}'$ that before was emitted is now absorbed and, conversely, that the wave $\mathbf{k}$ absorbed is now emitted. But the probabilities of the direct and inverse processes are equal; consequently, $w^{\sigma}_{\alpha,\sigma}$ ought to remain invariant with this substitution. Here, however, a slight refinement is called for. The momentum $\mathbf{p}$ in the argument of $w^{\sigma}_{\alpha,\sigma}$ refers to the initial particle momentum before scattering. For the inverse process this value differs from $\mathbf{p}$ by $\hbar(\mathbf{k}-\mathbf{k}')$. This disparity, however, is unimportant and may be ignored in a classical analysis.

Consequently, making the substitution $\mathbf{k}\rightleftarrows\mathbf{k}'$ in (7.21) for $\sigma=\sigma'$, we readily verify that the expression changes sign. This means that

$$\frac{\partial N^{\sigma}}{\partial t}=0, \tag{7.22}$$

i.e., the total number of quanta of type $\sigma$ in $\sigma\sigma$-scattering is conserved. This is easily understood when we realize that in each elementary $\sigma\sigma$-scattering event the number of quanta does not change.

As an example we cite one inference of the derived conservation laws. We see from (7.22) that in the induced scattering of Langmuir waves into Langmuir waves their total energy cannot change appreciably. Hence nonlinear interactions of this type for Langmuir waves cannot be true nonlinear absorption effects. In fact, the energy of the Langmuir waves depends only slightly on their wave number.

$$\Omega^{l}\simeq\omega_{0e}+\frac{3}{2}\frac{k^{2}v^{2}_{Te}}{\omega_{0e}};\quad \frac{k^{2}v^{2}_{Te}}{\omega_{0e}}\ll\omega_{0e}, \tag{7.23}$$

so that, by the conservation of $N^{l}$, the change in the total Langmuir wave energy $\Delta W^{l}$ must be small compared with their total energy $W^{l}$:

$$W^l = \int \hbar\omega^l N^l_{\mathbf{k}} \frac{d\mathbf{k}}{(2\pi)^3} \simeq \hbar\omega_{0e} N^l + \frac{3}{2}\hbar \int \frac{k^2 v_{Te}^2}{\omega_{0e}} N^l_{\mathbf{k}} \frac{d\mathbf{k}}{(2\pi)^3}. \qquad (7.24)$$

Therefore, nonlinear interactions of Langmuir waves can produce only a change in the wave spectrum, the total energy being left essentially unchanged. We recall that the decay processes of Langmuir into low-frequency waves have the same effect.

## 4. SOME GENERAL CONSEQUENCES OF THE KINETIC EQUATIONS FOR INDUCED SCATTERING

We now show that the induced scattering of waves by equilibrium plasma particles leads to the nonlinear transfer of waves toward lower frequencies in the spectrum. Assuming that $f^\alpha_{\mathbf{p}}$ depends only on the absolute value of the energy, we obtain

$$\frac{\partial f^\alpha_{\mathbf{p}}}{\partial \mathbf{p}} = \frac{\partial f^\alpha_{\mathbf{p}}}{\partial \varepsilon}\frac{\partial \varepsilon_{\mathbf{p}}}{\partial \mathbf{p}} = \mathbf{v}\frac{\partial f^\alpha_{\mathbf{p}}}{\partial \varepsilon}. \qquad (7.25)$$

By the conservation of energy

$$(\mathbf{k} - \mathbf{k}')\,\mathbf{v} = \Omega^\sigma_{\mathbf{k}} - \Omega^{\sigma'}_{\mathbf{k}'};$$

$$\frac{\partial N^\sigma_{\mathbf{k}}}{\partial t} = N^\sigma_{\mathbf{k}} \sum_{\alpha\sigma'} \int w^{\sigma'}_{\alpha\sigma}(\mathbf{p}, \mathbf{k}, \mathbf{k}')\, N^{\sigma'}_{\mathbf{k}'} \left(\Omega^\sigma_{\mathbf{k}} - \Omega^{\sigma'}_{\mathbf{k}'}\right) \frac{\partial f^\alpha_{\mathbf{p}}}{\partial \varepsilon} \frac{d\mathbf{p}\,d\mathbf{k}'}{(2\pi)^6}\hbar. \qquad (7.26)$$

If $\partial f^\alpha_{\mathbf{p}}/\partial\varepsilon < 0$, as in the case of an equilibrium Maxwellian particle distribution, then according to (7.26) the lower-frequency waves will grow. If, for example, $\sigma \neq \sigma'$, i.e., if scattering is accompanied by the conversion of one wave type into others, then waves are generated whose frequencies are lower than those of the waves exciting them. Thus, in an isotropic plasma during the conversion of Langmuir into transverse waves the frequencies of the transverse waves are lower than those of the Langmuir waves, and since the latter frequencies are close to $\omega_{0e}$, the transverse waves (for which $\omega > \omega_{0e}$) also have frequencies near $\omega_{0e}$.

But if $\sigma = \sigma'$, it is clear from (7.26) that the higher-frequency waves are damped, while the lower-frequency ones grow, i.e., a spectral transfer of the waves toward lower frequencies takes place. Suppose, for example, that we have a packet of $\sigma$-waves at time zero. As time passes, the mean frequency of this packet must decrease. This is nicely illustrated by using (7.26) to calculate the $\sigma$-wave energy:

$$\frac{\partial W^{\sigma}}{\partial t}=\int\frac{\hbar\omega_{\mathbf{k}}^{\sigma}}{(2\pi)^{3}}\frac{\partial N_{\mathbf{k}}^{\sigma}}{\partial t}d\mathbf{k}=\sum_{\alpha}\int N_{\mathbf{k}}^{\sigma}N_{\mathbf{k}'}^{\sigma}w_{\alpha,\,\sigma}^{\sigma}\Omega_{\mathbf{k}}^{\sigma}\left(\Omega_{\mathbf{k}}^{\sigma}-\Omega_{\mathbf{k}'}^{\sigma}\right)\times$$
$$\times\frac{\partial f_{\mathbf{p}}^{\alpha}}{\partial\varepsilon}\frac{d\mathbf{p}d\mathbf{k}d\mathbf{k}'\hbar^{2}}{(2\pi)^{9}}. \tag{7.27}$$

Making the substitution $\mathbf{k}\rightleftarrows\mathbf{k}'$, we obtain the expression (7.27), in which the factor $\Omega_{\mathbf{k}}^{\sigma}$ is replaced by $-\Omega_{\mathbf{k}}^{\sigma}$. Taking the half-sum of these expressions, we obtain

$$\frac{\partial W^{\sigma}}{\partial t}=\frac{1}{2}\sum_{\alpha}\int N_{\mathbf{k}}^{\sigma}N_{\mathbf{k}'}^{\sigma}w_{\alpha,\,\sigma}^{\sigma}\left(\Omega_{\mathbf{k}}^{\sigma}-\Omega_{\mathbf{k}'}^{\sigma}\right)^{2}\frac{\partial f_{\mathbf{p}}^{\alpha}}{\partial\varepsilon}\frac{d\mathbf{p}d\mathbf{k}d\mathbf{k}'}{(2\pi)^{9}}. \tag{7.28}$$

If $\partial f_{\mathbf{p}}^{\alpha}/\partial\varepsilon<0$, then $\partial W^{\sigma}/\partial t<0$, i.e., the wave energy decreases. The mean frequency of the packet is determined by the equation

$$\bar{\omega}=\frac{\int\Omega_{\mathbf{k}}^{\sigma}N_{\mathbf{k}}^{\sigma}\frac{d\mathbf{k}}{(2\pi)^{3}}}{\int N_{\mathbf{k}}^{\sigma}\frac{d\mathbf{k}}{(2\pi)^{3}}}=\frac{W^{\sigma}}{\hbar N^{\sigma}}. \tag{7.29}$$

As $N^{\sigma}$ is conserved, $d\bar{\omega}/dt<0$ for $\partial f_{\mathbf{p}}^{\alpha}/\partial\varepsilon<0$, as it was required to prove.

Another important consequence of (7.26) is the possibility of "true" nonlinear absorption in the scattering of high-frequency waves into low-frequency waves. Let $\Omega_{\mathbf{k}'}^{\sigma'}\ll\Omega_{\mathbf{k}}^{\sigma}$. Then

$$\frac{\partial N_{\mathbf{k}}^{\sigma}}{\partial t}=\sum_{\alpha}N_{\mathbf{k}}^{\sigma}\int w_{\alpha,\,\sigma}^{\sigma'}(\mathbf{p},\ \mathbf{k},\ \mathbf{k}')\,N_{\mathbf{k}'}^{\sigma'}\Omega_{\mathbf{k}}^{\sigma}\frac{\partial f_{\mathbf{p}}^{\alpha}}{\partial\varepsilon}\frac{d\mathbf{p}d\mathbf{k}'}{(2\pi)^{6}}. \tag{7.30}$$

For $\partial f_{\mathbf{p}}^{\alpha}/\partial\varepsilon<0$ the waves $\sigma$ are absorbed. They vanish in this case, i.e., the absorption indicated is "true" absorption. An example of this kind of nonlinear absorption in an isotropic plasma is found in the scattering of Langmuir waves with conversion into ion-acoustic modes ($\sigma=l$, $\sigma'=s$). If the intensity of the ion-acoustic waves is given and regarded as invariant, Eq. (7.30) describes the nonlinear damping rate for Langmuir waves in a plasma in which ion-acoustic waves are present. It is also clear that any other waves whose frequencies are less than the Langmuir waves produce an analogous effect of nonlinear absorption.

It is clear from the above proof that the results are valid not only for a Maxwellian particle distribution, but also for any other isotropic distributions with $\partial f/\partial \varepsilon > 0$.

If f is isotropic and $\partial f/\partial \varepsilon > 0$ in some definite energy interval, it must be borne in mind always that $f \to 0$ as $\varepsilon \to \infty$, so that there exists an energy interval in which $\partial f/\partial \varepsilon < 0$. With the existence of regions where $\partial f/\partial \varepsilon > 0$ and $\partial f/\partial \varepsilon < 0$, therefore, the direction of transfer is determined by whichever of these two energy intervals predominates in scattering. We recall that an analogous problem came up in the analysis of linear absorption, where it was demonstrated that for any isotropic f containing intervals with $\partial f/\partial \varepsilon > 0$ absorption will occur. The analogous theorem in the present situation would appear as follows: for any isotropic distribution of particles $\sigma\sigma$-scattering produces a transfer of waves toward lower frequencies, and $\sigma\sigma'$-scattering with $\Omega^{\sigma'} \ll \Omega^{\sigma}$ produces nonlinear absorption of the waves $\sigma$. This theorem applies to a broad class of particle and wave distributions.

## 5. INDUCED SCATTERING OF WAVES WITH BEAMS OF PARTICLES PRESENT IN THE PLASMA

As we saw earlier, when beams of particles are present in the plasma the balance between the processes of wave emission and absorption results in emission as the dominating process, i.e., the beam–plasma system becomes unstable. When this happens, instead of wave absorption in an isotropic plasma we get growth of the waves. Particle beams can similarly alter the direction of induced scattering processes. Thus, if $\sigma = \sigma'$, the presence of beams makes possible spectral transfer toward higher frequencies, while for $\sigma \neq \sigma'$, $\Omega^{\sigma} \gg \Omega^{\sigma'}$ the beams can cause growth of the waves $\sigma$.

For example, let the distribution function for the beam particles $f^{(1)}$ be the function $E = (\mathbf{p} - \mathbf{p}_1)^2/2m$, where $\mathbf{p}_1/m = \mathbf{v}_1$ is the beam velocity. Then

$$(\mathbf{k} - \mathbf{k}')\frac{\partial f_{\mathbf{p}}^{(1)}}{\partial \mathbf{p}} = (\mathbf{k} - \mathbf{k}')(\mathbf{v} - \mathbf{v}_1)\frac{\partial f^{(1)}}{\partial \varepsilon} =$$

$$= \left[\left(\Omega_{\mathbf{k}}^{\sigma} - \Omega_{\mathbf{k}'}^{\sigma'}\right) - (\mathbf{k} - \mathbf{k}')\,\mathbf{v}_1\right]\frac{\partial f^{(1)}}{\partial \varepsilon}. \tag{7.31}$$

If

$$(\mathbf{k}-\mathbf{k}')\,\mathbf{v}_1 > \left(\Omega_{\mathbf{k}}^{\sigma}-\Omega_{\mathbf{k}'}^{\sigma'}\right), \tag{7.32}$$

the direction of transfer is determined by $(\mathbf{k}-\mathbf{k}')\mathbf{v}_1$, i.e., for $\partial f/\partial \varepsilon < 0$ the transfer can go toward larger $|\mathbf{k}|$. In the case of normal dispersion $d\omega/dk > 0$, as in a fully ionized plasma, an increase in k in the absence of external magnetic fields always corresponds to an increase in $\omega_k$, i.e., transfer occurs in the opposite direction from that in the absence of a beam. This result has been obtained for Langmuir waves in [4].

For $\sigma = \sigma'$ condition (7.32) is analogous to the beam instability condition $v_1 > v_\varphi$ in the linear approximation. However, instead of $v_\varphi$ we are dealing with $(\omega - \omega')/|\mathbf{k}-\mathbf{k}'|$. Condition (7.32) therefore has entirely different physical consequences. For example, the implication of the beam instability condition $v_1 > v_\varphi > v_{Te}$ for Langmuir waves is that $v_1$ must be greater than $v_{Te}$, whereas according to (7.32) the spectral transfer of the Langmuir waves can change direction in the presence of beams whose velocity is much smaller than $v_{Te}$. This is related to the fact that $(\omega - \omega')/|\mathbf{k}-\mathbf{k}'|$ can be very small for large values of $\omega/k$ if the wave frequencies are very close to one another.

For Langmuir waves

$$\frac{\omega-\omega'}{|\mathbf{k}-\mathbf{k}'|} \simeq \frac{3}{2}\,\frac{(k-k')\,(k+k')\,v_{Te}^2}{|\mathbf{k}-\mathbf{k}'|\,\omega_{0e}} \sim 3\,\frac{v_{Te}^2}{v_\varphi} \ll v_{Te}. \tag{7.33}$$

This example shows that the direction of transfer of waves with large phase velocities can reverse. This is due to the fact that not all the plasma electrons take part in scattering for Langmuir waves with large phase velocities, but only those whose velocities are much smaller than the average thermal velocity. Therefore, the occurrence of beams with low velocities on the order of those of the particles participating in scattering causes a change in the direction of transfer.

As another example of how the direction of the process can be changed in the presence of beams we mention the induced scattering of Langmuir waves by electrons with conversion into ion-acoustic waves. In the absence of beams, this process results in nonlinear absorption of the Langmuir waves in the plasma with the generation of ion-acoustic modes. Nonlinear growth of the Langmuir waves is possible with beams present. This requires, according to (7.32), fulfillment of the condition $v_1 \gg \omega_{0e}/|\mathbf{k}-\mathbf{k}'|$. For the case of ion oscillations with $\omega_s \approx \omega_{0i}$, as we are now considering, $k' \gg 1/\lambda_{De}$, whereas the Langmuir waves satisfy $k \ll 1/\lambda_{De}$, i.e., $k' \gg k$, so that $v_1 \gg \omega_{0e}/k^s$. The maximum value of $k^s$ is

$$k^s \sim \frac{1}{\lambda_{Di}} = \frac{\omega_{0i}}{v_{Ti}} = \frac{\omega_{0e}\sqrt{m_e/m_i}}{v_{Te}\sqrt{m_e/m_i}\,\sqrt{T_i/T_e}},$$

i.e.,

$$v_1 \gg v_{Te}\sqrt{\frac{T_i}{T_e}}. \tag{7.34}$$

A more detailed analysis, taking account of the cancellation of $ls$-scattering processes by simultaneous emission of $l$- and s-waves, shows that the criterion is actually more stringent than (7.34), viz., $v_1 > v_{Te}$.

In order for the process to change direction, not only must the conditions (7.32) be satisfied for the beam velocity, but also a condition on the beam density. In particular, for the effects of scattering by the beam to dominate those of scattering by the plasma particles:

$$\frac{|k-k'|\,v_1 n_1}{T_1} \gg \frac{|\omega-\omega'|\,n_0}{T_0},$$

where $T_1$ and $T_0$ are the temperatures of the beam and plasma, and $n_1$ and $n_0$ are their respective densities. For Langmuir waves, for example, we have

$$\frac{n_1}{n_0} \gg \frac{|\omega-\omega'|}{|k-k'|\,v_1}\,\frac{T_1}{T_0} \approx \frac{3v_{Te}^2}{v_\varphi v_1}\,\frac{T_1}{T_0}.$$

This condition is usually not too restrictive.

## 6. SCATTERING MECHANISMS

Let us now examine the possible mechanisms of scattering by particles in plasma. There are two scattering mechanisms. One is conventional Thomson scattering. It is equally possible in vacuum. For Thomson scattering (which we also call Compton scattering) the $\sigma$-wave sets a plasma particle into oscillation, and the particle, acting as a dipole, then emits a wave $\sigma'$. Another scattering mechanism exists only in the plasma. Specifically, the wave $\sigma$ modulates the density or polarization of the plasma, i.e., it alters its electromagnetic properties. As a result, the presence of the wave causes the plasma to become weakly inhomogeneous and weakly nonstationary. Consequently, a particle executing uniform rectilinear motion in the plasma is capable of emitting electromagnetic waves [5-7]. Therefore, the indicated scattering occurs as follows: the wave $\sigma$ modulates the properties of the plasma, and a particle moving uniformly in the plasma emits a wave $\sigma$. We shall refer to this type of scattering henceforth as nonlinear, since a change in the electromagnetic properties of a plasma by a wave is a nonlinear effect.

Nonlinear scattering is also amenable to another straightforward treatment. For this we first explain how waves are emitted by a uniformly moving particle in a homogeneous plasma in the absence of the wave $\sigma$. A charge moving through a plasma produces an electromagnetic field around itself, polarizing the plasma by repelling particles of like charge and attracting particles of opposite charge. This creates a polarization cloud around the particle. Will this polarization cloud be transported together with the particle, or will it leave a polarization "trail"? This question is equivalent to the following: will the particle emit an electromagnetic wave or not? We note that outside the field source (outside the actual charge, i.e., in the plasma) every field can be expanded in plane waves, which are capable of propagating in the plasma. It is clear that if the phase velocity of these waves is smaller than the velocity of the moving charge, then roughly speaking the waves cannot keep pace with the charge and will become detached.

At this point we clarify the Cerenkov radiation mechanism in order to stress the part that it plays in the formation of the polarization "cloud" generated by a charge in a plasma.

Suppose the Cerenkov condition is not met, i.e., let the velocity of the moving charge be smaller than the phase velocities of the waves generating the polarization cloud. The charge will not radiate in this situation, and the polarization cloud will be transported together with it (at the velocity $\mathbf{v}$ of the charge). This happens when the plasma is homogeneous and isotropic. But if its properties are modified in space and time (even if very slightly), the charge, in going from one point to the next, is forced in effect to reconstruct its polarization cloud. The polarization of the plasma $p = \Sigma enr$ depends not only on the displacement r of the plasma particles caused by the external field (in this case the field of the charge), but also on the plasma density. Consequently, the polarization cloud generated by a charge moving uniformly in a straight line will vary in time and space and will therefore become a source of radiation of electromagnetic waves. It is clear from the above that only the interaction of the polarization created by the charge and a wave propagating in the plasma will be involved in nonlinear scattering. It is reasonable to assert in this case that nonlinear scattering is related to the radiation from a charge in a plasma modulated by an electromagnetic wave (as above), and that the wave is scattered by the charge of the screening polarization cloud. Both treatments are equivalent. The important thing is that the shielding cloud begins to oscillate as a result of interaction with the wave and thereby serves as a source of radiation.

We observe that the interaction of the two fields, the field of the screening cloud and the wave field, is a nonlinear effect; it would be absent if the superposition principle applied to the fields. Thus, the name given to the type of scattering in question is well founded.

The next issue is to examine the importance of the effect described by nonlinear scattering in plasma. It might seem at first glance that nonlinear scattering is nothing more than a small correction to the usual Compton scattering. Actually this is not so. It turns out in several instances that nonlinear scattering far outweighs Compton scattering. From the mere fact that the total screening charge is of the same magnitude as the scattering charge it is apparent that scattering by the screening charge can be of the same order as scattering by the main charge.

Let us see what type of scattering is the more important when a high-frequency wave is scattered by an ion. The ion produces a polarization cloud of electrons about itself. If the frequency of the scattering wave is high enough, the wave sets the electrons of the polarization cloud into oscillation, while the ion itself (because of its large mass) plays only a very small part in the scattering. Consequently, the scattering of high-frequency waves by ions is largely nonlinear. If the scattering charge is an electron, the oscillations of the polarization cloud can also be associated with the displacement of the polarization cloud electrons (they move in such a way, of course, that the screening charge has the opposite sign to that of the scattering electrons, i.e., an ion surplus is created). The oscillations of the scattering electron and polarization charge are in phase, and, since the charges are opposite, the nonlinear and Compton scattering effects oppose one another. In other words, for electrons the interference of the two types of scattering becomes very important. This example also shows that the scattering of high-frequency waves by ions can greatly exceed that by electrons [8].

Notice that the scattering condition (7.6) cannot always be met for ions whose velocities are of the order of the mean thermal velocity

$$v_{Ti} > \frac{\omega_1 - \omega_2}{|\mathbf{k}_1 - \mathbf{k}_2|}. \qquad (7.35)$$

In some cases, therefore, scattering by plasma ions is forbidden by the conservation laws. Under these conditions only scattering by plasma electrons is significant.

A few words are in order concerning the method of test particles. In the foregoing discussion we were concerned with an individual charged particle, aptly called a test particle. It is clear that a single particle only slightly distorts the overall distribution of the plasma particles through its field, hence the distribution may be regarded as if the test particle were not present. This is admissible, in particular, if the mean potential energy of the particle is much smaller than their mean kinetic energy, i.e., if $e^2/r \ll T$ or, equivalently, if $n\lambda_{De}^3 \gg 1$. If there are no isolated particles in the plasma, any particle may be treated as the test particle, i.e., for each particle the remainder constitute the screening polarization charge. Only in this case is the test particle concept proper.

The benefits of the test particle method have been shown recently in the analysis of the collision integral describing the collisions of plasma particles. We made use of the same basic idea at the beginning of this chapter to describe the "collisions" of particles and waves (those associated with induced scattering effects). The equations can be derived, of course, from the plasma kinetic equations by an expansion in the amplitudes of the interacting waves and subsequent phase averaging, as we have demonstrated for nonlinear decay interactions [1, 9–12]. Here we confine ourselves to the simple concept of induced scattering as first used in [8, 13].

## 7. COMPTON SCATTERING OF LONGITUDINAL WAVES

Suppose both waves involved in scattering are longitudinal. In this case the particle is acted upon only by the electric field of the wave, because the magnetic field is zero.* Let the field acting on the particle be represented by a set of plane longitudinal waves with spectrum $\omega = \Omega_{\mathbf{k}}$:

$$\mathbf{E}^{\sigma}(\mathbf{r},\ t) = \int \mathbf{E}^{\sigma}_{\mathbf{k}} e^{i\mathbf{k}\mathbf{r} - i\Omega_{\mathbf{k}}t}\, d\mathbf{k};\qquad \mathbf{E}_{\mathbf{k}} = E_{\mathbf{k}}\frac{\mathbf{k}}{k}. \tag{7.36}$$

The equation of motion of the charge in the wave field (7.36) is

$$\frac{d^2\mathbf{r}}{dt^2} = \frac{e}{m}\mathbf{E}^{\sigma}(\mathbf{r},\ t). \tag{7.37}$$

If we neglect the effect of the waves, the charge moves uniformly. Its velocity in this case is denoted by $\mathbf{v}_0$; we have $\mathbf{r}_0 = \mathbf{v}_0 t$. To find the field-perturbed oscillations of the charge, which are linear with respect to the field $\mathbf{E}$, as a first approximation we neglect the oscillations on the right-hand side of (7.37):

$$\mathbf{E}^{\sigma}(\mathbf{r},\ t) \simeq \mathbf{E}^{\sigma}(\mathbf{v}_0 t,\ t) = \int \mathbf{E}^{\sigma}_{\mathbf{k}} e^{-i(\Omega_{\mathbf{k}} - \mathbf{k}\mathbf{v}_0)t}\, d\mathbf{k}. \tag{7.38}$$

Here $\Omega'_{\mathbf{k}} = \Omega_{\mathbf{k}} - \mathbf{k}\mathbf{v}_0$ is the frequency of the field $\mathbf{E}_{\mathbf{k}}$, in a coordinate system in which the charge is at rest, i.e., the frequency seen by the moving charge (by the Doppler effect).

Thus, in this reference system the charge experiences the effect of the field in the approximation of (7.38) as a superposition

* The Lorentz force associated with the magnetic field of a wave is always small for nonrelativistic particles. However, it still needs to be included in the analysis of nonlongitudinal waves, because most of the scattering associated with the electric field of the wave is compensated by nonlinear scattering.

of time-periodic oscillations. Let us find the forced oscillations of the charge. Since the problem is linear, the oscillation frequencies of the charge will correspond with the external perturbation:

$$\mathbf{r}(t) = \int \mathbf{r}_{\mathbf{k}} e^{-i\Omega'_{\mathbf{k}} t}\, d\mathbf{k}.$$

For each of the components $\mathbf{r}_{\mathbf{k}}$ we obtain

$$-\Omega'^2_{\mathbf{k}} \mathbf{r}_{\mathbf{k}} = \frac{e}{m} \mathbf{E}^{\sigma}_{\mathbf{k}};$$

$$\mathbf{r}(t) = \int \frac{-e\mathbf{E}^{\sigma}_{\mathbf{k}}}{m(\Omega_{\mathbf{k}} - \mathbf{k}\mathbf{v}_0)^2} e^{-i(\Omega_{\mathbf{k}} - \mathbf{k}\mathbf{v}_0)t}\, d\mathbf{k} + \mathbf{v}_0 t. \tag{7.39}$$

We now find the intensity of the longitudinal wave emitted by the oscillating charge. For this we compute the work done by the radiation field on the charge. It follows from the conservation of energy in emission that the work done on the charge per unit time is equal to the radiation intensity:

$$W^{\sigma} = -\int \mathbf{E}_Q \mathbf{j}_Q\, d\mathbf{r}, \tag{7.39'}$$

where $\mathbf{E}_Q$ is the field produced in the plasma by the current of the charge Q.

$\mathbf{E}$ and $\mathbf{j}$ are conveniently represented as superimposed plane waves:

$$\mathbf{E}_Q = \int \mathbf{E}_{\mathbf{k}} e^{i\mathbf{k}\mathbf{r}}\, d\mathbf{k}; \quad \mathbf{j}_Q = \int \mathbf{j}_{\mathbf{k}} e^{i\mathbf{k}\mathbf{r}}\, d\mathbf{k}.$$

Hence

$$-W^{\sigma} = \int \mathbf{E}_{\mathbf{k}} \mathbf{j}_{\mathbf{k}'} e^{i(\mathbf{k}+\mathbf{k}')\mathbf{r}}\, d\mathbf{r}\, d\mathbf{k}\, d\mathbf{k}' = (2\pi)^3 \int \mathbf{E}_{-\mathbf{k}} \mathbf{j}_{\mathbf{k}}\, d\mathbf{k}.$$

Thus, to find the radiation intensity we need to know the current $\mathbf{j}_{\mathbf{k}}$ and the field $\mathbf{E}_{\mathbf{k}}$, created by the charge oscillating according to (7.39). Both can be expressed in terms of the charge density $\rho_{\mathbf{k}}$, generated in the plasma by the oscillating charge. From the equation of continuity

$$\operatorname{div} \mathbf{j} + \frac{\partial \rho}{\partial t} = 0$$

we infer

$$i(\mathbf{k}\mathbf{j}_{\mathbf{k}}) = -\frac{\partial \rho_{\mathbf{k}}}{\partial t}; \quad \rho = \int \rho_{\mathbf{k}} e^{i\mathbf{k}\mathbf{r}}\, d\mathbf{k},$$

and from Poisson's equation

$$\operatorname{div} \mathbf{D} = 4\pi\rho$$

we obtain

$$\varepsilon^l i\,(\mathbf{k}\mathbf{E_k}) = 4\pi\rho_{\mathbf{k}}. \tag{7.40}$$

Because the waves are longitudinal, $\mathbf{E_k} = \frac{\mathbf{k}}{k} E_{\mathbf{k}}$ and $\mathbf{E_{-k}} = \mathbf{E_k}$,

$$W^\sigma = (2\pi)^3 \int \frac{4\pi}{k^2 \varepsilon^{l*}} \overset{*}{\rho}_{\mathbf{k}} \frac{\partial \rho_{\mathbf{k}}}{\partial t}\, d\mathbf{k}. \tag{7.41}$$

The charge density $\rho_{\mathbf{k}}$ is found from (7.39):

$$\rho = e\delta(\mathbf{r} - \mathbf{r}(t)) = \frac{e}{(2\pi)^3} \int e^{i\mathbf{kr} - i\mathbf{kr}(t)}\, d\mathbf{k} = \int \rho_{\mathbf{k}} e^{i\mathbf{kr}}\, d\mathbf{k},$$

i.e.,

$$\rho_{\mathbf{k}} = \frac{e}{(2\pi)^3} e^{-i\mathbf{kr}(t)} \simeq \frac{e}{(2\pi)^3} e^{-i\mathbf{kv}_0 t} (1 - i\mathbf{kr}'(t)), \tag{7.42}$$

where $\mathbf{r}'(t)$ is the oscillation of the charge in a reference system in which it (neglecting the effect of the wave) is at rest. The first term of (7.42) corresponds to the charge density created by a uniformly moving particle and produces Cerenkov radiation. Since this radiation is presumed absent, only the second term of (7.42) need be included:

$$\rho_{\mathbf{k}} = \frac{ie^2}{(2\pi)^3 m} \int \frac{(\mathbf{kk}')\, E^\sigma_{\mathbf{k}'}}{(\Omega_{\mathbf{k}'} - \mathbf{k}'\mathbf{v}_0)^2 k'} e^{-i(\Omega_{\mathbf{k}'} - (\mathbf{k}' - \mathbf{k})\mathbf{v}_0)t}\, d\mathbf{k}'. \tag{7.43}$$

We now calculate the Compton scattering, neglecting nonlinear scattering. As already mentioned, this is not in general admissible, but the result of the calculation is important for comparing the true scattering cross section with the Compton cross section.

As implied by (7.40), the frequency involved in the dielectric constant $\varepsilon^l$ must correspond to the frequency in $\rho_{\mathbf{k}}$. Since $\rho_{\mathbf{k}}$ represents a set of monochromatic waves $e^{-i\Omega_{\mathbf{k}'} t}$, $\widetilde{\Omega}_{\mathbf{k}'} = \Omega_{\mathbf{k}'} - (\mathbf{k}' - \mathbf{k})\,\mathbf{v}_0$, allowance must be made for the fact that $\varepsilon^l$ has different values for different components, i.e.,

$$\frac{1}{\varepsilon^{l*}} \overset{*}{\rho}_{\mathbf{k}} \rightarrow \frac{-ie^2}{(2\pi)^3 m} \int \frac{E^{\sigma*}_{\mathbf{k}'} (\mathbf{kk}')\, e^{i\widetilde{\Omega}_{\mathbf{k}'} t}\, d\mathbf{k}'}{(\Omega_{\mathbf{k}'} - \mathbf{k}'\mathbf{v}_0)^2 k' \varepsilon^{l*} (\widetilde{\Omega}_{\mathbf{k}'}, \mathbf{k})}. \tag{7.44}$$

Substituting (7.44) and $d\rho_{\mathbf{k}}/dt$ into (7.41), we obtain

$$W^{\sigma} = \frac{4\pi}{(2\pi)^3 i} \int \frac{d\mathbf{k}\, d\mathbf{k}'\, d\mathbf{k}''\widetilde{\Omega}_{\mathbf{k}''}\,(\mathbf{k}, \mathbf{k}')(\mathbf{k}, \mathbf{k}'')\, E^{\sigma *}_{\mathbf{k}'} E^{\sigma}_{\mathbf{k}''} e^{i(\widetilde{\Omega}_{\mathbf{k}'} - \widetilde{\Omega}_{\mathbf{k}''})t}\, e^4}{k^2 k' k''\, (\Omega_{\mathbf{k}'} - \mathbf{k}'\mathbf{v}_0)^2\, (\Omega_{\mathbf{k}''} - \mathbf{k}''\mathbf{v}_0)^2\, \varepsilon^{l*}(\widetilde{\Omega}_{\mathbf{k}'}, \mathbf{k})\, m^2}. \tag{7.45}$$

We assume the wave fields $E^{\sigma}_{\mathbf{k}}$ are random, i.e.,

$$\langle E^{\sigma *}_{\mathbf{k}'} E^{\sigma}_{\mathbf{k}''} \rangle = |\, E^{\sigma}_{\mathbf{k}}\,|^2\, \delta\,(\mathbf{k}' - \mathbf{k}''). \tag{7.46}$$

As $W^{\sigma}$ is real, the result is proportional to $\operatorname{Im} \frac{1}{\varepsilon^{l*}(\widetilde{\Omega}_{\mathbf{k}'}, \mathbf{k})} = \frac{\operatorname{Im} \varepsilon^l}{|\operatorname{Re} \varepsilon^l|^2 + |\operatorname{Im} \varepsilon^l|^2}$. In the transmission range, where $\operatorname{Im} \varepsilon^l \to 0$, this expression vanishes unless $\operatorname{Re} \varepsilon^l \to 0$. Consequently, $\omega$ must be near $\Omega_{\mathbf{k}}$, i.e., the solution of the equation $\operatorname{Re} \varepsilon^l (\omega, \mathbf{k}) = 0$. Thus,

$$\operatorname{Re} \varepsilon^l \simeq (\omega - \Omega_{\mathbf{k}}) \left.\frac{\partial \varepsilon^l}{\partial \omega}\right|_{\omega = \Omega_{\mathbf{k}}}; \qquad \operatorname{Im} \frac{1}{\varepsilon^{l*}(\widetilde{\Omega}_{\mathbf{k}'}, \mathbf{k})} \simeq \frac{1}{\left.\frac{\partial \varepsilon^l}{\partial \omega}\right|_{\omega = \Omega_{\mathbf{k}}}} \frac{\gamma}{\gamma^2 + (\widetilde{\Omega}_{\mathbf{k}'} - \Omega_{\mathbf{k}})^2},$$

where

$$\gamma = \frac{\operatorname{Im} \varepsilon}{\partial \varepsilon / \partial \omega\, |_{\omega = \Omega_{\mathbf{k}}}}.$$

As $\gamma \to 0$ the result becomes

$$\operatorname{Im} \frac{1}{\varepsilon^{l*}(\widetilde{\Omega}_{\mathbf{k}'}, \mathbf{k})} \to \pi \frac{\delta\,(\widetilde{\Omega}_{\mathbf{k}'} - \Omega_{\mathbf{k}})}{\left.\frac{\partial \varepsilon^l}{\partial \omega}\right|_{\omega = \Omega_{\mathbf{k}}}}. \tag{7.47}$$

After substituting (7.46) and (7.47) into (7.45) we obtain the final result for the radiation intensity:

$$W^{\sigma} = \frac{1}{2\pi} \int d\mathbf{k}\, d\mathbf{k}'\, \Omega_{\mathbf{k}}\, |\, E^{\sigma}_{\mathbf{k}}\,|^2 \frac{(\mathbf{k}\mathbf{k}')^2}{k^2 k'^2} \frac{e^4 \delta\,(\Omega_{\mathbf{k}} - \Omega_{\mathbf{k}'} - (\mathbf{k} - \mathbf{k}')\,\mathbf{v}_0)}{(\Omega_{\mathbf{k}'} - \mathbf{k}'\mathbf{v}_0)^4\, m^2 \left.\frac{\partial \varepsilon^l}{\partial \omega}\right|_{\omega = \Omega_{\mathbf{k}}}}. \tag{7.48}$$

To find the probability of Compton scattering from this expression, we compare (7.48) with the expression for spontaneous scattering (7.14) deduced from simple balance considerations:

$$I^{\sigma} = \int \frac{\hbar \Omega_{\mathbf{k}}}{(2\pi)^3} \frac{\partial N^{\sigma}_{\mathbf{k}}}{\partial t}\, d\mathbf{k} = \sum_{\alpha} \int w^{\sigma'}_{\alpha,\, \sigma} N^{\sigma'}_{\mathbf{k}'} \Omega_{\mathbf{k}} f^{\alpha}_{\mathbf{p}} \frac{d\mathbf{p}\, d\mathbf{k}\, d\mathbf{k}'\hbar}{(2\pi)^9}.$$

This expression can be written as the sum of the emissions of the individual particles:

$$I^{\sigma} = \sum_{\alpha} \int W^{\sigma}_{\mathbf{p}} f^{\alpha}_{\mathbf{p}} \frac{d\mathbf{p}}{(2\pi)^3};$$

$$W^{\sigma}_{\mathbf{p}} = \int w^{\sigma'}_{\alpha,\sigma} N^{\sigma'}_{\mathbf{k}'} \Omega_{\mathbf{k}} \frac{d\mathbf{k}\, d\mathbf{k}' \hbar}{(2\pi)^6}. \tag{7.49}$$

By comparison with (7.48) we have

$$\frac{w^{\sigma'}_{\alpha,\sigma} N^{\sigma}_{\mathbf{k}'}}{(2\pi)^6} = \frac{(\mathbf{k}\mathbf{k}')^2}{k^2 k'^2} \frac{e^4 \delta(\Omega_{\mathbf{k}} - \Omega_{\mathbf{k}'} - (\mathbf{k} - \mathbf{k}')\mathbf{v}_{\alpha})}{2\pi m^2 (\Omega_{\mathbf{k}'} - \mathbf{k}'\mathbf{v}_{\alpha})^4 \left.\frac{\partial \varepsilon^l}{\partial \omega}\right|_{\omega=\Omega_{\mathbf{k}}}} |E^{\sigma}_{\mathbf{k}'}|^2 \hbar.$$

All that remains is to express $N^{\sigma}_{\mathbf{k}'}$ in terms of $|E^{\sigma}_{\mathbf{k}'}|$:

$$|E^{\sigma}_{\mathbf{k}'}|^2 = \frac{\hbar}{\pi^2} \frac{N^{\sigma}_{\mathbf{k}'}}{\left.\frac{\partial \varepsilon}{\partial \omega}\right|_{\omega=\Omega_{\mathbf{k}'}}}$$

i.e.,

$$w^{\sigma'}_{\alpha,\sigma} = \frac{4(2\pi)^3 e^4 \delta(\Omega_{\mathbf{k}} - \Omega_{\mathbf{k}'} - (\mathbf{k} - \mathbf{k}')\mathbf{v}_{\alpha})}{m^2 (\Omega_{\mathbf{k}} - \mathbf{k}\mathbf{v})^4 \left.\frac{\partial \varepsilon^l}{\partial \omega}\right|_{\omega=\Omega_{\mathbf{k}}} \left.\frac{\partial \varepsilon^l}{\partial \omega}\right|_{\omega=\Omega_{\mathbf{k}'}}} \frac{(\mathbf{k}\mathbf{k}')^2}{k^2 k'^2}. \tag{7.50}$$

The resulting equation demonstrates several unique characteristics of plasma scattering. To compare scattering in plasma and in vacuum, we present the expression for the probability of Compton scattering of transverse waves:

$$w^{t}_{\alpha,t} = \frac{2(2\pi)^3 e^4}{m^2 (\Omega_{\mathbf{k}} - \mathbf{k}\mathbf{v})^4} \left(1 + \frac{(\mathbf{k}\mathbf{k}')^2}{k^2 k'^2}\right) \frac{\Omega^2_{\mathbf{k}} \Omega^2_{\mathbf{k}'} \delta(\Omega_{\mathbf{k}} - \Omega_{\mathbf{k}'} - (\mathbf{k} - \mathbf{k}')\mathbf{v}_{\alpha})}{\left.\frac{\partial \varepsilon^t \omega^2}{\partial \omega}\right|_{\omega=\Omega_{\mathbf{k}}} \left.\frac{\partial \varepsilon^t \omega^2}{\partial \omega}\right|_{\omega=\Omega_{\mathbf{k}'}}}. \tag{7.51}$$

In the high-frequency limit $\omega \gg \omega_{0e}$, when plasma effects are unimportant, this expression describes ordinary Thomson scattering. The probability (7.51) can also be deduced in the same manner as (7.50).

Let us now show how the Thomson scattering cross section is obtained from (7.51) and compare (7.51) and (7.50). We begin with the latter. The main difference between (7.50) and (7.51) is that the factor $(\mathbf{k}\mathbf{k}')^2/k^2k'^2$ is involved for longitudinal waves, the factor $\frac{1}{2}[1 + (\mathbf{k}\mathbf{k}')^2/k^2k'^2]$ for transverse waves. These factors reflect the polarizations of the scattered waves. A longitudinal wave, for example, cannot be scattered perpendicularly to an incident longitudinal wave. This is because the longitudinal wave causes the charge to oscillate in its direction of propagation, whereas scattered longitudinal waves are stimulated only if the oscillations have a projection in the direction of scattering. The scattering is proportional to the square of the product of the unit polarization vectors $(\mathbf{e}_k\mathbf{e}_{k'})^2$, which is equal to $(\mathbf{kk}')^2/k^2k'^2$ for longitudinal waves and, after averaging over both directions of polarization, to $\frac{1}{2}[1 + (\mathbf{kk}')^2/k^2k'^2] = \frac{1}{2}(1 + \cos^2\theta)$ for transverse waves. A second difference lies in the factors $\frac{\omega^2}{\partial\varepsilon^t\omega^2/\partial\omega}$ and $\frac{1}{\partial\varepsilon^l/\partial\omega}$, which are of the same order of magnitude as the scattered-wave frequency. The longitudinal-wave frequencies in this case can only be close to $\omega_{0e}$, the transverse-wave frequency being always greater than $\omega_{0e}$.

These differences, although having serious consequences, are not fundamental. More important aspects emerge when comparing the scattering in vacuum and in plasma for Eq. (7.51). Under certain conditions the scattering of transverse waves in plasma is altogether different from that in a vacuum. For transverse waves this is true only in a narrow frequency range near $\omega_{0e}$. For Langmuir plasma waves whose frequencies are near $\omega_{0e}$ characteristic effects appear for practically the entire spectrum.

When scattering occurs in vacuum, $\Omega_k \simeq kc$, and the conservation law for scattering has the form $c(k - k') = (\mathbf{k} - \mathbf{k}')\mathbf{v}$. For nonrelativistic particles $k \approx k'$, i.e., only the direction of propagation and not the wave frequency changes on scattering. The scattering probability in this case is, according to (7.51),

$$w_{\alpha,\,t}^{t} = \frac{1}{2}\left(1 + \frac{(\mathbf{kk}')^2}{k^2k'^2}\right)\frac{(2\pi)^3 e^4}{m^2k^2c^3}\,\delta(k - k'), \tag{7.52}$$

and the scattering intensity is

$$W_{\mathbf{p}}^{\sigma} = \int \frac{e^4}{2m^2c^3}(1 + \cos^2\theta)\frac{N_{\mathbf{k}'}^{t}\hbar}{(2\pi)^3}\,kdk\delta(k - k')\,2\pi\sin\theta d\theta\,cd\mathbf{k}' =$$

$$= \frac{8\pi e^4}{3m^2c^3}\int\frac{k'dk'N_{\mathbf{k}'}^{t}\hbar}{(2\pi)^3} = \sigma_{\text{Thom}}W^t c,$$

where $\sigma_{\text{Thom}} = (8\pi/3)e^4/m^2c^4$ is the Thomson scattering cross section.

If the scattering occurs in plasma, the frequency change of the waves in scattering can be appreciable. For instance, for transverse waves having a spectrum $\Omega^t = \sqrt{\omega_{0e}^2 + c^2k^2}$, for $kc \ll \omega_{0e}$ we obtain $\Omega^t = \omega_{0e} + c^2k^2/2\omega_{0e}$ and, hence, $\Omega_k - \Omega_{k'} \simeq (c^2/2\omega_{0e})(k^2 - k'^2)$.

For sufficiently small k the difference $\Omega_{\mathbf{k}} - \Omega_{\mathbf{k}'}$ may be of order $(\mathbf{k}-\mathbf{k}')\mathbf{v}$, i.e., the fractional change of the wave frequency is not small. If, moreover, $\omega \gg kv$ and $\Omega_{\mathbf{k}} - \Omega_{\mathbf{k}'} \ll |\mathbf{k}-\mathbf{k}'| v$, the scattering probability becomes

$$w_{\alpha,t}^{t} = \frac{1}{2}\left(1 + \frac{(\mathbf{k}\mathbf{k}')^2}{k^2k'^2}\right)\frac{(2\pi)^3 e^4}{m^2\omega_{0e}^2}\,\delta\left((\mathbf{k}-\mathbf{k}')\,\mathbf{v}\right),$$

and the scattering intensity is of order

$$W_{\mathbf{p}}^{\sigma} \simeq 2\pi \int \frac{N_{k'}\,dk'}{(2\pi)^3}\,\frac{\hbar e^4}{m^2\omega_{0e}}\,\frac{k_{\perp}^2}{v} \simeq \frac{2\pi e^4}{m^2}\,\frac{k_{\perp}^2}{v\omega_{0e}^2}\,W^t. \tag{7.53}$$

The scattering cross section (7.53) differs from the usual Thomson cross section and, in particular, depends on the particle velocity (the scattering of waves by a set of plasma particles depends on their velocity distribution).

Turning now to the scattering of longitudinal waves, for Langmuir waves we have

$$\Omega_{\mathbf{k}} - \Omega_{\mathbf{k}'} = \frac{3v_{Te}^2}{\omega_{0e}}\left(k^2 - k'^2\right) = \frac{3v_{Te}^2\,(k+k')\,(k-k')}{\omega_{0e}},$$

which is of order

$$3v_{Te}^2\left(\frac{1}{v_{\varphi}} + \frac{1}{v'_{\varphi}}\right)(k-k') \ll v_{Te}\,(k-k'),$$

since $v_{\varphi} \gg v_{Te}$. In other words, even scattering by thermal electrons changes the wave frequency. This is in fact the frequency shift responsible for the nonlinear spectral transfer discussed earlier.

It must also be pointed out that induced scattering processes have a number of special characteristics that distinguish them from spontaneous scattering. For instance, we have shown that induced processes are proportional to $(\mathbf{k}-\mathbf{k}')\,\partial f/\partial \mathbf{p}$, which is equal to $(\mathbf{k}-\mathbf{k}')\mathbf{v}\;\partial f/\partial \varepsilon$ for isotropic particles.

For the induced scattering of high-frequency transverse waves described by the Thomson probability (7.52) this is zero, since the mean value of $(\mathbf{k}-\mathbf{k}')\mathbf{v} = 0$ for an isotropic particle distribution. This implies that induced scattering is strongly dependent on the corrections $(\mathbf{k}-\mathbf{k}')\mathbf{v}$ to the frequency difference $\Omega_{\mathbf{k}} - \Omega_{\mathbf{k}'}$, small as they may be. Thus we must take account of the first terms in the expansion of the corresponding expressions in $(\mathbf{k}-\mathbf{k}')\mathbf{v}$:

$$\delta(\Omega_{\mathbf{k}} - \Omega_{\mathbf{k}'} - (\mathbf{k} - \mathbf{k}')\mathbf{v}) = \delta(\Omega_{\mathbf{k}} - \Omega_{\mathbf{k}'}) - (\mathbf{k} - \mathbf{k}')\mathbf{v}\delta'(\Omega_{\mathbf{k}} - \Omega_{\mathbf{k}'}). \tag{7.53'}$$

It is easily demonstrated that the induced scattering contains the derivative of the spectral density of the waves with respect to frequency. Despite the fact that only the second term of (7.53'), which is small compared with the first, is involved in the induced scattering, induced effects may greatly exceed spontaneous scattering effects, because they are proportional to the next higher power of the intensity of the waves.

It is clear from the above example, first that induced scattering differs qualitatively from the spontaneous form and, second, that the Doppler corrections to the frequency of the incident and scattered waves can be important even when they are small, $(\mathbf{k} - \mathbf{k}')\mathbf{v} \ll \Omega_{\mathbf{k}} - \Omega_{\mathbf{k}'}$. All of the above implications of the conservation laws apply equally to nonlinear scattering, to be discussed below. We note that the Doppler frequency corrections occur not only in the scattering conservation laws, but also in the probability (7.50) through the quantity

$$\frac{1}{(\Omega - \mathbf{kv})^4} \simeq \frac{1}{\Omega^4} + \frac{4\mathbf{kv}}{\Omega^5}.$$

It turns out that they are very significant through the almost complete cancelation of the principal term $1/\Omega^4$ by nonlinear scattering.

## 8. NONLINEAR SCATTERING OF LONGITUDINAL WAVES

We now set out to analyze the nonlinear scattering of longitudinal waves into longitudinal waves. For this we need to know the density $\rho_k$ of the charges screening the scattering charge and oscillating in the field of the scattered wave. As we have seen, nonlinear scattering is associated with the nonlinear interaction of the propagating waves and the polarization created in the plasma by the charge.

The plasma charges surrounding the scattering charge are acted upon simultaneously by two forces, one from the wave field, the other from the moving charge. As in the analysis of Compton scattering, we may, in the expression for the force acting on the screening charges in this case, neglect the oscillations of the scattering charge, i.e., regard its field as that which arises from a uniform rectilinear motion of the scattering charge. This field has been computed above [first term of (7.42) and (7.40)]:

$$\rho_{\mathbf{k}} = \frac{e}{(2\pi)^3} e^{-i\mathbf{k}\mathbf{v}_0 t}; \quad \mathbf{E}_{\mathbf{k}} = \frac{\mathbf{k} 4\pi e\, e^{-i\mathbf{k}\mathbf{v}_0 t}}{ik^2 \varepsilon^l(\mathbf{k}\mathbf{v}_0, \mathbf{k})(2\pi)^3}. \tag{7.54}$$

Here we have allowed for the fact that the field frequency $\omega$ is $\mathbf{k}\mathbf{v}_0$ and substituted this value into $\varepsilon^l$. Then the field of the charge is

$$\mathbf{E}^Q = \int \mathbf{E}_{\mathbf{k}} e^{i\mathbf{k}\mathbf{r}}\, d\mathbf{k} = \int \frac{\mathbf{k} 4\pi e}{ik^2 (2\pi)^3} \frac{d\mathbf{k}}{\varepsilon^l(\mathbf{k}\mathbf{v}_0, \mathbf{k})} e^{i\mathbf{k}\mathbf{r} - i\mathbf{k}\mathbf{v}_0 t}. \tag{7.55}$$

On the other hand, the field of the wave is

$$\mathbf{E}^\sigma = \int \mathbf{E}^\sigma_{\mathbf{k}} e^{-i\Omega_{\mathbf{k}} t + i\mathbf{k}\mathbf{r}}\, d\mathbf{k}. \tag{7.56}$$

The total field due to the wave and the charge is

$$\mathbf{E} = \mathbf{E}^Q + \mathbf{E}^\sigma = \sum_\lambda \int E_{\mathbf{k}\lambda} e^{-i\Omega_{\mathbf{k}\lambda} t + i\mathbf{k}\mathbf{r}} \frac{\mathbf{k}}{k}\, d\mathbf{k}, \tag{7.57}$$

where

$$\lambda = 1,2;\ \Omega_{\mathbf{k},1} = \mathbf{k}\mathbf{v}_0;\ \Omega_{\mathbf{k},2} = \Omega^\sigma_{\mathbf{k}};\ E_{\mathbf{k},1} = \frac{4\pi e}{ik(2\pi)^3 \varepsilon^l(\mathbf{k}\mathbf{v}_0, \mathbf{k})};\ E_{\mathbf{k},2} = E^\sigma_{\mathbf{k}}. \tag{7.58}$$

The motion of the plasma charges in a field equal to the sum of $\mathbf{E}^Q$ and $\mathbf{E}^\sigma$ is described by the kinetic equation

$$\frac{\partial f_{\mathbf{p}}}{\partial t} + \mathbf{v}\frac{\partial f_{\mathbf{p}}}{\partial \mathbf{r}} + \frac{e}{m}\mathbf{E}\frac{\partial f_{\mathbf{p}}}{\partial \mathbf{v}} = 0. \tag{7.59}$$

Since $\mathbf{E}$ is assumed weak, the solution of (7.59) can be sought by expanding f with respect to the field:

$$f = f_0 + f^{(1)} + f^{(2)} + \ldots,$$

where $f_0$ is independent of the field, $f^{(1)}$ is proportional to the first power of the field, $f^{(2)}$ to the second, etc.:

$$\frac{\partial f^{(1)}_{\mathbf{p}}}{\partial t} + \mathbf{v}\frac{\partial f^{(1)}_{\mathbf{p}}}{\partial \mathbf{r}} = -\frac{e}{m}\mathbf{E}\frac{\partial f_{\mathbf{p},0}}{\partial \mathbf{v}}; \tag{7.60}$$

$$\frac{\partial f^{(2)}_{\mathbf{p}}}{\partial t} + \mathbf{v}\frac{\partial f^{(2)}_{\mathbf{p}}}{\partial \mathbf{r}} = -\frac{e}{m}\mathbf{E}\frac{\partial f^{(1)}_{\mathbf{p}}}{\partial \mathbf{v}}. \tag{7.61}$$

We need to know $f^{(2)}$, as it contains the desired nonlinear effect proportional to both the charge field $E^Q$ and the wave field $E^\sigma$.

Note that the linear perturbation induced by the field E is easily found from (7.60) if, corresponding to the form of the perturbing field (7.57), $f^{(1)}$ is sought in the form

$$f_{\mathbf{p}}^{(1)} = \sum_{\lambda} \int f_{\mathbf{k}\lambda}^{(1)} e^{-i\Omega_{\mathbf{k}\lambda}t + i\mathbf{kr}}\, d\mathbf{k}.$$

This at once gives

$$i\,(\Omega_{\mathbf{k}\lambda} - \mathbf{kv})\, f_{\mathbf{k}\lambda}^{(1)} = \frac{e}{m} E_{\mathbf{k}\lambda} \frac{\mathbf{k}}{k} \frac{\partial f_0}{\partial \mathbf{v}}\,. \tag{7.62}$$

Using (7.62), we write the right-hand side of (7.61), which "generates" $f^{(2)}$, in the form

$$\frac{e}{m}\mathbf{E}\frac{\partial f_{\mathbf{p}}^{(1)}}{\partial \mathbf{v}} = \frac{e^2}{m^2}\sum_{\lambda\lambda'}\int \mathbf{E}_{\mathbf{k}\lambda} e^{-i\Omega_{\mathbf{k}\lambda}t + i\mathbf{kr}} \frac{\partial}{\partial \mathbf{v}} \frac{1}{i\,(\Omega_{\mathbf{k}'\lambda'} - \mathbf{k}'\mathbf{v})} \times \mathbf{E}_{\mathbf{k}'\lambda'} \frac{\partial f_0}{\partial \mathbf{v}} e^{-i\Omega_{\mathbf{k}'\lambda'}t + i\mathbf{k}'\mathbf{r}} d\mathbf{k}\, d\mathbf{k}'.$$

In order not to burden the analysis unnecessarily by including terms not used later, we confine our discussion to the nonlinear scattering of waves satisfying the relation $\omega \gg kv_T$ by particles whose velocities are of the same order or less than the thermal velocities of the plasma particles. Then of the two terms $\lambda' = 1, 2$ the first ($\lambda' = 1$) contains $1/(\mathrm{kv}_0 - \mathrm{k'v})$, which is of the order $1/kv_T$, and the second ($\lambda' = 2$) contains $1/(\Omega_{\mathbf{k}} - \mathbf{kv}) \sim 1/\Omega \ll 1/kv_T$. Neglecting the term $\lambda' = 2$, in $E_{\mathbf{k}'\lambda'}$ we are left with the term corresponding to the charge field. Since we are concerned only with the effect in which the charge and wave fields participate, only $\lambda = 2$ need be retained. Denoting $\mathbf{k} + \mathbf{k}' = \mathbf{k}_1$, we obtain

$$\frac{e}{m}\mathbf{E}\frac{\partial f_{\mathbf{p}}^{(1)}}{\partial \mathbf{v}} = \frac{e}{m}\int \left(\mathbf{E}\frac{\partial f_{\mathbf{p}}^{(1)}}{\partial \mathbf{v}}\right)_{\mathbf{k}_1} e^{i\mathbf{k}_1\mathbf{r}}\, d\mathbf{k}_1;$$

$$\frac{e}{m}\left(\mathbf{E}\frac{\partial f_{\mathbf{p}}^{(1)}}{\partial \mathbf{v}}\right)_{\mathbf{k}_1} = \frac{e^2}{m^2}\int \mathbf{E}_{\mathbf{k}'}^{\sigma} e^{-i\,(\Omega_{\mathbf{k}'}^{\sigma} + (\mathbf{k}_1 - \mathbf{k}')\mathbf{v}_0)t} \frac{\partial}{\partial \mathbf{v}} \frac{1}{i(\mathbf{k}_1 - \mathbf{k}')(\mathbf{v}_0 - \mathbf{v})} \mathbf{E}_{\mathbf{k}_1 - \mathbf{k}'} \frac{\partial f_0}{\partial \mathbf{v}} d\mathbf{k}';$$

$$\widetilde{\Omega}_{\mathbf{k}'} = \Omega_{\mathbf{k}'}^{\sigma} - (\mathbf{k}' - \mathbf{k}_1)\,\mathbf{v}_0.$$

Expanding $f^{(2)}$ in plane waves

$$f^{(2)} = \int f_{\mathbf{k}_1}^{(2)} e^{i\mathbf{k}_1\mathbf{r}}\, d\mathbf{k}_1,$$

we obtain

$$\frac{\partial f_{\mathbf{k}_1}^{(2)}}{\partial t} + i\,(\mathbf{k}_1\mathbf{v})\, f_{\mathbf{k}_1}^{(2)} = -\frac{e}{m}\left(\mathbf{E}\frac{\partial f_{\mathbf{p}}^{(1)}}{\partial \mathbf{v}}\right)_{\mathbf{k}_1}. \tag{7.63}$$

Since the perturbation [right-hand side of (7.63)] represents the superposition of perturbation components with frequencies $\widetilde{\Omega}_{\mathbf{k}'}$, the forced motion of the screening charges as described by $f^{(2)}$ is also a superposition of components with frequencies $\tilde{\Omega}_{k'}$:

$$f^{(2)}_{k_1} = \int f^{(2)}_{k_1 k'} e^{-i\tilde{\Omega}_{k'} t}\, d\mathbf{k}'. \tag{7.64}$$

Hence

$$f^{(2)}_{k_1 k'} = -\frac{e^2}{m^2} \frac{1}{\tilde{\Omega}_{k'} - \mathbf{k}_1\mathbf{v}} \mathbf{E}^{\sigma}_{k'} \frac{\partial}{\partial \mathbf{v}} \frac{1}{(\mathbf{k}_1 - \mathbf{k}')(\mathbf{v}_0 - \mathbf{v})} \mathbf{E}_{k_1 - k', 1} \frac{\partial f_0}{\partial \mathbf{v}}.$$

Thus we have found that the variation of the distribution function for the screening electrons represents oscillations with frequencies coinciding with those of the scattering charge [see (7.43)]. From (7.64) it is a simple matter to determine the charge density associated with the oscillations of the screening charge:

$$\rho_{k_1} = e\int f^{(2)}_{k_1} \frac{d\mathbf{p}}{(2\pi)^3} = \int \frac{e f^{(2)}_{k_1 k'}\, d\mathbf{p}}{(2\pi)^3} e^{-i\tilde{\Omega}_{k'} t}\, d\mathbf{k}' = \int \rho_{k_1 k'} e^{-i\tilde{\Omega}_{k'} t}\, d\mathbf{k}',$$

where

$$\rho_{kk'} = \int e f^{(2)}_{kk'} \frac{d\mathbf{p}}{(2\pi)^3} =$$

$$= -\frac{e^3}{m^2} E^{\sigma}_{k'} \int \frac{\mathbf{k}'/k'}{\tilde{\Omega}_{k'} - \mathbf{k}\mathbf{v}} \frac{\partial}{\partial \mathbf{v}} \frac{1}{(\mathbf{k} - \mathbf{k}')(\mathbf{v}_0 - \mathbf{v})} \mathbf{E}_{k-k', 1} \frac{\partial f_0}{\partial \mathbf{v}} \frac{d\mathbf{p}}{(2\pi)^3}. \tag{7.65}$$

This expression can be simplified by integrating by parts. We recognize that

$$\frac{\mathbf{k}'}{k'} \frac{\partial}{\partial \mathbf{v}} \frac{1}{\tilde{\Omega}_{k'} - \mathbf{k}\mathbf{v}} \frac{\mathbf{k}'}{k'} \simeq \frac{\mathbf{k}}{(\Omega_{k'} - \mathbf{k}\mathbf{v})^2} \simeq \frac{\mathbf{k}'\mathbf{k}}{k'\Omega^2_{k'}},$$

because $\tilde{\Omega}_{k'} = \Omega_{k'} - (\mathbf{k}' - \mathbf{k})\mathbf{v}_0 \simeq \Omega_{k'}$ from the underlying assumption that $kv_T \ll \Omega_k$. Inserting the charge field (7.58) $\mathbf{E}_{k-k',1}$ into (7.65), we obtain

$$\rho_{kk'} = \frac{e^4 4\pi}{m^2 i (2\pi)^3} \frac{(\mathbf{k}'\mathbf{k}) E^{\sigma}_{k'}}{k'\Omega^2_{k'} |\mathbf{k} - \mathbf{k}'|^2 \varepsilon^l((\mathbf{k} - \mathbf{k}')\mathbf{v}_0, \mathbf{k} - \mathbf{k}')} \times$$

$$\times \int \frac{\mathbf{k} - \mathbf{k}'}{(\mathbf{k} - \mathbf{k}')(\mathbf{v}_0 - \mathbf{v})} \frac{\partial f_0}{\partial \mathbf{v}} \frac{d\mathbf{p}}{(2\pi)^3}. \tag{7.66}$$

This integral can be expressed in terms of the plasma permittivity

$$\varepsilon^l_e(\omega, \mathbf{k}) = 1 + \frac{4\pi e^2}{mk^2} \int \frac{\mathbf{k}\frac{\partial f_0}{\partial \mathbf{v}}}{\omega - \mathbf{k}\mathbf{v}} \frac{d\mathbf{p}}{(2\pi)^3}. \tag{7.66'}$$

The subscript e is included here to stress the fact that this is the electronic part of the dielectric constant of the plasma. By elementary deduction Eq. (7.66) is derived from (7.62) if $f_{\mathbf{k}}^{(1)}$ is used to compute $\rho_{\mathbf{k}} = e \int f_{\mathbf{k}}^{(1)}\, d\mathbf{p}/(2\pi)^3$ and Poisson's equation $\rho_{\mathbf{k}} = (1/4\pi)(\varepsilon^l - 1)\, ik\ E_{\mathbf{k}}$ is used.

Note that both electrons and ions can contribute to the screening effect $\rho_{\mathbf{kk}'}$, and the general expression of the type (7.66') must include summation over all plasma charges. We are interested mainly in longitudinal Langmuir waves, for which the ions cannot take part in the screening process or, more precisely, their contribution is negligible. We shall give a detailed analysis of the role of ions and electrons later. Observing that $\rho_{\mathbf{kk}}$ contains $1/m^2$, we see that the numerator of (7.66) includes only the electronic part of the dielectric constant:

$$\rho_{\mathbf{kk}'} = \frac{e^2(\mathbf{kk}')}{im(2\pi)^3 k'\Omega_{\mathbf{k}'}^2}\, \frac{\varepsilon_e^l[(\mathbf{k}-\mathbf{k}')\mathbf{v}_0, \mathbf{k}-\mathbf{k}'] - 1}{\varepsilon^l[(\mathbf{k}-\mathbf{k}')\mathbf{v}_0, \mathbf{k}-\mathbf{k}']} E_{\mathbf{k}'}^{\sigma}. \tag{7.67}$$

It is important that the denominator (7.67) contain the total dielectric constant, including both the electron and the ion contributions:

$$\varepsilon^l = \varepsilon_e^l + \varepsilon_i^l - 1 = 1 + (\varepsilon_e^l - 1) + (\varepsilon_i^l - 1).$$

Therefore

$$\varepsilon_e^l - 1 = \varepsilon^l - \varepsilon_i^l;\quad \frac{\varepsilon_e^l - 1}{\varepsilon^l} = 1 - \frac{\varepsilon_i^l}{\varepsilon^l}. \tag{7.68}$$

We now compare the charge density of the oscillating screening cloud with that of the scattering particle (7.43):

$$\rho_{\mathbf{kk}'}^{\mathrm{Comp}} = \frac{-e^2}{i(2\pi)^3 m}\, \frac{(\mathbf{kk}')E_{\mathbf{k}'}^{\sigma}}{k'(\Omega_{\mathbf{k}'} - \mathbf{kv}_0)^2}.$$

Since the Doppler corrections $kv/\Omega_{\mathbf{k}}$ to the frequency as seen by the moving charge are small, we have

$$\rho_{\mathbf{kk}'}^{\mathrm{Comp}} = \frac{-e^2}{i(2\pi)^3 m}\, \frac{(\mathbf{kk}')}{k'\Omega_{\mathbf{k}'}^2}\left(1 + \frac{2\mathbf{kv}_0}{\Omega_{\mathbf{k}'}}\right) E_{\mathbf{k}'}^{\sigma}.$$

But the charge density of the screening cloud is

$$\rho_{\mathbf{kk}'}^{\mathrm{nlr}} \simeq \frac{e^2 E_{\mathbf{k}'}^{\sigma}}{i(2\pi)^3 m}\, \frac{(\mathbf{kk}')}{k'\Omega_{\mathbf{k}'}^2}\left(1 - \frac{\varepsilon_i^l((\mathbf{k}-\mathbf{k}')\mathbf{v}_0, \mathbf{k}-\mathbf{k}')}{\varepsilon^l((\mathbf{k}-\mathbf{k}')\mathbf{v}_0, \mathbf{k}-\mathbf{k}')}\right). \tag{7.68'}$$

We see that if $\varepsilon^l \gg \varepsilon_i^l$ and the Doppler corrections to the Compton scattering are neglected, the nonlinear scattering completely cancels the Compton scattering:

$$\rho_{\mathbf{kk}'}^{\mathrm{nlr}} + \rho_{\mathbf{kk}'}^{\mathrm{Comp}} = \frac{-e^2 E_{\mathbf{k}'}^{\sigma}}{i(2\pi)^3 m} \frac{(\mathbf{kk}')}{k'\Omega_{\mathbf{k}'}^2}\left(\frac{2\mathbf{kv}_0}{\Omega_{\mathbf{k}'}} + \frac{\varepsilon_i^l((\mathbf{k}-\mathbf{k}')\mathbf{v}_0, \mathbf{k}-\mathbf{k}')}{\varepsilon^l((\mathbf{k}-\mathbf{k}')\mathbf{v}_0, \mathbf{k}-\mathbf{k}')}\right). \quad (7.69)$$

It becomes clear, therefore, that the Doppler corrections to the Compton scattering can play a vital role, like the contribution of the ions in the screening of scattering.

We now write the total scattering probability, taking account of both the Compton and the nonlinear scattering and their interference. It is obvious from a comparison of (7.69) and (7.68) that $1/(\Omega_{\mathbf{k}} - \mathbf{kv}_0)^2$ must be replaced in the final expression (7.50) by

$$\frac{1}{\Omega_{\mathbf{k}}^2}\left(\frac{2\mathbf{kv}_0}{\Omega_{\mathbf{k}}} + \frac{\varepsilon_i^l(\Omega_{\mathbf{k}} - \Omega_{\mathbf{k}'}, \mathbf{k}-\mathbf{k}')}{\varepsilon^l(\Omega_{\mathbf{k}} - \Omega_{\mathbf{k}'}, \mathbf{k}-\mathbf{k}')}\right).$$

Here we have allowed for the fact that in scattering, owing to the conservation laws, $(\mathbf{k}-\mathbf{k}')\mathbf{v}_0 = \Omega_{\mathbf{k}} - \Omega_{\mathbf{k}'}$. We thus obtain the result [8, 14]

$$w_{el}^{l'} = \frac{4(2\pi)^3 e^4 (\mathbf{kk}')^2 \delta(\Omega_{\mathbf{k}} - \Omega_{\mathbf{k}'} - (\mathbf{k}-\mathbf{k}')\mathbf{v})}{m^2\Omega_{\mathbf{k}}^4 \left.\frac{\partial\varepsilon^l}{\partial\omega}\right|_{\omega=\Omega_{\mathbf{k}}} \left.\frac{\partial\varepsilon^l}{\partial\omega}\right|_{\omega=\Omega_{\mathbf{k}'}} k^2 k'^2} \left|\frac{2(\mathbf{kv})}{\Omega_{\mathbf{k}'}} + \frac{\varepsilon_i^l(\Omega_{\mathbf{k}} - \Omega_{\mathbf{k}'}, \mathbf{k}-\mathbf{k}')}{\varepsilon^l(\Omega_{\mathbf{k}} - \Omega_{\mathbf{k}'}, \mathbf{k}-\mathbf{k}')}\right|^2. \quad (7.70)$$

The resulting probability describes scattering by plasma electrons. For scattering by ions the Compton scattering becomes negligible compared with the nonlinear scattering. Retaining only (7.68'), we at once write the probability of scattering by ions:

$$w_{il}^{l'} = \frac{4(2\pi)^3 e^4 (\mathbf{kk}')^2 \delta(\Omega_{\mathbf{k}} - \Omega_{\mathbf{k}'} - (\mathbf{k}-\mathbf{k}')\mathbf{v})}{m^2\Omega_{\mathbf{k}}^4 \left.\frac{\partial\varepsilon^l}{\partial\omega}\right|_{\omega=\Omega_{\mathbf{k}}} \left.\frac{\partial\varepsilon^l}{\partial\omega}\right|_{\omega=\Omega_{\mathbf{k}'}} k^2 k'^2} \left|\frac{\varepsilon_e^l(\Omega_{\mathbf{k}} - \Omega_{\mathbf{k}'}, \mathbf{k}-\mathbf{k}') - 1}{\varepsilon^l(\Omega_{\mathbf{k}} - \Omega_{\mathbf{k}'}, \mathbf{k}-\mathbf{k}')}\right|^2. \quad (7.71)$$

These equations, generally speaking, apply only to scattering of high-frequency longitudinal (Langmuir) waves; this arises from the theoretically trivial simplifying assumptions included in the analysis.

## 9. NONLINEAR INTERACTION OF LANGMUIR WAVES IN SCATTERING BY PLASMA ELECTRONS

As an example let us consider the nonlinear interaction of two Langmuir waves, including only the Doppler corrections to the Compton scattering and neglecting $\varepsilon_1^l/\varepsilon^l$ compared with $\mathbf{kv}/\Omega_{\mathbf{k}'}$. We concern ourselves with the effects produced by scattering from electrons in a Maxwellian plasma, i.e., we assume

$$f_{\mathbf{p}} = (2\pi)^{3/2} \frac{n}{(m_e v_{Te})^3} e^{-v^2/2v_{Te}^2} = f_{p_x} f_{p_y} f_{p_z} n, \tag{7.72}$$

where

$$f_{p_x} = \frac{\sqrt{2\pi}}{m_e v_{Te}} e^{-v_x^2/2v_{Te}^2}; \quad \int f_{p_x} \frac{dp_x}{2\pi} = 1. \tag{7.72'}$$

We recall that the equation describing nonlinear interaction may be written as

$$\frac{\partial N_{\mathbf{k}}^l}{\partial t} = N_{\mathbf{k}}^l \int N_{\mathbf{k}_1}^l \frac{d\mathbf{k}_1}{(2\pi)^3} w_{el}^l (\mathbf{kk}_1) (\Omega_{\mathbf{k}}^l - \Omega_{\mathbf{k}_1}^l) \frac{\partial f}{\partial \varepsilon} \frac{d\mathbf{p}}{(2\pi)^3} \hbar,$$

or because

$$\frac{\partial f}{\partial \varepsilon} = -\frac{f}{m_e v_{Te}^2}; \quad \Omega_{\mathbf{k}} - \Omega_{\mathbf{k}_1} = \frac{3}{2} \frac{k^2 - k_1^2}{\omega_{0e}} v_{Te}^2,$$

$$\frac{\partial N_{\mathbf{k}}^l}{\partial t} = N_{\mathbf{k}}^l \int N_{\mathbf{k}_1}^l \frac{d\mathbf{k}_1}{(2\pi)^3} w_{el}^l (\mathbf{kk}_1) \frac{3(k_1^2 - k^2)}{2\omega_{0e} m_e} f_{\mathbf{p}} \frac{d\mathbf{p}}{(2\pi)^3} \hbar. \tag{7.73}$$

Since $\partial\varepsilon^l/\partial\omega \simeq 2/\omega_{0e}$ and $\Omega^l \simeq \omega_{0e}$, the scattering probability according to (7.70) is

$$w_{el}^l = \frac{4(2\pi)^3 e^4}{m_e^2 \omega_{0e}^2} \frac{(\mathbf{kk}_1)^2}{k^2 k_1^2} \frac{(\mathbf{kv})^2}{\omega_{0e}^2} \delta\left(\frac{3}{2} \frac{(k^2 - k_1^2) v_{Te}^2}{\omega_{0e}} - (\mathbf{k} - \mathbf{k}_1)\mathbf{v}\right).$$

Without loss of generality we align the $p_x$ axis with $\mathbf{k} - \mathbf{k}_1$, i.e., we let $(\mathbf{k} - \mathbf{k}_1)\mathbf{v} = |\mathbf{k} - \mathbf{k}_1| v_x$. From the conservation of energy we have

$$v_x = \frac{3}{2} \frac{v_{Te}^2 (k + k_1)}{\omega_{0e}} \frac{k - k_1}{|\mathbf{k} - \mathbf{k}_1|}, \tag{7.74}$$

and, since $\frac{k-k_1}{|\mathbf{k}-\mathbf{k}_1|}<1$, we have $\frac{k+k_1}{\omega_{0e}}v_{Te}\ll 1$ and $v_x \ll v_{Te}$, i.e., only those particles whose velocity components parallel to the difference wave vector $\mathbf{k}-\mathbf{k}_1$ of the two interacting waves are much smaller than the mean thermal velocity take part in the scattering process. Consequently, we may argue that these velocity components are approximately zero, i.e.,

$$\mathbf{kv} = k_y v_y + k_z v_z;$$

$$(\mathbf{kv})^2 = k_y^2 v_y^2 + k_z^2 v_z^2 + 2k_y k_z v_y v_z.$$

Because particles with arbitrary $v_y$ and $v_z$ take part in the scattering, for every particle with a given $v_y v_z$ there is another for which the product $v_y v_z$ has the opposite sign. Hence

$$(\mathbf{kv})^2 \to k_y^2 v_y^2 + k_z^2 v_z^2 = (k_y^2 + k_z^2)\, v_{Te}^2.$$

Here we have used the facts that since the directions of y and z have equal weight, the mean value of $v_z^2$ is equal to the mean value of $v_y^2$, and for a Maxwellian distribution both equal $v_{Te}^2$. We note that $k_y^2 + k_z^2 = k_\perp^2$, where $k_\perp$ is the component of $\mathbf{k}$ perpendicular to $\mathbf{k}-\mathbf{k}_1$,

$$k_\perp^2 = k^2 - k_x^2 = k^2 - \frac{(\mathbf{k}(\mathbf{k}-\mathbf{k}_1))^2}{|\mathbf{k}-\mathbf{k}_1|^2} = \frac{[\mathbf{k}(\mathbf{k}-\mathbf{k}_1)]^2}{|\mathbf{k}-\mathbf{k}_1|^2} = \frac{[\mathbf{kk}_1]^2}{|\mathbf{k}-\mathbf{k}_1|^2}.$$

Consequently, we obtain an approximate expression for the probability:

$$w_{el}^{l} \simeq \frac{4\,(2\pi)^3 e^4}{m_e^2\omega_{0e}^4}\,\frac{(\mathbf{kk}_1)^2}{k^2k_1^2}\,\frac{[\mathbf{kk}_1]^2}{|\mathbf{k}-\mathbf{k}_1|^2}\,\delta\,(|\mathbf{k}-\mathbf{k}_1|\,v_x)\,v_{Te}^2. \tag{7.75}$$

To obtain the desired expression describing the nonlinear interaction it suffices to integrate (7.73) over the particle momenta, taking account of the normalization (7.72'), and $(4\pi)^2e^4/m_e^2 = \omega_{0e}^4/n^2$,

$$\int \delta\,(|\mathbf{k}-\mathbf{k}_1|\,v_x)\, f_{p_x}\frac{dp_x}{2\pi} = \frac{m_e f_0}{|\mathbf{k}-\mathbf{k}_1|\,2\pi} = \frac{1}{\sqrt{2\pi}v_{Te}\,|\mathbf{k}-\mathbf{k}_1|}.$$

we obtain [1, 4, 8, 14]

$$\frac{\partial N_{\mathbf{k}}^l}{\partial t} = \frac{3v_{Te}}{2m_e n\omega_{0e}}\,N_{\mathbf{k}}^l\int\frac{N_{\mathbf{k}_1}^l}{(2\pi)^{5/2}}\,\frac{(\mathbf{kk}_1)^2}{k^2k_1^2}\,\frac{[\mathbf{kk}_1]^2}{|\mathbf{k}-\mathbf{k}_1|^3}\,(k_1^2 - k^2)\,d\mathbf{k}_1\hbar. \tag{7.76}$$

Two characteristic features of the interaction described by (7.76) need to be brought to attention:

**1.** The interaction vanishes for waves with mutually perpendicular or parallel directions of propagation. This follows from the particular approximation invoked by expanding the interaction in terms of the parameter $v_{Te}/v_{\varphi}$. In order to clarify the interaction of mutually perpendicular and parallel waves it is necessary to include higher-order terms in the parameter $v_{Te}/v_{\varphi}$. This shows that, although (7.76) does not now vanish, the interaction is still much weaker than for nonparallel nonperpendicular waves.

**2.** The interaction (7.76) describes the same spectral transfer of waves as implied by general considerations. If, for example, at $t = 0$ the wave distribution represents a wave packet of width $\Delta k$ about $\mathbf{k} = \mathbf{k}_0$, (7.76) determines the change in the number of waves outside $\Delta k$, provided only that "seed" waves exist outside $\Delta k$. This change is determined to a first approximation by the integral factor (7.76), in which the main contribution comes from the initial packet

$$\frac{\partial N_{\mathbf{k}}^{l}}{\partial t} = \gamma N_{\mathbf{k}}^{l} \qquad (7.77)$$

where $\gamma$ depends on the intensity of the initial packet.

If in (7.77) the absolute value of k is smaller than all those for the waves in the packet $\Delta k$, then $\gamma > 0$, but if it is larger, then $\gamma < 0$. In other words, waves having frequencies higher than those in the packet are damped. Because their initial intensity is small, the damping of these waves due to nonlinear effects is of no interest.

But waves having frequencies lower than the packet $\Delta k$ grow. Initially their growth rate may be regarded as independent of time. As soon as energy on the order of the wave energy of the initial packet is transferred to the growing wave, it is necessary to take account of the dissipation of the waves in the original packet, because, according to the general theorems, the total number of quanta is conserved in induced scattering effects. The characteristic transfer time may be estimated from (7.77), disregarding the change in intensity of the initial packet. If the angle $\theta$ between the vectors $\mathbf{k}$ and $\mathbf{k}_1$ is of order unity and $k_1 - k = \Delta k \lessdot k$, then according to (7.76)

$$\gamma \simeq \frac{3\sqrt{2\pi}}{2}\,\omega_{0e}\left(\frac{\Delta k}{\omega_{0e}}v_{Te}\right)\left(\frac{k_0}{\omega_{0e}}v_{Te}\right)^2\frac{W^l}{nm_e v_{Te}^2}. \tag{7.78}$$

Here

$$W^l = \int\frac{\omega^l N_{\mathbf{k}}^l d\mathbf{k}}{(2\pi)^3}\,\hbar \simeq \omega_{0e}\int\frac{N_{\mathbf{k}}^l d\mathbf{k}}{(2\pi)^3}\,\hbar$$

is the energy density of the Langmuir waves, and

$$\frac{W^l}{nm_e v_{Te}^2} = \frac{W^l}{nT_e}$$

is the value of $W^l$ relative to the thermal energy density of the plasma electrons.

It follows from (7.78) that the smaller $\Delta k$ is, the less efficient will be the transfer. With increasing $\Delta k$ the rate grows. Let us estimate what value of $\Delta k$ corresponds to the maximum rate. For $k \ll k_0$ the characteristic increment has the form

$$\gamma \approx \omega_{0e}\left(\frac{k_0}{\omega_{0e}}v_{Te}\right)\left(\frac{k}{\omega_{0e}}v_{Te}\right)^2\frac{W^l}{nm_e v_{Te}^2} = \omega_{0e}\frac{v_{Te}}{v_\varphi^0}\left(\frac{v_{Te}}{v_\varphi}\right)^2\frac{W^l}{nm_e v_{Te}^2}, \tag{7.79}$$

i.e., for the transfer of waves into the interval of phase velocities much larger than the phase velocities $v_\varphi^0$ of the waves in the initial packet, the rate falls off with increasing $v_\varphi$. Therefore, the most efficient transfer occurs for $\Delta k$ of the same order as k, when the transfer rate is of order

$$\gamma \simeq \omega_{0e}\left(\frac{v_{Te}}{v_\varphi}\right)^3\frac{W^l}{nm_e v_{Te}^2}. \tag{7.80}$$

It is important to note that even for relatively small values of $W^l$ the increment (7.80) can greatly exceed the characteristic collision frequency of the plasma particles with each other:

$$\gamma \gg \nu_{\text{eff}} = \frac{1}{3}\frac{\omega_{0e}}{(2\pi)^{3/2}}\frac{1}{n\lambda_{De}^3}L,$$

where L is the Coulomb logarithm ($n\lambda_{De}^3 \gg 1$).

The Langmuir-wave phase velocities increase through nonlinear transfer. Since transfer is most efficient for $\Delta k$ of the order of k, it must proceed, so to speak, in "relay" fashion. If each leg or step in the "relay" transfer of oscillation energy were small, the transfer would be differential (as we find out later on, this happens, for example, with ion-acoustic or transverse waves). In this case, however, the "relay" step $\Delta k$ is of the order of k, and transfer is described by an integral equation. The important thing to realize, nevertheless, is that not just any k smaller than $k_0$ is excited; rather there is a maximum probability of transfer by a step $\Delta k$ of the order of k. The "relay" nature of the transfer process roughly corresponds to a linear time growth of some mean phase velocity of the waves in the packet, $\overline{v}_{\varphi} = v_{\varphi}^0(1 + \gamma t)$.

**3.** It is not necessary for the effective nonlinear interaction of two waves that both waves have large intensity. It is enough if one of them is intense. The other wave can increase its intensity, beginning with very small values, by nonlinear interaction with the first. This effect is highly typical of induced scattering processes and follows from Eq. (7.76).

It may be inferred from the above that nonlinear transfer must be accompanied by isotropization of the distribution of the oscillatory modes. Let us suppose then that the waves of a given packet lie in a definite range of directions. Low-intensity "seed" waves with $k < k_0$ generally exist for all directions. Consequently, during transfer those waves whose directions differ from that of the waves in the initial spectrum (and, of course, of lower frequencies) will be the ones to grow. The characteristic isotropization times of the waves are of the same order as the spectral transfer times. We note that the effects of isotropization of plasma oscillations are very significant in the analysis of particle acceleration by plasma oscillations.

**4.** Generally speaking, nonlinear transfer can be accompanied by narrowing of the wave packet. Let us imagine that "seed" waves are absent outside the spread $\Delta k$ of the packet. Then the wave intensity outside the packet does not increase, and the nonlinear transfer inside the packet causes the waves to pile up toward lower values of k, i.e., the packet is constricted. Now let there be "seed" waves outside $\Delta k$. Let us see which happens more quickly, transfer inside the packet, causing it to narrow, or transfer outside the packet, resulting in a significant change in the mean phase velocities of the waves. In general this depends on the distributions of the waves inside $\Delta k$, as well as their magnitude. If $\Delta k$ is of the order of k, the characteristic transfer inside the packet will be the more efficient. For a narrow packet, $\Delta k \ll k_0$, transfer outside the packet is the more efficient, and this does not translate the waves to a definite value of k, but to different values. The fastest rate in this case occurs for $k - k_0$ of the order of $k_0$. Therefore, transfer can cause the packet to flatten out. It is conceivable that under certain conditions the initial wave packet could periodically constrict and expand, moving toward larger values of $v_\varphi$.

We should note that as the spectral transfer is also accompanied by changes in direction, these periodic oscillations are also possible in the angular distribution of the waves. For the induced scattering of ion-acoustic waves by ions, which is very similar to the scattering of Langmuir waves by electrons, the oscillations in the angular and spectral distributions have been investigated in [15], and spectral transfer for the scattering of isotropic Langmuir waves by ions has been studied in [16].

Several different effects contribute to the broadening of the spectrum. We have seen, in particular, how the nonlinear effects of Langmuir waves decaying into ion-acoustic waves produce the effects of diffusion of Langmuir waves in a sound field accompanied by broadening of the Langmuir-wave spectrum. Spectral broadening effects are also caused by four-plasmon decay interactions, collisional effects, etc.

Under real conditions the range of possible values of k normally has a lower limit, for example, if the system has a dimension $a$, $k_{min} \sim 1/a$. Consequently, the end result of transfer could be grouping of the waves about $k_{min}$. The values of $k_{min} \sim 1/a$ correspond to $v_{\varphi max} = \omega_{0e}/k_{min} \simeq \omega_{0e} a$. Only for very low-density plasmas do the values $v_{\varphi max}$ become of the same order as $v_{Te}$. Thus, for $a \sim 1$ cm and $v_\varphi \sim 10^{-3}c$ we have $n \sim 10^6$ cm$^{-3}$. For $n \sim 10^{13}$ and $a = 1$ cm we have $v_{\varphi max} \approx 10c$, i.e., $v_{\varphi max}$ is much greater than the velocity of light. The application of (7.76) is limited, however, to small phase velocities

$$\frac{v_\varphi}{v_{Te}} \ll \left(\frac{3m_i}{m_e}\right)^{1/5}, \tag{7.81}$$

which will be explained presently. We merely point out that the efficiency of spectral transfer when (7.81) is not fulfilled is greater than that considered above. We also note that one consequence of

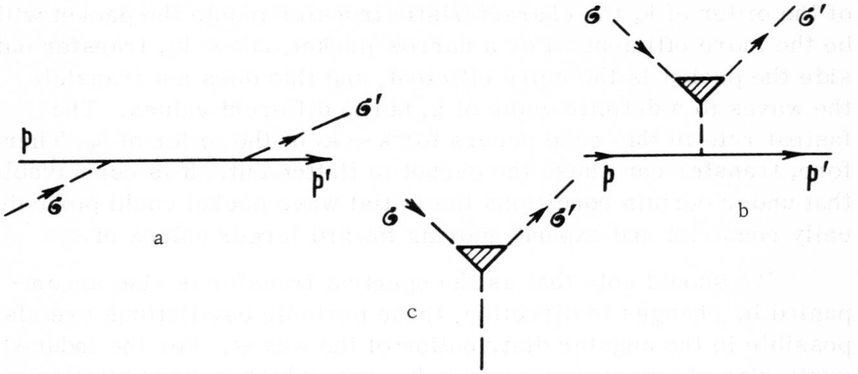


Fig. 31. Graphical representation of the mechanisms of scattering by particles: a) Compton scattering; b) nonlinear scattering; c) decay element describing nonlinear scattering.

the nonlinear interactions discussed might be the appearance of fundamental modes with wavelengths of the same order as the characteristic dimension of the system, a result which could be experimentally checked.

The example quoted, which has been explored in some detail, serves as an illustration for the analysis of more complex interactions associated with induced scattering. In the ensuing section, therefore, we merely review these interactions and highlight their physical consequences.

## 10. SCATTERING PROBABILITIES

Compton and nonlinear scattering are easy to represent in a clear graphical form. As we are well aware, Compton scattering corresponds to the absorption of one wave by a particle and the emission of another wave by the same particle, as illustrated in Fig. 31a. The dashed lines indicate the absorbed and emitted waves, the solid lines the scattering particle. In a plasma, however, as explained above, the nonlinear decay interaction of three waves is possible, as illustrated in Fig. 31c. Through the existence of this type of interaction a new type of scattering can occur, which is also nonlinear; the particle can "emit" a wave, which is then capable of absorbing a scattered wave and emitting a scattered wave by decay interaction (Fig. 34b). The intermediate wave that was emitted by the particle is often called a virtual wave. As a matter of fact, it need not be an ordinary propagating wave satisfying the dispersion relation $\omega = \Omega(k)$. The conservation of energy and momentum need be fulfilled only for the initial and final state, which correspond to a particle of momentum p and wave $\sigma$, and a particle of momentum p' and wave $\sigma'$, respectively. The decay element of Fig. 31c, as explained earlier, is determined by the nonlinear plasma current. Consequently, nonlinear scattering can be properly interpreted if the nonlinear plasma current is known [17]. As for Compton scattering, its analysis is also straightforward, as it reduces to a calculation of the oscillations of a charge in a wave and the emission of waves due to those oscillations [17].

A general method is given in Appendix 3 for computing the scattering probabilities, and a summary list is presented for the scattering probabilities of waves in isotropic plasma.

## Literature Cited

1. B. B. Kadomtsev and V. I. Petviashvili, "Weakly turbulent plasma in a magnetic field," Zh. Éksp. Teor. Fiz., 43:2234 (1962).
2. A. Gailitis and V. N. Tsytovich, "Radiation of transverse electromagnetic waves in the scattering of charged particles by plasma waves," Zh. Éksp. Teor. Fiz., 46:1726 (1964).
3. A. Gailitis and V. N. Tsytovich, "Radiation in the scattering of charged particles by electromagnetic waves in an isotropic plasma," Zh. Éksp. Teor. Fiz., 46:1455 (1964).
4. V. P. Silin and L. M. Gorbunov, "Nonlinear interaction of plasma waves," Zh. Éksp. Teor. Fiz., 47:200 (1964).
5. M. L. Ter-Mikaelyan and A. D. Gazazyan, "Resonance effects associated with radiation in a layered medium," Zh. Éksp. Fiz., 39:1963 (1961).
6. K. A. Barsukov and B. M. Bolotovskii, "Radiation of fast particles in a nonstationary inhomogeneous medium," Zh. Éksp. Teor. Fiz., 45:303 (1963).
7. Ya. B. Fainberg and N. A. Khizhnyak, "Energy losses of a charged particle in transmission through a layered dielectric," Zh. Éksp. Teor. Fiz., 32:883 (1957).
8. A. Gailitis and V. N. Tsytovich, "Theory of the nonlinear interaction of transverse and longitudinal plasma waves," Izv. Vuzov, Radiofizika, 7:119 (1964).
9. A. A. Galeev, V. I. Karpman, and R. Z. Sagdeev, "Many-particle aspects of the theory of a turbulent plasma," Yadernyi Sintez, 5:20 (1965).
10. L. I. Rudakov, "Problems in the nonlinear theory of the oscillations of an inhomogeneous plasma," Zh. Éksp. Teor. Fiz., 48:1372 (1965).
11. V. P. Silin, "Kinetic theory of the interaction of plasma waves," Zh. Prikl. Mekhan. i Tekh. Fiz., 1:31 (1964).

12. V. N. Al'tshul' and V. I. Karpman, "Kinetics of waves in a weakly turbulent plasma," Zh. Éksp. Teor. Fiz., 47:1552 (1964).
13. L. M. Kovrizhnykh and V. N. Tsytovich, "Interaction of transverse and longitudinal waves in a plasma," Zh. Éksp. Teor. Fiz., 46:2212 (1964).
14. L. M. Kovrizhnykh (Kovrischnich), Raport interne EVR-CEA, F. C.-258 Fontanay-aux-Roses (Scin.), France (1964).
15. I. A. Akhiezer, "Theory of turbulence in a two-temperature plasma," Zh. Éksp. Teor. Fiz., 47:2269 (1964).
16. L. M. Kovrizhnykh, "Plasmon interaction effects," Zh. Éksp. Teor. Fiz., 49:237 (1965).
17. V. N. Tsytovich and A. B. Shvartsburg, "Theory of nonlinear wave interaction in a magnetoactive anisotropic plasma," Zh. Éksp. Teor. Fiz., 49:797 (1965).

*Chapter VIII*

# Induced Scattering of Waves in an Isotropic Plasma

## 1. NONLINEAR INTERACTIONS OF LANGMUIR WAVES IN SCATTERING BY ELECTRONS

We have earlier considered the nonlinear interaction of Langmuir waves in connection with the Doppler corrections to the induced Thomson scattering of waves by plasma electrons and we have shown that the characteristic time for the most effective transfer by $\Delta k$ of the order of k, and $\theta \sim 1$, leads to the following estimate:

$$\frac{1}{\tau} \simeq \omega_{0e} \frac{W^{l}}{nT_{e}} \left( \frac{v_{\mathrm{T}e}}{v_{\varphi}} \right)^{3} \tag{8.1}$$

This interaction has been obtained in [2-4].

As mentioned, Eq. (8.1) has a limited range of applicability (strictly speaking, it applies only in a plasma with infinitely massive ions, $m_i \to \infty$). Here we are interested in the interaction of Langmuir waves, which cannot be described by (7.76). We shall first investigate the interaction of a one-dimensional wave spectrum with $\mathbf{k}_1 \parallel \mathbf{k}$, as well as $\mathbf{k}_1 \perp \mathbf{k}$, for which (7.76) goes to zero. This interaction is described by the next higher-order terms in the small parameter $(v_{Te}/v_{\varphi})$. Of course, it is meaningful to consider such effects only when the wave vectors are strictly parallel or strictly perpendicular. For an electron plasma ($m_i = \infty$) we have for $\mathbf{k} \parallel \mathbf{k}_1$ [2, 3, 5-7]

$$\frac{\partial N^l_{\mathbf{k}_1}}{\partial t} = N^l_{\mathbf{k}} \int N^l_{\mathbf{k}_1} \frac{27\hbar}{8\,(2\pi)^{5/2}} \frac{k_1^2 - k^2}{|k_1 - k|} \frac{v_{Te}}{n\omega_{0e} m_e} \frac{k^2 k_1^2 v_{Te}^2}{\omega_{0e}^2} d\mathbf{k}_1, \tag{8.2}$$

and for $\mathbf{k} \perp \mathbf{k}_1$ the interaction differs from (8.2) by a factor of 1/9. The characteristic transfer time is smaller than (8.1) roughly by a factor $(v_{Te}/v_\varphi)^2$. The interaction is described by (8.2) when

$$\theta \ll \theta_0 = \frac{v_{Te}}{v_\varphi}, \tag{8.3}$$

where $\theta$ is the angle between $\mathbf{k}$ and $\mathbf{k}_1$ or its complement. As the interaction of one-dimensional spectra is relatively weak, the more likely process is isotropization with transfer of waves into the angular region $\theta > \theta_0$.

We now consider the role of the ion contribution to the polarization in nonlinear scattering, as described by the term $\varepsilon_i/\varepsilon$ in the scattering probability. We emphasize that these corrections do not describe scattering by ions, but by the screening electron charge and the part played by the ions in this screening. The only way in which the ions can affect the screening appreciably is through cancellation of the Compton and nonlinear scattering in the first approximation. Let us assess the conditions under which the influence of the ions on the screening of scattering by electrons begins to prevail over the Doppler corrections to the Compton scattering. It is important to note that this assessment is best made in the case when scattering by ions is impossible, otherwise, as we shall see later, the scattering by electrons will generally represent a small effect. In order for ion scattering to be possible, it is required that the velocity component of the scattering ion in the direction $\mathbf{k} - \mathbf{k}_1$, i.e., $v_i = (\omega - \omega_1)/|\mathbf{k} - \mathbf{k}_1|$, be smaller than $v_{Ti}$ (otherwise, for a Maxwellian distribution the number of scattering ions will be exponentially small.

If we assume that ion scattering is small:

$$\omega - \omega_1 \gg |\mathbf{k} - \mathbf{k}_1| v_{Ti}. \tag{8.4}$$

Moreover, as shown, for electron scattering

$$\omega - \omega_1 \ll |\mathbf{k} - \mathbf{k}_1| v_{Te}. \tag{8.5}$$

Under these conditions the approximate expressions for the dielectric constants for electrons and ions are

$$\varepsilon_i^l \simeq -\frac{\omega_{0i}^2}{(\omega - \omega_1)^2}; \quad \varepsilon_e^l = \frac{\omega_{0e}^2}{(\mathbf{k} - \mathbf{k}_1)^2 v_{Te}^2}. \tag{8.6}$$

For $k_1 \sim k$ and $k - k_1 \sim k$, $\varepsilon_i^l$ is of order $(m_e/m_i)(v_\varphi/v_{Te})^4$, and for $v_\varphi/v_{Te} \ll \sqrt{m_i/m_e}$, $\varepsilon^l$ is of order $(v_\varphi/v_{Te})^2$, i.e., $\varepsilon_i/\varepsilon \sim (m_e/m_i)\cdot(v_\varphi/v_{Te})^2$ is of order $v_{Te}/v_\varphi$ for

$$\frac{v_\varphi}{v_{Te}} \sim \left(\frac{m_i}{m_e}\right)^{1/3}. \tag{8.7}$$

Recognizing, in addition, that $v_\varphi/v_{Te} \gg 1$, we see that for a hydrogen plasma, $m_i/m_e \sim 2\cdot 10^3$, the estimate (8.1) applies over a narrow range: $1 \ll v_\varphi/v_{Te} \ll 10$. For heavy-gas plasma, for example, a cesium plasma, (8.1) applies to a wider range. If we neglect the Doppler corrections, the nonlinear interaction of electrons for $(m_i/m_e)^{1/3} \ll v_\varphi/v_{Te} \ll (m_i/m_e)^{1/2}$ is described approximately by

$$\frac{\partial N_{\mathbf{k}}^l}{\partial t} = N_{\mathbf{k}}^l \int N_{\mathbf{k}_1}^l d\mathbf{k}_1 \frac{2}{27} \frac{m_e^2}{m_i^2} \left(\frac{\omega_{0e}}{v_{Te}}\right)^5 \frac{(\mathbf{k} - \mathbf{k}_1)^4 (\mathbf{k}\mathbf{k}_1)^2 \hbar}{|\mathbf{k} - \mathbf{k}_1| \left(k_1^2 - k^2\right)^3 n m_e v_{Te} (2\pi)^{5/2} k^2 k_1^2}, \tag{8.8}$$

and for $v_\varphi/v_{Te} \gg (m_i/m_e)^{1/2}$ by

$$\frac{\partial N_{\mathbf{k}}^l}{\partial t} = N_{\mathbf{k}}^l \int N_{\mathbf{k}_1}^l \frac{3\omega_{0e}\left(k_1^2 - k^2\right)(\mathbf{k}\mathbf{k}_1)^2 \hbar}{8(2\pi)^{5/2} n m_e v_{Te} |\mathbf{k} - \mathbf{k}_1| k_1^2 k^2} d\mathbf{k}_1. \tag{8.9}$$

The role of ions in the screening of electron scattering has been described in [8], in which the nonlinear interaction (8.8) is derived for a one-dimensional spectrum ($\mathbf{k} \parallel \mathbf{k}_1$), and in [9] for the threedimensional case.

It is important to note that the interaction (8.9), unlike (7.76), depends only weakly on the directions $\theta$ of the interacting waves for $\theta \lesssim 1$ and does not vanish for the one-dimensional case $\mathbf{k} \parallel \mathbf{k}_1$. The characteristic transfer time under conditions such that Eq. (8.9) is valid and $k - k_1$ is of the same order as $k$ becomes

$$\frac{1}{\tau} \sim \omega_{0e} \frac{W^l}{n m_e v_{Te}^2} \frac{v_{Te}}{v_\varphi}, \tag{8.10}$$

which is many times $[(v_\varphi/v_{Te})^2$-fold] greater than (8.1).

Notice that the polarization effects of ions are particularly significant in the investigation of one-dimensional spectra. The following general expression for the nonlinear interaction in one-dimensional spectra [10] is equally suitable for waves with phase velocities of the same order as or greater than the velocity of light:

$$\frac{\partial N_{\mathbf{k}}^l}{\partial t} = N_{\mathbf{k}}^l \int N_{\mathbf{k}_1}^l d\mathbf{k}_1 \frac{3\omega_{0e}\hbar}{8(2\pi)^{5/2}} \frac{k_1^2 - k^2}{|\mathbf{k}_1 - \mathbf{k}|\, nm_e v_{Te}} \times$$

$$\times \left\{ \frac{v_{Te}^4}{c^4} + \left| \frac{v_{Te}^2}{c^2} + \frac{1 - \varepsilon_i^l(\omega - \omega_1, \mathbf{k} - \mathbf{k}_1)}{\varepsilon^l(\omega - \omega_1, \mathbf{k} - \mathbf{k}_1)} + \frac{3kk_1}{(\mathbf{k} - \mathbf{k}_1)^2 \varepsilon^l(\omega - \omega_1, \mathbf{k} - \mathbf{k}_1)} \right|^2 \right\}. \tag{8.11}$$

The result (8.2) follows from (8.11) only for $v_\varphi/v_{Te} \ll (m_i/m_e)^{1/4}$. If this inequality is not satisfied, the nonlinear interaction of one-dimensional waves is of the same order as for multidimensional waves, i.e., (8.10). Only nonlinear scattering through a virtual longitudinal wave was included above. Scattering by a virtual transverse wave is strictly forbidden in the one-dimensional case, and (8.10) describes the interaction of an arbitrary one-dimensional spectrum.

## 2. NONLINEAR SCATTERING BY ELECTRONS THROUGH VIRTUAL TRANSVERSE WAVES

Nonlinear scattering through virtual transverse waves can be significant only for Langmuir waves with phase velocity much larger than the velocity of light. We have seen, however, that nonlinear interactions at small phase velocities transfer waves very efficiently into this region. The wave lengths of the plasma cannot exceed the dimension $a$ of the system: $v_{\varphi\,\max} \simeq \omega_{0e}a$. For reasonable values of $a$ and n (plasma density) $v_{\varphi\,\max}$ can be $\sim 10 - 10^4$ c.

The investigation of the interaction of these long-wave modes, whose wavelength approaches the dimensions of the system, might be of interest because spectral transfer causes oscillatory energy to be concentrated in the long-wave region of the spectrum. Nonlinear scattering through a virtual transverse wave was first analyzed in [11] for the scattering of longitudinal waves into transverse waves. For the nonlinear interaction of Langmuir waves with each other the analysis of scattering through a virtual transverse wave has been carried out in [12] for $m_i = \infty$. There the related effects were comparable with (8.1), rather than with (8.10), which is valid for large phase velocities. Moreover, the polarization effects of ions can also affect scattering through a virtual transverse wave.

We now present a result that describes all three types of interaction (Compton scattering, scattering through a longitudinal wave, and scattering through a transverse wave) when the weak interactions corresponding to one-dimensional spectra are ignored [10]:

$$\frac{\partial N_{\mathbf{k}}^{l}}{\partial t} = N_{\mathbf{k}}^{l} \int N_{\mathbf{k}_1}^{l} d\mathbf{k}_1 \frac{3\omega_{0e}(\mathbf{k}_1^2 - \mathbf{k}^2)\hbar}{8(2\pi)^{5/2}|\mathbf{k}-\mathbf{k}_1| n m_e v_{Te}} \left\{ \frac{[\mathbf{k}\mathbf{k}_1]^2 v_{Te}^2}{k^2 k_1^2 \omega_{0e}^2 |k_-|^2} \times \right.$$

$$\left. \times \left| 2(\mathbf{k}\mathbf{k}_1) + \frac{\omega_-^2 \left(\varepsilon_e^t(\omega_-, \mathbf{k}_-) - 1\right) 2(\mathbf{k}\mathbf{k}_1) - \omega_{0e}^2 k_-^2}{k_-^2 c^2 - \omega_-^2 \varepsilon^t(\omega_-, \mathbf{k}_-)} \right|^2 + \frac{(\mathbf{k}\mathbf{k}_1)^2}{k^2 k_1^2} \left| \frac{\varepsilon_i^l(\omega_-, \mathbf{k}_-)}{\varepsilon^l(\omega_-, \mathbf{k}_-)} \right|^2 \right\};$$

$$\omega_- = \omega - \omega_1; \quad \mathbf{k}_- = \mathbf{k} - \mathbf{k}_1. \tag{8.12}$$

The term containing the denominator $k_-^2 c^2 - \omega_-^2 \varepsilon^t$ corresponds to scattering through a virtual transverse wave. It is clear from the above expression that scattering through a virtual transverse wave is possible only for waves whose wave vectors are not strictly parallel. Let us find when scattering through a transverse wave is decisive, assuming that the angle between k and $k_1$ is not small ($\theta \sim 1$). Scattering through a transverse wave can be large only when the denominator $k_-^2 c^2 - \omega_-^2 \varepsilon^t$ is small. If k and $k_1$ have the same order of magnitude, then for $\theta \sim 1$ the quantity $k_-$ is of the same order as k, whereas $\omega_- \sim (3k\Delta k/\omega_0) v_{Te}^2$ ($\Delta\mathbf{k} = |\mathbf{k}| - |\mathbf{k}_1|$). Since we are neglecting scattering by ions, we have $\omega_- \gg |k_-| v_{Ti}$, or

$$\Delta k \gg \frac{v_{Ti}}{v_{Te}} \frac{v_\varphi}{v_{Te}} k.$$

With this approximation the transverse dielectric constant $\varepsilon^t$ has the form

$$\varepsilon_e^t = \frac{i}{3} \sqrt{\frac{\pi}{2}} \left(\frac{v_\varphi}{v_{Te}}\right)^3 \frac{k}{\Delta k}; \quad \varepsilon_i^t = -\frac{m_e}{9m_i} \left(\frac{v_\varphi}{v_{Te}}\right)^4 \left(\frac{k}{\Delta k}\right)^2.$$

Under the conditions

$$\frac{\Delta k}{k} \gg \frac{v_\varphi}{v_{Te}} \frac{m_e}{9m_i} \tag{8.13}$$

electrons provide the main contribution to $\varepsilon^t$. It is now possible to deduce in approximate form the nonlinear interaction for the special cases $k_-^2 c^2 \gg \omega_-^2 \varepsilon^t(\omega_-, \mathbf{k}_-)$ and $k_-^2 c^2 \ll \omega_-^2 \varepsilon^t(\omega_-, \mathbf{k}_-)$, which under the given conditions are equivalent to $\frac{\Delta k}{k} \gg \frac{c^2}{v_\varphi v_{Te}}$ and $\frac{\Delta k}{k} \ll \frac{c^2}{v_\varphi v_{Te}}$.

If $\Delta k/k \gg c^2/v_\varphi v_{Te}$ and (8.13) holds, we obtain the interaction in the form

$$\frac{\partial N_{\mathbf{k}}^l}{\partial t} = N_{\mathbf{k}}^l \int \frac{2}{3} \frac{\omega_{0e}}{(2\pi)^{7/2}} \frac{[\mathbf{k}\mathbf{k}_1]^2}{k_1^2 - k^2} \frac{|\mathbf{k} - \mathbf{k}_1|^3}{n m_e v_{Te}} \frac{N_{\mathbf{k}_1}}{k_1^2 k^2} d\mathbf{k}_1 \hbar. \tag{8.14}$$

It has been assumed here that $v_\varphi \gg c$, because only under these conditions can (8.14) be greater than scattering via a longitudinal wave.

In fact, the characteristic time for spectral transfer (8.14) is of order

$$\frac{1}{\tau} \simeq \frac{k}{\Delta k} \frac{W^l}{n m_e v_{Te}^2} \frac{v_{Te}}{v_\varphi} \omega_{0e}, \tag{8.15}$$

whereas for scattering via a longitudinal wave when $\varepsilon_i^l \ll \varepsilon_e^l$, i.e.,

$$\frac{\Delta k}{k} \gg \frac{v_\varphi}{v_{Te}} \sqrt{\frac{m_e}{9m_i}},$$

it is of order

$$\frac{1}{\tau} \simeq \left(\frac{k}{\Delta k}\right)^3 \frac{W^l}{n m_e v_{Te}^2} \left(\frac{v_\varphi}{v_{Te}}\right)^3 \omega_{0e} \left(\frac{m_e}{9m_i}\right)^2. \tag{8.16}$$

By comparison we obtain the conditions for (8.14) to dominate:

$$\frac{\Delta k}{k} \gg \left(\frac{v_\varphi}{v_{Te}}\right)^2 \frac{m_e}{9m_i}. \tag{8.17}$$

Because $\Delta k/k < 1$ we have $v_\varphi/v_{Te} \ll \sqrt{9m_i/m_e}$, and, as $\Delta k/k \gg c^2/v_\varphi v_{Te}$, it follows that $v_\varphi \gg c^2/v_{Te}$, i.e.,

$$v_{Te} \gg \left(\frac{m_e}{9m_i}\right)^{1/4} c. \tag{8.18}$$

The last condition shows that scattering via a transverse wave is significant only for sufficiently high plasma temperatures. Equation (8.18) involves the electron temperature, which can become fairly high in some cases, for example, in turbulent heating. Notice that at high temperatures particle collision effects, which clearly could become pronounced for large $v_\varphi$, become negligibly small. If $\frac{\Delta k}{k} \ll \frac{v_\varphi}{v_{Te}} \sqrt{\frac{m_e}{9m_i}}$ scattering via a transverse wave can exceed that via a longitudinal wave for $\Delta k \ll k$, $v_\varphi \gg c^2/v_{Te}$, and $v_{Te} \gg (m_e/9m_i)^{1/2} c$. We shall not give the expression derived in [12] for the interaction when $\Delta k/k \ll c^2/v_\varphi v_{Te}$ and in which the ion contribution to the transverse dielectric constant was not included because, in addition to the temperature condition, which in this case is $v_{Te} \gg (m_e/9m_i)^{1/3} c$, we must also satisfy $c \ll v_\varphi \ll (9m_i/m_e)^{1/4}$, which shows that this interaction is valid only in a very narrow range of phase velocities.

Given the opposite inequality to (8.13), the ions assert themselves in polarization effects. Scattering via a transverse wave can predominate in this case if $1 \ll v_\varphi/c \ll \sqrt{m_i/m_e}$, $v_\varphi \gg c^2/v_{Te}$, i.e., $v_{Te} \gg \sqrt{m_e/m_i}/c$. The approximate form of the interaction agrees with that deduced in [12] (neglecting the ion contribution to $\varepsilon^t$). This is because for $\omega_-^2 \varepsilon^t \gg k_-^2 c^2$ the interaction turns out to be identical to:

$$\frac{\partial N_{\mathbf{k}}^l}{\partial t} = N_{\mathbf{k}}^l \int \frac{3\omega_{0e}^3 \left(k_1^2 - k^2\right) [\mathbf{k}\mathbf{k}_1]^2 v_{Te} N_{\mathbf{k}_1}^l d\mathbf{k}_1 \hbar}{8(2\pi)^{5/2} |\mathbf{k} - \mathbf{k}_1|^3 n m_e k^2 k_1^2 c^4}. \tag{8.19}$$

The characteristic transfer time is

$$\frac{1}{\tau} \simeq \omega_{0e} \frac{\Delta k}{k} \frac{W^l}{n m_e v_{Te}^2} v_{Te}^3 v_\varphi / c^4. \tag{8.20}$$

In conclusion we stress two points: first, only at sufficiently high plasma temperatures can scattering be significant; second, the interaction is appreciable only for

$$v_{\varphi\,\max} > v_\varphi \gg c^2/v_{Te}, \quad \text{i.e.,} \quad v_{Te} \gg c^2/v_{\varphi\,\max} = c^2/\omega_0 a.$$

For small densities and relatively small system dimensions the interaction can well prove to be of no consequence.

## 3. INDUCED SCATTERING OF LANGMUIR WAVES BY IONS

In order that plasma ions can scatter Langmuir waves, the following inequality must hold:

$$\frac{3\Delta k}{k} < \frac{v_{Ti}}{v_{Te}} \frac{v_\varphi}{v_{Te}}. \tag{8.21}$$

This is a straightforward result of the conservation of energy in scattering. For $\Delta k \sim k$ it reduces to

$$T_i > T_e 9 \frac{m_i}{m_e} \left(\frac{v_{Te}}{v_\varphi}\right)^2. \tag{8.22}$$

For a nonisothermal plasma with strongly heated electrons it is always possible to have small enough $T_i$ that the conditions (8.21) and (8.22) are not fulfilled and scattering by ions is not significant. Even for $T_i \sim T_e$, however, Langmuir waves with large phase velocities [for (8.22) $v_\varphi/v_{Te} \sim \sqrt{9m_i/m_e}$] and small $\Delta k/k$ [see (8.21)] interact mainly through ion scattering. As a matter of fact, induced ion scattering, if allowed by the conservation laws, normally exceeds that by electrons. This follows from the fact that, first, Compton scattering by ions is negligible compared to nonlinear scattering and there is no cancellation of the two scattering mechanisms [11]; second, induced scattering is proportional to the derivative $\partial f/\partial p$, which for ions with $T_i \ll T_e$ is much smaller than for electrons.

Recognizing that $(\varepsilon_e^l - 1)/\varepsilon^l$, which occurs in the scattering probability for $\omega_- \ll k_- v_{Ti}$, is of order $1/(1 + T_e/T_i)$, we obtain the following expression for the nonlinear interaction of Langmuir waves during induced scattering by ions [3, 8, 9, 13]:

$$\frac{\partial N_{\mathbf{k}}^l}{\partial t} = N_{\mathbf{k}}^l \frac{3}{8} \frac{\omega_{0e} T_e/T_i}{n m_e v_{Ti} (1 + T_e/T_i)^2} \int \frac{\hbar N_{\mathbf{k}_1}^l}{(2\pi)^{5/2}} \frac{(\mathbf{k}\mathbf{k}_1)^2}{k^2 k_1^2} \frac{k_1^2 - k^2}{|\mathbf{k} - \mathbf{k}_1|} d\mathbf{k}_1. \tag{8.23}$$

The characteristic transfer time is to order of magnitude

$$\frac{1}{\tau} \simeq \omega_{0e} \frac{\Delta k}{k} \frac{W^l}{n m_e v_{Te}^2} \frac{v_{Te}}{v_{Ti}} \frac{v_{Te}}{v_\Phi} \frac{T_e/T_i}{(1 + T_e/T_i)^2}. \tag{8.24}$$

In comparing the nonlinear interaction (8.23) with nonlinear scattering by electrons, one must bear in mind that the latter has the following form for conditions (8.21) and (8.22) [8]:

$$\frac{\partial N_{\mathbf{k}}^{l}}{\partial t} = N_{\mathbf{k}}^{l} \int N_{\mathbf{k}_1}^{l} d\mathbf{k}_1 \frac{3\omega_{0e}\left(k_1^2 - k^2\right)(T_e/T_i)^2 (\mathbf{k}\mathbf{k}_1)^2 \hbar}{(1 + T_e/T_i)^2\, 8\, (2\pi)^{5/2}\, |\mathbf{k} - \mathbf{k}_1|\, n m_e v_{Te} k^2 k_1^2}. \tag{8.25}$$

The characteristic interaction time (8.25) is larger than (8.23) for $T_e/T_i \ll m_i/m_e$. Under these conditions scattering by ions predominates over the nonlinear interaction associated with the Doppler corrections to Compton scattering by electrons (8.1) for

$$\frac{v_\varphi}{v_{Te}} \gg \frac{(1 + T_e/T_i)}{(T_e/T_i)^{3/4}} \left(\frac{m_e}{m_i}\right)^{1/4}. \tag{8.26}$$

This is automatically satisfied for $T_e \gg T_i$, $T_e \ll T_i m_i/m_e$ since $v_\varphi \gg v_{Te}$. In the case $T_e \ll T_i$, however, only for $T_i \ll T_e \cdot (m_i/m_e)^{1/3}$ are there no restrictions on the above conditions. For large $T_i$, on the other hand, the interaction of waves with small $v_\varphi$ near $v_{Te}$ cannot depend on scattering by ions. It is essential to realize that for waves with $\Delta k \sim k$ the relations (8.26) and (8.22) are incompatible if $T_i \gg T_e$. It is only meaningful to consider small $\Delta k$, therefore, as long as (8.21) is not violated.

So far spectral transfers by electrons and ions have been compared for identical values of $\Delta k$. There is another kind of comparison that is useful, however, namely, to consider $\Delta k \lessgtr k$, for which ion scattering is impossible, assuming that (8.22) is violated. Under these conditions "step-wise" scattering by ions is still possible. In each step $\Delta k \ll \Delta k_0$ and satisfies (8.21). This type of scattering must be compared with the allowed single transfer of energy by electrons.

It follows from (8.21) that even for small values of $T_i$ there can be small enough $\Delta k$ that the nonlinear interaction will be determined by ions. However, the characteristic time for spectral transfer increases with decreasing $\Delta k$. Consequently, the greatest nonlinear interaction occurs for the maximum values of $\Delta k$ permitted by the inequality (8.21):

$$\frac{1}{\tau} \simeq \frac{\omega_{0e}}{3} \frac{W^l}{n m_e v_{Te}^2} \frac{T_e/T_i}{(1 + T_e/T_i)^2}. \tag{8.27}$$

This time corresponds to transfer by an amount $\Delta k \ll \Delta k_0$, whereas the time for transfer by $\Delta k \sim \Delta k_0$ is $(\Delta k_0/\Delta k)$ times greater, i.e., is of the order*

$$\frac{1}{\tau} \simeq \omega_{0e} \frac{W^l}{n m_e v_{Te}^2} \left(\frac{v_{Ti}}{v_{Te}} \frac{v_\varphi}{v_{Te}}\right)^2 \frac{T_e/T_i}{(1+T_e/T_i)^2} \frac{k^2}{\Delta k_0^2}. \tag{8.28}$$

For $v_\varphi/v_{Te} \ll (m_i/m_e)^{1/3}$ (8.28) must be compared with (8.1), whereupon a criterion is obtained for ion scattering to predominate:

$$\frac{v_\varphi}{v_{Te}} \gg \left(1+\frac{T_e}{T_i}\right)^{2/5} \left(\frac{3m_i}{m_e}\right)^{1/5}. \tag{8.29}$$

The analogous criterion for $v_\varphi/v_{Te} \gg (m_i/m_e)^{1/2}$ has the form*

$$\frac{v_\varphi}{v_{Te}} \gg \left(1+\frac{T_e}{T_i}\right)^{2/3} \left(\frac{3m_i}{m_e}\right)^{1/3}. \tag{8.30}$$

It is important to bear in mind that for one-dimensional spectra, when all interacting waves have the same direction, the criterion delimiting scattering by ions is contained in (8.22). The characteristic transfer time in this case,

$$\frac{1}{\tau_{1-d}} \simeq \omega_{0e} \frac{W^l}{n m_e v_{Te}^2} \frac{v_{Te}}{v_{Ti}} \frac{v_{Te}}{v_\varphi} \frac{T_e/T_i}{(1+T_e/T_i)^2} \tag{8.31}$$

does not increase with decreasing $\Delta k$. In the process analogous to "step-wise" energy transfer by small $\Delta k$ steps, however, in the one-dimensional case the direction of the waves reverses in each step [8].

## 4. COMPARISON OF THE ENERGY TRANSFER BY INDUCED SCATTERING AND BY DECAY OF LANGMUIR WAVES INTO ION-ACOUSTIC WAVES

Both in induced scattering processes and in the decay of Langmuir waves into low-frequency waves the total number of

* One must, of course, bear in mind that the above estimates are very crude. Only a solution of the integral equation (8.25) with $\exp(-\omega_-^2/2k_-^2 v_{Ti}^2)$ on the right-hand side can provide precise information about the evolution of the spectrum and the "step-wise" transfer.

Langmuir waves remains unchanged, i.e., spectral transfer occurs. In the absence of strong low-frequency waves decay transfer is in the direction of larger phase velocities. However, in contrast with scattering, it is bounded by a maximum phase velocity equal to

$$v_{\varphi\,\max} = v_{Te}\sqrt{9\,m_i/m_e}. \tag{8.32}$$

The energy transfer of waves with $v_\varphi > v_{\varphi\,\max}$ is wholly determined by induced scattering. It is necessary, therefore, to compare both processes for $v_\varphi < v_{\varphi\max}$.

Let us make a crude comparison ignoring the qualitative differences involved in these processes. Using the diffusion approximation for decays and assuming that the width of the Langmuir spectrum satisfies the conditions

$$\Delta k \gg k^s \sqrt{\frac{9\,m_i}{m_e}\frac{v_{Te}}{v_\varphi^l}} \tag{8.33}$$

and $k^s \ll \omega_{0e}/3v_{Te}\sqrt{m_e/m_i}$, we obtain the following estimate of the characteristic time for transfer by an amount $\Delta k$ of order k due to decays:

$$\frac{1}{\tau} \simeq \omega_{0e}\frac{W^l}{nm_e v_{Te}^2}\left(\frac{v_\varphi}{v_{Te}}\right)^2\left(\frac{m_e}{9m_i}\right), \tag{8.34}$$

which is greater than that for electron scattering (8.28) if $T_e \gg T$; for the smallest values of $\Delta k$ allowed by (8.33) the estimate of transfer due to decays is

$$\frac{1}{\tau} \simeq \omega_{0e}\frac{W^l}{nm_e v_{Te}^2}. \tag{8.35}$$

The decay processes considered in this section are possible only if $T_e \gg T_i$, in which case they become more important than ion scattering.

## 5. INDUCED SCATTERING OF ION-ACOUSTIC WAVES

The interaction of ion-acoustic waves differs markedly from that of Langmuir waves.

First, the phase velocities of the waves have an upper limit $v_s = v_{Te}\sqrt{m_e/m_i}$, and transfer toward smaller $v_\varphi$ takes oscillatory energy into the region where collisional absorption becomes increasingly important. Therefore, this type of transfer can serve as a mechanism for the absorption of oscillations generated by plasma instability.

Second, in the acoustic branch the wave number dependence on oscillation frequency causes the transfer process, in which the number of quanta is conserved, to be accompanied by a significant reduction in the energy of the ion-acoustic oscillations, unlike for the Langmuir-wave case.

Third, the main participants in the scattering process are plasma ions, and on the acoustic branch the transfer of energy can have only a "step-wise" character; in other words, it can be approximately described by differential equations (transfer to wave numbers having similar magnitudes). This is directly implied by the conservation of energy in scattering:

$$(k-k')\,v_s = (k-k')\,v_{Ti}\sqrt{T_e/T_i} < |\mathbf{k}-\mathbf{k}'|\,v_{Ti}, \tag{8.36}$$

i.e., $\Delta k/|\mathbf{k}-\mathbf{k}'| \ll \sqrt{T_i/T_e}$. In the case of small angles $\theta$ between interacting waves $\Delta k$ must be especially small, $\Delta k/k \ll \theta\sqrt{T_i/T_e}$. Yet for short-wave ion-acoustic oscillations, $k^2\lambda_{De}^2 \gg 1$, whose frequencies are close to $\omega_{0i}$, the interaction can be of an integral character (transfer by an amount $\Delta k$ of order k). This follows from the dispersion curve for these modes. Thus, for $1 \ll k^2\lambda_{De}^2 \ll \sqrt{T_e/T_i}$ we have $\omega - \omega' \simeq \Delta k/k^3\lambda_{De}^3\omega_{0i}$, and for $\Delta k \sim k$ we have $\omega - \omega' < |\mathbf{k}-\mathbf{k}'|v_{Ti}$, provided $k^2\lambda_{De}^2 > (T_e/T_i)^{1/3}$. For $\sqrt{T_e/T_i} \ll k^2\lambda_{De}^2 < T_e/T_i$ the transfer process can always be integral.

The differential form of nonlinear interaction for the acoustic branch of the spectrum has been derived in [14, 15] and later in [16]. It follows from the above that in terms of direction the transfer process has an integral character, a fact that is most clearly elucidated in [16]. However, we must lay particular stress on the approximate nature of the differential form of nonlinear interaction. This is apparent just from the fact that the resulting equations for the nonlinear interaction begin to depend on the boundary conditions in wave number space and can become many-valued. Actually the equations can only have a differential character in a limited wave number range. Outside that range the transfer is integral.

As the phase velocities of ion-acoustic waves exceed the mean thermal velocity of the ions, their scattering by ions is very similar to that of Langmuir waves by electrons. The nonlinear scattering in this case need include only the ion contribution. We have $(\varepsilon_i^l - 1)/\varepsilon^l \sim 1/(1 + T_i/T_e) \approx 1$. The Doppler corrections to the Compton scattering are described by lower-order terms in the small parameter $T_i/T_e$; they are larger than $\sqrt{T_i/T_e}$. The scattering probability is described by the approximate expression [cf. (7.75)]

$$w_p^{s,s}(\mathbf{k}, \mathbf{k}_1) \simeq \frac{[\mathbf{k}\mathbf{k}_1]^2 (\mathbf{k}\mathbf{k}_1)^2}{|\mathbf{k}_-|^2 k^2 k_1^2} \delta(\omega_- - \mathbf{k}_- \mathbf{v}) \pi \frac{v_{Ti}^2}{n^2}. \tag{8.37}$$

From this we readily obtain expressions for the nonlinear interaction for the entire range of ion-acoustic waves:

$$\frac{\partial N_{\mathbf{k}}^s}{\partial t} = -\hbar N_{\mathbf{k}}^s \int N_{\mathbf{k}_1}^s \frac{d\mathbf{k}_1}{(2\pi)^2} \frac{(\mathbf{k}\mathbf{k}_1)^2}{k^2 k_1^2} \frac{T_i}{m_i^2 n} \frac{[\mathbf{k}\mathbf{k}_1]^2}{\sqrt{2\pi}} \frac{\omega_-}{v_{Ti}^3 |\mathbf{k}_-|^3} \exp\left(-\frac{\omega_-^2}{2v_{Ti}^2 k_-^2}\right). \tag{8.38}$$

In the acoustic part of the spectrum, using

$$-\frac{\omega}{\sqrt{2\pi}\, v_{Ti}^3 k^3} \exp\left(-\frac{\omega^2}{2v_{Ti}^2 k^2}\right) \to \delta'(\omega) \tag{8.39}$$

we readily obtain the result of [14]:

$$\frac{\partial N_{\mathbf{k}}^s}{\partial t} = N_{\mathbf{k}}^s \int N_{\mathbf{k}_1}^s \frac{d\mathbf{k}_1}{(2\pi)^2} \frac{(\mathbf{k}\mathbf{k}_1)^2}{k^2 k_1^2} \frac{T_i}{m_i^2 n} [\mathbf{k}\mathbf{k}_1]^2 \delta'(\omega_-), \tag{8.40}$$

which is equivalent to the differential form (for $\theta \ll 1$ in [16])

$$\frac{\partial N_{k\Omega}^s}{\partial t} = N_{k\Omega}^s k^2 \frac{\partial}{\partial k} \int \frac{T_i \sin^2 2\theta}{T_e 16\pi^2 m_i n} N_{k\Omega_1}^s k^4 \, d\Omega_1. \tag{8.41}$$

Here $\theta$ is the angle between $\mathbf{k}_1$ and $\mathbf{k}$, and $d\boldsymbol{\Omega}_1$ is the solid angle subtended by $\mathbf{k}_1$. It should be emphasized that for small angles, where (8.41) tends to zero, as well as for $\theta \to \pi/2$, the next-higher terms of the expansion in $T_i/T_e$ become important. Then (8.41) holds only for $\theta \gg T_i/T_e$. The characteristic transfer time for (8.41) with $\theta \sim 1$ is of the order

$$\frac{1}{\tau} \simeq \omega^s \frac{W^s}{nm_e v_{Te}^2} \frac{k}{\Delta k} \frac{T_i}{T_e}. \tag{8.42}$$

In the spectral range corresponding to ion waves, $\exp\left(-\frac{\omega_-^2}{2v_{Ti}^2 k_-^2}\right) \simeq 1$, the interaction turns out to be similar to the nonlinear interaction of Langmuir waves:

$$\frac{\partial N_{\mathbf{k}}^s}{\partial t} = -\hbar N_{\mathbf{k}}^s \int N_{\mathbf{k}_1}^s \frac{d\mathbf{k}_1}{(2\pi)^{3/2}} \frac{\omega_- [\mathbf{k}\mathbf{k}_1]^2}{v_{Ti} n m_i |\mathbf{k}_-|^3} \frac{(\mathbf{k}\mathbf{k}_1)^2}{k^2 k_1^2}. \tag{8.43}$$

## 6. INDUCED SCATTERING OF LANGMUIR WAVES WITH CONVERSION INTO ION-ACOUSTIC WAVES

This type of interaction plays an important part in plasma, particularly in the development of instabilities, because it is a possible mechanism for the absorption of Langmuir waves. The laws of energy and momentum conservation in scattering provide verification of the fact that the interaction of $l$-waves with sound waves ($\omega = k^s v_s$) is exponentially small in a plasma without particle beams. As a matter of fact, it follows from

$$\omega^l \pm \omega^s = (\mathbf{k}^l \pm \mathbf{k}^s)\,\mathbf{v} \tag{8.44}$$

that the projection of the velocity on $\mathbf{k} \pm \mathbf{k}^s$, which is equal to $v_\parallel \simeq \omega_{0e}/|\mathbf{k}^l \pm \mathbf{k}^s|$, is much larger than $v_{Te}$ for $k^l \ll 1/\lambda_{De}$, $k^s \ll 1/\lambda_{De}$, i.e., those particles involved in this type of scattering must have a velocity greatly exceeding the mean electron thermal velocity. In the ion oscillation region $k^s \gg k^l$, and $\omega_{0e}/k^s \gg \omega_{0i}/k^s \gg v_{Ti}$, i.e., ions cannot take part in scattering. Because $\omega_{0e}/k^s \ll v_{Te}$, for $\omega^s \simeq \omega_{0i}$ the scattering is determined by the electrons.

Thus, if decay interactions are allowed only for the acoustic branch of the ion-acoustic waves, then scattering by electrons is allowed only for the ion resonance branch. In this respect the two processes complement one another. The nonlinear absorption of Langmuir waves is described by

$$\frac{\partial N_{\mathbf{k}}^{l}}{\partial t} = - N_{\mathbf{k}}^{l} \int \frac{d\mathbf{k}^{s}}{2} \frac{N_{\mathbf{k}^{s}}^{s} \omega_{0i} \omega_{0e}}{(2\pi)^{5/2} n m_{e} v_{Te}^{2}} \frac{(\mathbf{k}\mathbf{k}^{s})^{2}}{(kk^{s})^{2}} \frac{\omega_{0e}\hbar}{|\mathbf{k}^{s}| v_{Te}}. \tag{8.45}$$

The characteristic damping rate can be estimated from the relation

$$\gamma = \frac{1}{\tau} = \pi \frac{W^{s}}{n m_{e} v_{Te}^{2}} \omega_{0e} \frac{1}{k^{s} \lambda_{De}}, \tag{8.46}$$

where $W^s$ is the ion oscillation energy. The maximum damping corresponds to $k^s$ of order $1/\lambda_{De}$, the minimum damping to a value smaller by $\sqrt{T_i/T_e}$. The variation of the number of ion-acoustic waves is described by the expression

$$\frac{\partial N_{\mathbf{k}^{s}}^{s}}{\partial t} = N_{\mathbf{k}^{s}}^{s} \int N_{\mathbf{k}}^{l} \frac{d\mathbf{k}}{(2\pi)^{5/2}} \frac{\omega_{0i}\omega_{0e}}{n m_{e} v_{Te}^{2}} \left(\frac{\mathbf{k}\mathbf{k}^{s}}{kk^{s}}\right)^{3} \frac{k\omega_{0e}\hbar}{(k^{s})^{2} v_{Te}}. \tag{8.47}$$

It is important to consider the dependence of the sign of this effect (8.47) (generation or absorption) on the relative orientation of **k** and $\mathbf{k}^s$ (which may be parallel or antiparallel). For a given sign of (8.45), this possibility is due to the fact that in this process the plasma particles absorb part of the energy and momentum of the waves.

## 7. INDUCED SCATTERING OF TRANSVERSE WAVES INTO TRANSVERSE WAVES

We do not propose to discuss in detail the effects of induced scattering through virtual transverse waves, as they are generally important only for very high plasma temperatures and for waves whose frequencies very precisely coincide with $\omega_{0e}$. At this point we merely examine the conditions for which the process becomes ordinary Compton scattering. Analysis shows that this requires that the difference frequency of the interacting waves $\omega_{-} = \omega_1 - \omega_2$ fulfills the following conditions:

$$\omega_{-} \gg |\mathbf{k}_{-}| v_{Te}, \quad \omega_{-} \gg |\mathbf{k}_{-}| v_{Ti}; \quad \omega_{-} \gg \omega_{0e}. \tag{8.48}$$

For $\omega_{-} \gg \omega_{0e}$, of course, at least one of the interacting waves must have a frequency much above $\omega_{0e}$. Under these conditions the first two inequalities of (8.48) are usually satisfied.

As already emphasized, induced scattering contains information different from that of spontaneous scattering; in particular, it is determined by the Doppler corrections, which are negligible for spontaneous scattering. The transfer of energy in this case becomes differential and is described by the equation

$$\frac{\partial N_{\mathbf{k}}^t}{\partial t} = \frac{\omega_{0e}^4 N_{\mathbf{k}}^t \hbar}{8(2\pi)^2 nm_e} \int N_{\mathbf{k}_1}^t \left(1 + \frac{(\mathbf{k}\mathbf{k}_1)^2}{k^2 k_1^2}\right) \frac{(\mathbf{k}-\mathbf{k}_1)^2}{\omega\omega_1} \delta'(\omega - \omega_1)\, d\mathbf{k}_1. \tag{8.49}$$

The characteristic transfer time has the following estimate:

$$\frac{1}{\tau} \sim \frac{W^t}{nm_e c^2} \omega_{0e} \frac{\omega_{0e}^3}{\omega^3} \left(\frac{\omega}{\Delta\omega}\right)^2. \tag{8.50}$$

The differential description of transfer, Eq. (8.49), as in the case of acoustic oscillations, can only be approximated under conditions such that the so-called physically infinitesimal $\Delta\omega$ meets the requirement $\Delta\omega \gg \omega_{0e}$. We note in this connection that, for example, scattering by ions cannot be represented analytically in differential form, even approximately. If the opposite conditions to (8.48) are satisfied, nonlinear scattering begins to be important, and compensation between the Compton and nonlinear scattering sets in. For electron scattering the Doppler corrections to the Compton scattering become negligible in comparison with the ion contribution to nonlinear scattering. Scattering by electrons leads to the following expression for the nonlinear interaction:

$$\frac{\partial N_{\mathbf{k}}^t}{\partial t} = N_{\mathbf{k}}^t \int N_{\mathbf{k}_1}^t \frac{-\hbar\omega_- \omega_{0e}^4 \left| \dfrac{\varepsilon_i^l(\omega_-, \mathbf{k}_-)}{\varepsilon^l(\omega_-, \mathbf{k}_-)} \right|^2 \left(1 + \dfrac{(\mathbf{k}\mathbf{k}_1)^2}{k^2 k_1^2}\right)}{8(2\pi)^{5/2} |\mathbf{k}_-| v_{Te} \omega^t \omega_1^t nm_e v_{Te}^2}\, d\mathbf{k}_1. \tag{8.51}$$

The fastest transfer for transverse waves is characterized by the time given by

$$\frac{1}{\tau} \sim \frac{W^t}{nm_e v_{Te}^2} \left(\frac{\omega_{0e}}{\omega^t}\right)^3 \omega_{0e} \frac{\Delta\omega}{|\mathbf{k} - \mathbf{k}_1| v_{Te}}. \tag{8.52}$$

When $\omega_- \ll |\mathbf{k}_-| v_{Ti}$ the wave interaction is determined solely by nonlinear scattering by ions:

$$\frac{\partial N^t_{\mathbf{k}}}{\partial t} = N^t_{\mathbf{k}} \int N^t_{\mathbf{k}_1} \frac{-\hbar\omega_- \left(1 + \frac{(\mathbf{k}\mathbf{k}_1)^2}{k^2 k_1^2}\right) \omega_{0e}^4}{\left(1 + \frac{T_e}{T_i}\right)^2 8 (2\pi)^{5/2} |\, \mathbf{k}_- \,| v_{Ti} \omega^t \omega_1^t n m_i v_{Ti}^2} d\mathbf{k}_1. \tag{8.53}$$

The characteristic time of the process (8.53) under these conditions is smaller than for (8.51) when $T_e < T_i m_i/m_e$. Note that the condition $\omega_- \ll |\, \mathbf{k}_- |v_{Ti}$ can be fulfilled for high frequencies $\omega \gg \omega_{0e}$, where it shows that the interaction in question describes transfer by small $\Delta\omega$:

$$\frac{\Delta\omega^t}{\omega^t} \ll \Delta\theta\, v_{T\alpha}. \tag{8.54}$$

The last condition demands that the scattering angle be fairly large for even a slight change in the quantum frequency (energy). This scattering is almost elastic and is accompanied by isotropization of the transverse-wave distribution. It is interesting that the transfer of transverse waves can in this case be more efficient than Compton transfer (8.49).

It follows at once from a comparison of (8.49) and (8.51) that for $\Delta\theta \sim 1$ transfer by an amount $\Delta\omega$ of order $\omega$ by "step-wise" transfer in small steps $\Delta\omega$ which satisfy $\omega_- \ll k_- v_{Te}$, is $(c/v_{Te})$ times as efficient as "step-wise" transfer in steps $\Delta\omega$ satisfying $\omega_- \gg k_- v_{Te}$ and $\Delta\omega \gg \omega_{0e}$. We note also that the transfer of high-frequency transverse waves $\omega \gg \omega_{0e}$ can occur through the effects considered above, involving the decay of transverse into Langmuir and transverse waves.

## 8. INDUCED SCATTERING OF LANGMUIR WAVES INTO TRANSVERSE WAVES

This effect is of interest to problems involving emissions from cosmic and laboratory plasma. From the radiation intensity of transverse waves, in particular, one can assess the intensities of plasma oscillations, etc.

It is readily seen that in scattering by plasma particles Langmuir waves can be converted into transverse waves whose frequencies are close to the electron plasma frequency $\omega_{0e}$. This is directly implied by the conservation of energy in scattering:

$$\Omega^t - \Omega^l = (\mathbf{k}^t - \mathbf{k}^l)\,\mathbf{v} \simeq \frac{(k^t)^2c^2}{2\omega_{0e}} - \frac{3}{2}\frac{(k^l)^2v_{Te}^2}{\omega_{0e}}. \tag{8.55}$$

The transfer is toward lower frequencies. In the presence of strong Langmuir and weak transverse oscillations, conversion to transverse waves becomes the dominant factor. Since $\omega^t < \omega^l$ we now have

$$ck^t < \omega_{0e}\sqrt{3}\,\frac{v_{Te}}{v_{\varphi}}. \tag{8.56}$$

In the case of intense transverse waves with frequencies $\omega \sim \omega_{0e}$, on the other hand, it is possible to have excitation of plasma waves whose phase velocities satisfy the inequality

$$\frac{v_{\varphi}}{v_{Te}} > \left(\frac{2(\omega^t - \omega_{0e})}{\omega_{0e}}\right)^{1/2}. \tag{8.57}$$

Here we write down the equations describing the generation of transverse waves from Langmuir waves, as they can be used without difficulty to derive the equations for the generation of Langmuir waves by transverse waves. Neglecting scattering through a virtual transverse wave, we have [10] ($\omega_- = \Omega^t_{k_2} - \Omega^l_{k_1}$; $\mathbf{k}_- = \mathbf{k}_2 - \mathbf{k}_1$)

$$\frac{\partial N^t_{\mathbf{k}_2}}{\partial t} = -N^t_{\mathbf{k}_2}\int d\mathbf{k}_1 N^l_{\mathbf{k}_1}\frac{\omega_{0e}^2\omega_-\hbar}{4m_e n v_{Te}^3(2\pi)^{5/2}|\mathbf{k}_-|}\left\{\frac{[\mathbf{k}_1\mathbf{k}_2]^2}{k_1^2k_2^2}\left[\left|\frac{\varepsilon_i^l(\omega_-,\mathbf{k}_-)}{\varepsilon^l(\omega_-,\mathbf{k}_-)}\right|^2 + \right.\right.$$
$$\left.\left. + \frac{v_{Te}^2}{\omega_{0e}^2}\frac{k_1^2k_2^2}{|\mathbf{k}_-|^2} + \frac{2v_{Te}^2(\mathbf{k}_2\mathbf{k}_-)(\mathbf{k}_1\mathbf{k}_2)}{\omega_{0e}^2|\mathbf{k}_-|^2} - \frac{2v_{Te}^2(\mathbf{k}_1\mathbf{k}_2)^2}{\omega_{0e}^2|\mathbf{k}_-|^2}\right] + \frac{2v_{Te}^2}{k_1^2\omega_{0e}^2}(\mathbf{k}_1\mathbf{k}_2)^2\right\}. \tag{8.58}$$

The first term of (8.58) describes the polarization effect associated with scattering by the screening cloud. As for Langmuir waves, the role of the ions in this screening is very important. The other terms describe the Doppler corrections to the Compton scattering. In particular, as $m_i \to \infty$ (8.58) goes over the interaction obtained in [11].

Analysis shows that these results have a narrow range of application:

$$\frac{v^l_{\varphi}}{v_{Te}} < \left(\frac{9m_i}{m_e c}v_{Te}\right)^{1/3} \tag{8.59}$$

and describe conversion into transverse waves, which is much less efficient than transfer into longitudinal waves, which violates (8.59). We therefore give an approximate equation that is valid when (8.59) fails, but when the following condition is met:

$$\frac{v_\varphi}{v_{Te}} \ll \sqrt{\frac{9m_i}{m_e}}, \tag{8.60}$$

namely,

$$\frac{\partial N^t_{\mathbf{k}_2}}{\partial t} = -N^t_{\mathbf{k}_2} \int N^l_{\mathbf{k}_1} d\mathbf{k}_1 \frac{\omega_{0e}^2 \,|\, \mathbf{k}_- \,|^3 v_{Te} \,[\mathbf{k}_1\mathbf{k}_2]^2}{4\omega_-^3 m_e n \,(2\pi)^{5/2} k_1^2 k_2^2} \frac{m_e^2}{m_i^2} \hbar. \tag{8.61}$$

Hence the characteristic time for $l \to$ t-transformation for $k_2 \sim k_1 v_{Te}$ is

$$\frac{1}{\tau} \simeq \frac{W^l}{n m_e v_{Te}^2} \left(\frac{m_e}{m_i}\right)^2 \left(\frac{v_\varphi}{v_{Te}}\right)^3. \tag{8.62}$$

Comparing (8.62) with the characteristic interaction time for longitudinal waves (8.1) for $v_\varphi/v_{Te} \ll (m_i/m_e)^{1/3}$ shows that (8.62) cannot become of the same order as the longitudinal-wave transfer time except at the very limit of its applicability. If $v_\varphi/v_{Te} \gg \sqrt{9m_i/m_e}$, the $l \to$ t-conversion is described by

$$\frac{\partial N^t_{\mathbf{k}}}{\partial t} = -N^t_{\mathbf{k}} \int d\mathbf{k}_1 N^l_{\mathbf{k}_1} \frac{\omega_{0e}^2 \omega_- \,[\mathbf{k}_1\mathbf{k}_2]^2 \,\hbar}{4 m_e n v_{Te}^3 \,(2\pi)^{5/2} \,|\, \mathbf{k}_- \,|\, k_1^2 k_2^2}. \tag{8.63}$$

The characteristic time for the process when $k_2 \sim k_1 v_{Te}$ is of order

$$\frac{1}{\tau} \sim \frac{W^l}{n m_e v_{Te}^2} \omega_{0e} \frac{v_{Te}}{v_\varphi}, \tag{8.64}$$

which corresponds to that for the nonlinear interaction of Langmuir waves.

Thus, we find that for $v_\varphi/v_{Te} > (m/me)^{1/3}$ the nonlinear conversion of Langmuir waves into transverse waves by scattering on electrons has the same order as the non-linear transfer of Langmuir waves through their mutual interaction. The conversion by scattering on ions is also very important.

## Literature Cited


1. A. Gailitis, L. M. Gorbunov, L. M. Kovrizhnykh, V. V. Pustovalov, V. P. Silin, and V. N. Tsytovich, "Elementary processes involving the nonlinear interaction of charged particles with a plasma and the equations for a weakly turbulent plasma," survey paper presented at the Conference on Phenomena in Ionized Gases, Belgrade, 1965; FIAN Preprint A-136 (1965).
2. B. B. Kadomtsev, Plasma turbulence, Academic Press, New York (1965).
3. L. M. Gorbunov and V. P. Silin, "Nonlinear interaction of plasma waves," Zh. Éksp. Teor. Fiz., 47:200 (1964).
4. A. Gailitis and V. N. Tsytovich, "Theory of the nonlinear interaction of transverse and longitudinal plasma waves," Izv. Vuzov, Radiofizika, 7:1190 (1964).
5. P. Sturrock, "Nonlinear effects in an electron plasma," Proc. Roy. Soc. A242:277 (1957).
6. W. Drummond and D. Pines, "Nonlinear stability of plasma oscillations," Nuclear Fusion, Suppl. Part 3, p. 1049 (1962).
7. V. D. Shapiro, Nonlinear Theory of the Interaction of Charged Particle Beams with a Plasma, Author's Abstract of Dissertation, OIYaI LVÉ (1962).
8. V. N. Tsytovich and V. D. Shapiro, "Nonlinear stabilization of beam plasma instabilities," Yadernyi Sintez, No. 5 (1965).
9. L. M. Kovrizhnykh [Kovrischnich], Raport Interne EUR-CEA, F. C. -258 Fontenay-aux-Roses (Scin.), France (1964).
10. V. N. Tsytovich, "Nonlinear generation of plasma waves by a beam of transverse waves (I)," Zh. Tekh. Fiz., 34:773 (1965).
11. A. Gailitis and V. N. Tsytovich, "Emission of transverse electromagnetic waves in the scattering of charged particles by plasma waves," Zh. Éksp. Teor. Fiz., 46:1726 (1964).
12. L. M. Gorbunov, V. V. Pustovalov, and V. P. Silin, "Nonlinear interaction of electromagnetic waves in a plasma," Zh. Éksp. Teor. Fiz., 47:1437 (1964).
13. A. A. Galeev, V. N. Karpman, and R. Z. Sagdeev, "Many-particle aspects of the theory of a turbulent plasma," Yadernyi Sintez, 5:20 (1965).
14. B. B. Kadomtsev and V. I. Petviashvili, "Weakly turbulent plasma in a magnetic field," Zh. Éksp. Teor. Fiz., 43:2234 (1962).
15. V. I. Petviashvili, "Ion-acoustic oscillations excited by an electron current," Dokl. Akad. Nauk SSSR, 153:1295 (1963).
16. I. A. Akhiezer, "Theory of nonlinear motions of a nonequilibrium plasma," Zh. Éksp. Teor. Fiz., 47:952 (1964).

*Chapter IX*

# Some Applications of Induced Wave Scattering in Plasma

## 1. CONVERSION OF LONGITUDINAL INTO TRANSVERSE WAVES

The interest in the conversion of longitudinal waves into transverse waves has arisen largely through astrophysics. There are a number of mechanisms which lead to the excitation of plasma oscillations under astrophysical conditions. Among these are excitation by strong nonlinear waves, magnetohydrodynamic waves, and particle beams. These modes of excitation occur frequently on the sun, particularly in chromospheric flares. However, evidence of plasma oscillations can be obtained only by the study of electromagnetic waves arriving on the earth. The importance of plasma effects on the radiation from the sun was first postulated by Shklovskii [1]. Later Ginzburg and Zheleznyakov [2, 3] were the first to analyze the mechanisms of longitudinal–transverse wave conversion (see also [4, 5]). Today, thanks to the detailed analysis of nonlinear conversion, it is possible to treat this problem in more precise terms. It is important to recognize, in addition to nonlinear effects, the significant contribution of surface effects associated with wave scattering by plasma inhomogeneities, surface irregularities, etc., all of which have been investigated by Tidman [4]. Normally these are considerably smaller than the nonlinear effects, which occur throughout the volume. The conversion length is important. If it is very small, i.e., if the conversion is very efficient, then, over paths whose lengths are such that energy on the order of the longitudinal wave energy is converted into transverse waves, the inverse transfer becomes important, and transverse radiation is inhibited. This means that transverse waves are radiated from a thickness for which $W^t$ becomes of the same order as $W^l$. It is important, therefore, how this length L relates to the dimensions of the object and the length characterizing the nonuniformity of the plasma distribution, etc.

The plasma radiation mechanism is currently being increasingly applied to astrophysical investigations. In addition to the radio emission from the sun there are the radio emission from the Earth's radiation belts, the radio emission from remnants of supernova stars, and finally emission from quasars [6]. For this reason the characteristic length for nonlinear conversion of longitudinal into transverse waves has become an important problem.

Experimental investigations on beam–plasma interactions [7-12] have demonstrated that the development of beam instability is accompanied by intense plasma radiation at frequencies $\sim \omega_{0e}$, etc. Consequently, the study of longitudinal–transverse wave conversion plays a most important part in the interpretation of laboratory experiments.

If we analyze some of the results obtained relevant to the nonlinear interaction of transverse and longitudinal waves, we arrive at the following qualitative conclusions:

**1.** As mentioned earlier, it is typical of induced scattering that waves may grow outside the region in which they originally existed at a relatively low initial level. We have explored this situation for longitudinal waves. When applied to the conversion of Langmuir into transverse waves this statement implies that transverse waves can grow from a small "seeding" level. The nonlinear effects associated with induced scattering govern the conversion process only in the event of small spontaneous scattering. References [2, 5] investigate mainly the conversion of waves by scattering from thermal fluctuations, i.e., spontaneous scattering. Such fluctuations may be neglected if the intensity of the "seed" transverse waves, though small, are nevertheless significantly larger than the thermal fluctuation level of the transverse waves. Consequently, the "seed" radiation must be above thermal. Of course, weak superthermal radiation is often observed in the astrophysical environment as a consequence of the emission of superthermal particles (cosmic rays accelerated in a plasma) and from other dynamic processes producing nonthermal radiation. The "seed" radiation can also develop as the result of spontaneous scattering.

One important consideration is relevant at this point, however. A possible factor contributing to induced scattering processes lies in the spontaneous scattering terms in the scattering probability which are neglected. Strictly speaking, therefore, the minimum "seeding" intensity of transverse radiation such that spontaneous scattering may be safely neglected must be assessed in each specific case by comparing the effects of spontaneous and induced scattering. As an example let us make such a comparison for the conversion of Langmuir waves on ions, this being an effective conversion mechanism. Since the number of "seed" quanta* is $N^t = T_{eff}/\hbar\omega_0$, the condition we seek is

$$T_{\text{eff}} \gg T_i \left(\frac{v_\varphi}{v_{Te}}\right)^2 \quad \text{for} \quad \Omega^t - \Omega^l \simeq \frac{k^2 v_{Te}^2}{\omega_{0e}}, \tag{9.1}$$

and for an ultimately large difference $\Omega^t - \Omega^l$, such that $\omega_- \approx |k_-| v_{Ti}$,

$$T_{\text{eff}} > T_i v_\varphi / v_{Ti}. \tag{9.2}$$

From the actual approximate expressions obtained for the scattering cross sections we can derive convenient expressions to estimate the spontaneous conversion.†

**2.** When investigating the conversion of plasma waves into transverse waves by induced scattering it is impossible, in general, to ignore the longitudinal-wave transfer effects engendered by the nonlinear interaction of such waves. In fact, the characteristic conversion time depends very strongly on the phase velocities of

* Remember that the frequency of the generated transverse waves is near $\omega_{0e}$.

† The "seed" radiation mentioned above can be produced by spontaneous processes, provided there are no other radiation sources in the frequency range investigated. The induced conversion actually determines the optical thickness, i.e., the characteristic length within which the buildup (or absorption) of electromagnetic waves occurs. An important direct consequence of the above equations is that there is always a region of electromagnetic wave growth (amplification) in induced scattering.

the Langmuir waves. Consequently, the study of particular stationary Langmuir-wave spectra which occur as a result of transfer processes tending to increase $v_\varphi$, as well as the inverse processes, would be rather important for a more precise estimate of the characteristic conversion times. Processes tending to reduce $v_\varphi$ should occur in the presence of beams, low-frequency oscillations, etc.

If there is no basis for postulating the existence of such spectra, we must take account of the increase in $v_\varphi$ for longitudinal waves and the reduction of their conversion to transverse waves in the estimates (8.62) and (8.64)*.

3. To determine the conversion lengths corresponding to induced scattering we use the estimates obtained earlier for the conversion time $\tau$ and invoke the equation

$$L = c\tau. \tag{9.3}$$

We note that, in addition to induced scattering, there can be a significant contribution from the decay of Langmuir waves into transverse and ion-acoustic waves; this effect becomes significant for plasma waves with phase velocities less than $v_{\varphi \max} = v_{Te}\sqrt{9m_i/4m_e}$ and for $\Delta k^l \simeq k^l$ is approximately described by:

$$\frac{1}{\tau} \simeq \frac{W^l}{nm_e v_{Te}^2}\left(\frac{4m_e}{9m_i}\right)^{1/2}\frac{v_\varphi}{v_{Te}}\,\omega_{0e}, \tag{9.4}$$

This effect can even exceed induced scattering by ions if $v_\varphi$ is of the order of $v_{\varphi \max}$. However, it decreases for $v_\varphi > v_{\varphi \max}$. The characteristic lengths for these processes described by (9.3) can be very short when $W^l$ is large. Efficient conversion therefore requires the plasma density to vary considerably within the length L and the radiated transverse waves not to be subsequently absorbed by the inverse processes when their frequency becomes greater than $\omega_{0e}$.†

---

* New information on the spectra of longmuir waves may be found in [23-26].

† Effects involving the conversion of plasma into electromagnetic waves have been investigated in [22], where a numerical solution of the nonlinear integral equations is used to find the spectra of plasma and electromagnetic waves stimulated by fast particle beams.

Finally, when low-frequency ion-acoustic and Langmuir waves are both present, the radiation intensity of transverse waves produced by coalescence with frequencies near $\omega_{0e}$ is proportional to the product of the intensities of the Langmuir ($W^l$) and ion-acoustic ($W^s$) waves (Chap. V). In the merging of two Langmuir waves transverse radiation is generated at frequency $2\omega_{0e}$ with intensity proportional to the square of the Langmuir-wave intensity (Chap. V).

## 2. STABILIZATION OF BEAM INSTABILITIES BY INDUCED SCATTERING OF LANGMUIR WAVES INTO LANGMUIR WAVES

The efficiency of interaction of charged particle beams with a plasma is a problem in urgent need of investigation. On the one hand, there is experimental evidence that beams of particles can under certain conditions give up a large part of their energy [8, 12]. On the other hand, there is doubt whether the interaction of beams with plasma is always efficient. The aim of nonlinear theory should be to indicate those plasma and beam parameters for which efficient collisionless energy transfer occurs. This problem, naturally, is important not only for laboratory experiments, but also for astrophysical applications. For example, it is well known that chromospheric flares are accompanied by the emission of particle streams from the sun, which clearly give rise to plasma waves. Also, many of the eruptive processes so common in outer space often lead to beam instabilities. Finally, the heating of the interstellar gas by cosmic rays represents an important problem [13].

As we have seen, the quasilinear approximation predicts a very efficient beam–plasma interaction (about 50 to 30% energy transfer). We have already shown, however, that decay processes, for example, can under certain conditions stabilize a beam instability. We recall that beam instability is attended by the generation of oscillations in a definite phase velocity interval $\Delta v_\varphi$ roughly corresponding to the beam particle velocities. The transfer of oscillations from the interval $\Delta v_\varphi$ to one in which they are not generated could stabilize beam instability. This kind of transfer is effected both by decay, discussed above, and by induced scattering.

We can show, for example, that in the induced scattering of Langmuir waves by ion waves can transfer rapidly out of the generation interval, causing the energy fed from the beam into plasma oscillations to be only a small fraction of its initial energy [14], i.e., resulting in low-efficiency interaction of the beam with the plasma.

For the moment we outline the qualitative side of the picture, referring to [14] for the pertinent details. To be specific, we shall assume that

$$T_i > T_e \frac{m_i}{m_e} \left(\frac{v_{Te}}{v_1}\right)^2, \tag{9.5}$$

where $v_1$ is the mean initial beam velocity. The plasma waves generated by the beam have phase velocities of order $v_1$. With (9.5) satisfied, the induced scattering of the Langmuir waves generated is determined by the ions [see (8.23)]. According to the estimate (8.24) the transfer of oscillation energy from the generation interval by induced scattering is

$$\frac{dW^l}{dt} = -\frac{W^l W_1^l}{n_0 m_e v_{Ti} v_1} \frac{T_e/T_i}{(1 + T_e/T_i)^2} \omega_{0e}, \tag{9.6}$$

where $W^l$ and $W_1^l$ represent the oscillation energy inside and outside the interval $\Delta v$.* The generation of waves by the beam is described by the quasilinear growth rate

$$\frac{dW^l}{dt} = \gamma W^l; \quad \gamma \approx \frac{n_1}{n_0} \omega_{0e} \left(\frac{v_1}{\Delta v}\right)^2. \tag{9.7}$$

Wave growth in the interval $\Delta v$ occurs only if the generation (9.7) exceeds the energy outflow (9.6), i.e., if

$$W_1^l < n_1 m_e v_{Ti} v_1 \left(\frac{v_1}{\Delta v}\right)^2 \frac{T_i}{T_e} \left(1 + \frac{T_e}{T_i}\right)^2. \tag{9.8}$$

The energy $W_1^l$ was produced by transfer from the interval of $W^l$, in other words, the energy transmitted by the beam into the plasma.

---

* We have assumed here that only scattering in the direction of the beam is significant (one-dimensional model). This means that $\Delta k / k v_\varphi$ in (8.24) must be replaced by $1/v_1 + 1/v_\varphi$ (on the order of $1/v_1$), because $v_\varphi > v_1$. For a multidimensional model, on the other hand, the most efficient transfer occurs for $\Delta k \sim k$, i.e., the estimate applies equally to this case.

Let us define the efficiency of interaction of a beam with a plasma as the ratio of the energy transmitted by the beam into oscillation modes to the initial beam energy. For an electron beam

$$\eta = \frac{W_1^l}{n_1 m_e v_1^2}. \tag{9.9}$$

From (9.8) we obtain

$$\eta = \eta_0 = \frac{v_{Ti}}{v_1}\left(\frac{v_1}{\Delta v}\right)^2 \frac{T_i}{T_e}\left(1 + \frac{T_e}{T_i}\right)^2. \tag{9.10}$$

Notice that condition (9.5) is also satisfied for $T_i < T_e$ for beams having large velocities and small electron temperatures. Since $v_{Ti} \ll v_1$, even for $v_0/\Delta v \sim 10^2$ $[v_1/\Delta v > (n/n_1)^{1/3}]$ we can still have $\eta \ll 1$.

The stabilization condition can be deduced directly from (9.8) by considering the fact that the energy generated by a beam of width $\Delta v$ is $W^l \simeq n m_e v_1 \Delta v$. This condition takes the form

$$\eta_0 \ll \frac{\Delta v}{v_1}, \tag{9.11}$$

which is always much smaller than unity for $\Delta v/v_1 \ll 1$.

Notice that condition (9.5) is not always necessary for stabilization by ion scattering. Violation of (9.5) does not forbid scattering by ions, although this implies a change in direction of the Langmuir wave. In the one-dimensional case, in particular, the Langmuir wave reverses its direction. Estimates show that stabilization is also possible in this case, but the stability criterion is somewhat more stringent [14]:

$$\frac{27}{4}\sqrt{\frac{\pi}{2}\frac{m_i}{m_e}}\left(\frac{v_{Te}}{v_1}\right)^3\left(\frac{v_{Te}}{\Delta v}\right) \ll \left(1 + \frac{T_e}{T_i}\right)^{-2}. \tag{9.12}$$

The conditions for stabilizing ion beams are considerably more lax than for electron beams. For instance, with condition (9.5) the efficiency $\eta_0$ for an ion beam is smaller than for an electron beam in the ratio $m_i/m_e$. An approximate graph showing the variation of the wave intensity in the generation region is given in Fig. 32a [14]. The reason that $W^l$ becomes zero is that after the transfer of a sufficient fraction of the wave energy into the off-

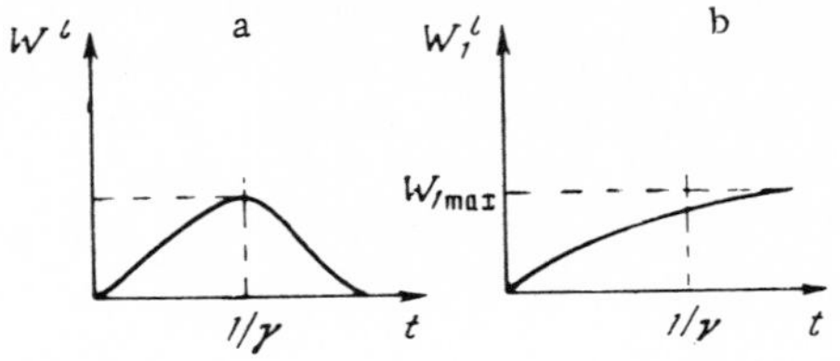


Fig. 32. Nonlinear stabilization of beam instabilities: a) Oscillation energy versus time in the beam; b) oscillation energy versus time outside the beam; $W_{max}$ is much smaller than the beam energy.

resonance interval [the opposite inequality to (9.8) being fulfilled] any newly generated wave is induced into the off-resonance interval faster than the beam particles can undergo a change of state. In the resonance interval, therefore, the waves are damped. Off-resonance waves stop growing, because they originate from waves in the resonance interval, which are no longer generated (Fig. 32b) Consequently, after becoming slightly diffuse, the beam eventually stops spreading.

It is also clear that if plasma oscillations complying with the opposite inequality to (9.8) were initially present in the off-resonance interval, the beam is stable from the very beginning. This situation is possible if the plasma has been inadequately prepared for the beam experiments, i.e., if oscillations were present in it prior to the injection of the beam.

The stabilization of beam instabilities offers promising opportunities for the acceleration of particles in a plasma [14, 15].*

## 3. ACCELERATION OF ELECTRONS DURING THE DEVELOPMENT OF BEAM INSTABILITY

During the transmission of electron beams through plasma one often observes electrons with velocities greater than those of the injected beam electrons [12]. The same effect is encountered

* Instability stabilization effects have been used in [22] to interprete the radio emission from the sun caused by beams.

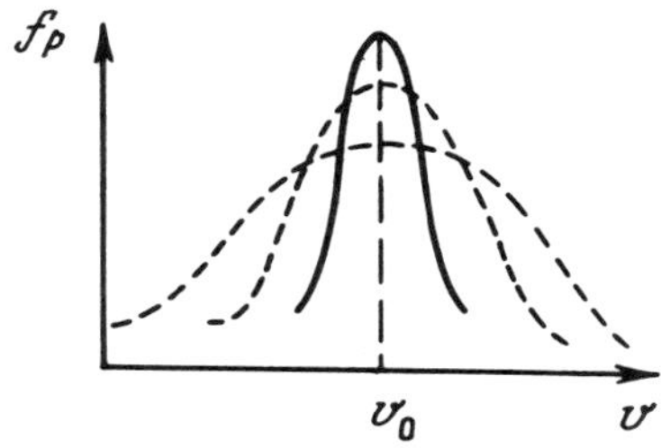


Fig. 33. Change of the beam particle distribution function due to interaction with oscillations, allowing for induced scattering effects.

in ion injection into magnetic traps, when the ions develop cyclotron instability. We suggest the following interpretation of these effects.

Let us suppose that $\eta_0$ is of order $\Delta v/v$, i.e., that the time for the oscillations to build up sizable phase velocities is of the same order as their generation time. Langmuir oscillations are generated in the interval where $\partial f/\partial v > 0$ (Fig. 33). According to quasilinear theory, particle diffusion occurs only in this interval, causing a local reduction of the derivative and the occurrence of particles whose energies are smaller than the initial energy of the beam particles. Under conditions such that $\eta_0 \sim \Delta v/v$ the Langmuir oscillations succeed during the generation time in filling the entire interval $\Delta v_\varphi = \Delta v$, i.e., they are ultimately found in the interval $\partial f/\partial v < 0$. The diffusion of beam particles through oscillations now leads to a reduction in $\partial f/\partial v$, i.e., to the appearance of accelerated particles.

It is important to note that this is not the only possible effect in which induced scattering promotes particle acceleration. As demonstrated above, induced scattering also leads to isotropization of the generated oscillations, while interactions with isotropic oscillations are always accompanied by an increase in the mean particle energy. In essence this effect is similar to the one-dimensional increase of the phase velocities. In fact, the increase in the phase velocities of the waves in the one-dimensional case enables them to interact with particles whose velocities are greater than $v_1$. The same effect is produced by isotropization, since a wave moving at an angle with respect to a particle can interact with the latter if its phase velocity is smaller than the particle velocity ($\omega = kv$, i.e., $v_\varphi = v \cos \theta$).

Here, however, there is a significant difference, namely, in that the condition $\eta_0 \sim \Delta v/v$ is by no means necessary. All that matters is that the isotropization time be comparable with the total quasilinear relaxation time. In this case waves having phase velocities much smaller than $v_1$ and occurring in the last stages of the development of beam instability can become isotropic. In general, however, waves having low phase velocities become isotropic only through electron scattering, i.e., a less efficient process than scattering by ions. It should be possible, therefore, to investigate waves whose phase velocities already satisfy the condition for ion scattering.

Suppose that nonlinear effects do not prevent beam instability. Then the condition that isotropization can occur during the quasilinear stage is

$$v_1 > v_{Ti} \frac{T_i}{T_e}\left(1 + \frac{T_e}{T_i}\right)^2. \tag{9.13}$$

External magnetic fields, which could influence isotropization appreciably, have not been included in this estimate, which is derived directly from (9.6). If condition (9.13) is satisfied, then the characteristic time for the beam particles to increase $\Delta v$ to the order of $v_1$ is of the same order as the quasilinear relaxation time.

## 4. ON THE STABILIZATION OF BEAM INSTABILITIES BY INDUCED SCATTERING OF LANGMUIR INTO ION-ACOUSTIC WAVES

In a nonisothermal plasma with a large ratio $T_e/T_i$ the decay of Langmuir waves by the excitation of ion oscillations becomes appreciable. This effect is described by the induced scattering of Langmuir oscillations with conversion into ion-acoustic waves. It can occur only if $k^s \geq 1/\lambda_{De}$. On the other hand, if the Landau damping of ion oscillations by ions is to be small, we must have $k^s < 3/\lambda_{Di}$, i.e., $\lambda_{Di} < 3\lambda_{De}$ or $T_i < 10T_e$. Strictly speaking, in order for the ion oscillation branch to be well defined we require $T_i \ll 10T_e$. Let us here confine our discussion to the one-dimensional case. It follows from (8.47) that nonlinear conversion can in this case* generate only ion-acoustic waves propagating in the

* This assertion, like the interaction relation (8.47), is valid for $v_1/v_{Te} \lessdot \sqrt{9m_i/m_e}$.

direction of the particle beam. The characteristic generation time can be estimated by allowing for the fact that the fastest growth occurs for smallest $k^s$, whose values in this instance are of order $1/\lambda_{De}$:

$$\gamma = \omega_{0e} \frac{W^l}{n_0 m_e v_{Te}^2} \frac{v_{Te}}{v_1} \sqrt{\frac{m_e}{m_i}}. \tag{9.14}$$

The Landau damping is of order $\gamma \sim \omega_{0i}\sqrt{m_e/m_i}$, i.e., the growth of the ion oscillations requires

$$W^l > n_0 m_e v_{Te}^2 \frac{v_1}{v_{Te}} \sqrt{\frac{m_e}{m_i}}. \tag{9.15}$$

On the other hand, for electron scattering not to prevent instability, we must have

$$\frac{n_1}{n_0} \left(\frac{v_1}{\Delta v}\right)^2 \omega_{0e} > \omega_{0e} \frac{W^l}{n_0 m_e v_{Te}^2} \frac{v_{Te}}{v_1},$$

or

$$W^l < n_1 m_e v_{Te}^2 \left(\frac{v_1}{\Delta v}\right)^2 \frac{v_1}{v_{Te}}. \tag{9.16}$$

The inequalities (9.15) and (9.16) are compatible only if

$$\frac{n_1}{n_0} > \left(\frac{\Delta v}{v_1}\right)^2 \sqrt{\frac{m_e}{m_i}}. \tag{9.17}$$

Recognizing that the energy $W^l$ has been produced by the beam, i.e., is of order $n_1 m_e \Delta v v_1$, we obtain from (9.15)

$$\frac{\Delta v}{v_1} > \frac{n_0}{n_1} \frac{v_{Te}}{v_1} \sqrt{\frac{m_e}{m_i}}.$$

The last inequality is compatible with (9.17) if

$$\frac{n_1}{n_0} \gtrsim \left(\frac{v_{Te}}{v_1}\right)^{2/3} \left(\frac{m_e}{m_i}\right)^{1/2}. \tag{9.18}$$

If $n_1/n_0$ is of the same order as (9.18), then $\Delta v/v > (v_{Te}/v_1)^{1/3}$. On the other hand, $\Delta v \ll v_1$, otherwise the beam will lose a sizable portion of its energy, i.e., $(v_1/v_{Te})^{1/3} \gg 1$, whereas from (9.15) $v_1/v_{Te} \ll (m_i/m_e)^{1/2}$, since $W^l \ll n_0 m_e v_{Te}^2$. Consequently, $n_1/n_0$ must be at least ten times the right-hand side of (9.18).

These estimates show that although the stabilization of beam instability by $ls$-scattering is still possible, it nevertheless requires the fulfillment of several severe demands on the parameters of the plasma and beam. We also note that in our estimates of $ll$-scattering we ignored ion scattering, which is admissible only for very low ion temperatures:

$$(T_e / T_i)^2 \gg \left(\frac{v_1}{v_{Te}}\right)^3 \left(\frac{v_1}{\Delta v}\right)^2 \frac{m_e}{m_i} \frac{4}{27} \sqrt{\frac{2}{\pi}}. \tag{9.19}$$

This condition is inferred from a comparison of scattering by electrons and ions allowing for the fact that the most probable ion scattering event for small $T_i/T_e$ is one in which the Langmuir wave changes its direction of propagation. This turns out to be a differential interaction and may be estimated from the relation

$$\frac{\partial N_k^l}{\partial t} = N_k^l \frac{4}{27} \frac{1}{(2\pi)^{5/2}} \frac{1}{n_0 m_i} \frac{\omega_{0e}^4}{v_{Te}^4} \left(1 + \frac{T_e}{T_i}\right)^{-2} \frac{\partial N_k^l}{\partial k} \hbar. \tag{9.20}$$

Assuming $\Delta k/k \approx \Delta v/v_1$ and $k = \omega_0/v_1$, we obtain the estimate (9.19). Substituting for $\Delta v$ from (9.17), we obtain

$$\left(\frac{T_i}{T_e}\right)^2 \ll \frac{n_1}{n_0} \left(\sqrt{\frac{m_i}{m_e}} \frac{v_{Te}}{v_1}\right)^3.$$

If this is not satisfied, ion scattering becomes important. Notice that $ls$-scattering is effective for low beam velocities, whereas the same is true of $ll$-scattering for high beam velocities. For $v_1$ of the same order as $v_{Te}$, however, the inequality (9.15) is no longer easily satisfied, as $W^l \ll n_1 m_e v_{Te}^2$.

The growth of ion-acoustic waves in $ls$-scattering can be efficient under conditions such that the Landau damping of s-waves is suppressed by the quasilinear effect. This is possible if intense ion oscillations are first excited in the plasma. An analogous situation was examined earlier for the interaction of electron beams with a plasma in which Langmuir waves had been excited. If their energy is greater than the beam energy, then, as demonstrated there, the generation of waves by the beam is not particularly important. Here we find a similar problem involving instead ion oscillations. Owing to nonlinear effects they are coupled with Langmuir oscillations. Let us estimate what the intensity of the ion-acoustic oscillations must be for the beam to be stable with respect to the generation of Langmuir waves. For $k^s \sim 1/\lambda_{De}$ it follows from (8.46) that

$$\gamma \sim \frac{W^s}{n_0 m_e v_{Te}^2}\,\omega_{0e} > \frac{n_1}{n_0}\,\omega_{0e}\left(\frac{v_1}{\Delta v}\right)^2,$$

i.e., the stabilization criterion is

$$W^s > n_1 m_e v_1^2 \left(\frac{v_{Te}}{\Delta v}\right)^2. \tag{9.21}$$

The stabilization mechanism in this case is not nonlinear transfer, but corresponds instead to nonlinear absorption. In some cases when the initial $W^s$ is small and nonlinear stabilization by *ll*-scattering is prohibited, the beam relaxes quasilinearly, losing considerable energy, and only afterwards are the Langmuir waves absorbed by the mechanism discussed here.

## 5. STABILITY OF RELATIVISTIC BEAMS WITH RESPECT TO LANGMUIR-WAVE GENERATION

We have investigated three mechanisms for the nonlinear stabilization of beam instability. Interestingly enough, the stability criteria, especially for *ll*-scattering, are easily met for relativistic beams, i.e., these beams can be stable in a plasma. The estimates given above for nonrelativistic beams must be made a little more precise for the present case. Relativistic beams generate waves with a maximum growth rate for those whose phase velocity is close to the velocity of light. Such waves, however, owing to induced scattering, are rapidly transferred to an interval in which the phase velocities are greater than that of light. While the velocity of light has no special significance with regard to phase velocities, it has for particles, because the latter cannot have velocities greater than that of light. Consequently, the waves are transferred into a range of phase velocities where they cannot make resonant interactions with the beam particles. In place of the criterion (9.5) we have the milder condition

$$v_{Ti} > \frac{3}{2}\frac{v_{Te}^2}{c}\left(1 - \frac{v_\varphi^2}{c^2}\right),$$

and the fraction of the energy lost by a relativistic electron beam is

$$\eta_e \simeq \frac{v_{Ti}}{c}\frac{T_i}{T_e}\left(1+\frac{T_e}{T_i}\right)^2\left(\frac{c}{\Delta v}\right)^2. \tag{9.22}$$

The stability conditions are especially easily satisfied for relativistic ion beams [16]:

$$\eta_i = \left(\frac{m_e}{m_i}\right)^2 \eta_e. \tag{9.23}$$

Such beams can also generate waves having nonrelativistic phase velocities. The direction of generation is almost perpendicular to the beam. For this reason the growth rate for the generation of nonrelativistic waves is much smaller than that for the generation of relativistic waves. Nonrelativistic waves are transferred more slowly into the range $v_\varphi > c$, but the growth rate is reduced more than the rate of transfer. Consequently, the losses of relativistic beams through the generation of nonrelativistic waves turns out to be even smaller than (9.22).

We should mention the instability associated with anisotropy of the distribution function. For beams with large velocities and a definite minimum velocity spread, this instability is unimportant. The necessary criterion for relativistic beams is

$$\frac{\Delta v_\perp}{c} > \left(\frac{n_1}{n_0}\right)^{1/2}. \tag{9.24}$$

Finally, if the relativistic beam is closely monoenergetic, the Langmuir waves they generate have a rather narrow spread $\Delta\omega$, and the nonlinear hydrodynamic instabilities of the Langmuir waves should compete with the $ll$-scattering. However, since the energy of the $l$-waves produced by $ll$-scattering is small, the required criterion need not be too strict.

The concept that relativistic beams may be stable, as we have shown, can have important implications.

**1.** An interesting problem concerns the interaction of cosmic rays with interstellar plasma. Simple estimates show that the linear growth rates for the beam instability are very great. They result in a short (on the cosmic scale) interaction time between cosmic rays and the plasma [17]. Nevertheless, cosmic rays are ion beams which satisfy the stabilization conditions with considerable room to spare. Therefore, the isotropization of cosmic rays through beam instability is often impossible. The most probable process is their isotropization through quasilinear particle scattering by the plasma oscillations, as analyzed above.

**2.** The stability of relativistic beams can be important for the acceleration of particles in a plasma, the creation of plasma configurations involving relativistic particles, etc.

**3.** Major importance lies in the problem of particle interaction in intersecting relativistic beams, in which the relative energy of the colliding particles is greatly augmented. The investigation of plasma instabilities should be highly relevant to this problem.

## 6. INTERACTION EFFICIENCY OF NONRELATIVISTIC BEAMS WITH PLASMA

Low-velocity beams can interact with plasma very efficiently. For instance, if the beam velocity is much smaller than $v_{Te} \cdot (3m_i/m_e)^{1/5}$ and ion scattering is weak, the nonlinear interaction is described by (8.1). In this case wave transfer can occur only after the full development of the quasilinear stage. In fact, substituting for $W^l$ a quantity of order $n_1 m_e v_1^2$ in (8.1) for the quasilinear stage, we obtain

$$\frac{1}{\tau} = \omega_{0e} \frac{n_1}{n_0} \frac{v_{Te}}{v_1}, \tag{9.25}$$

which is much less than the quasilinear growth rate $\gamma \sim \omega_{0e} n_1/n_0$ $(v_1 \gg v_{Te})$.

If the beam velocity is less than $v_{Te}$, its quasilinear relaxation in the plasma is associated with the generation of ion-acoustic oscillations, and, as one might easily verify, nonlinear wave interaction does not occur until after the quasilinear relaxation. In this stage, which we could call the nonlinear stage, spectral transfer becomes important. For example, it can alter the oscillation spectra created in the quasilinear stage, resulting in wave absorption, etc.

## 7. QUASISTATIONARY SPECTRA OF ACOUSTIC OSCILLATIONS EXCITED BY AN ELECTRON CURRENT

Here we describe qualitatively the results of [21]. Consider an electron current in a partially ionized plasma containing both ions and neutral gas atoms. Let the electron drift velocity be greater than the velocity of sound. The generation of sound reduces $\partial f_e/\partial v$ in the electron distribution until the distribution function acquires a plateau or, more accurately, a near plateau, since the collisions of electrons with neutral atoms, while tending to restore a Maxwellian distribution, increase $\partial f_e/\partial v$, whereas quasilinear effects tend to reduce it. The outcome is a small value of $\partial f_e/\partial v$, where the instability is not completely eliminated, because the collisional damping of the sound waves is determined by the collision frequency of the ions with neutral gas atoms, which is much smaller than that between the electrons and neutral particles. The growth rate (without collisions) is approximately given by (Chap. VI):

$$\gamma \simeq \frac{m_i}{2} k v_s \left( v_s^2 \frac{\partial f_e}{\partial v} \frac{1}{n} \right); \quad v_s = v_{Te} \sqrt{\frac{m_e}{m_i}}. \tag{9.26}$$

The derivative $\partial f_e/\partial v$ can be estimated by equating the electron–neutral collision rate to the rate of diffusion by the acoustic oscillations:

$$\partial f / \partial v \simeq \frac{\nu_e v_1 f}{D}; \quad D \simeq \frac{e^2}{m_e^2} \frac{W^s v_s \langle k \rangle}{\omega_{0i}^2} 4\pi^2,$$

where $\nu_e$ is the characteristic electron–neutral collision frequency, D is the diffusion coefficient for the sound waves, $\langle k \rangle$ is some average value of the acoustic wave number, and $v_1$ is the electron drift velocity.

If the wave distribution is such that the greatest contribution to $\int k\omega^s N_k \, dk$ is for k of the order $k_{max} \sim 1/\lambda_{De}$, while $W^s$ is of the order of the electron drift energy $n m_e v_s^2$ and $v_1 \sim v_s$, then

$$D \simeq \sqrt{\frac{m_e}{m_i}} v_{Te}^2 \omega_{0e} \;; \quad \frac{\partial f}{\partial v} \sim \frac{\nu_e}{\omega_{0e}} \frac{n}{v_{Te}^2}, \quad \text{i.e.,} \quad \gamma \simeq \frac{k v_s}{\omega_{0e}} \nu_e.$$

A very important feature is the decrease in the growth rate with decreasing k and the fact that for small k it becomes of the same order as the collision frequency:

$$\nu_i \simeq n v_{Ti} \sigma = \frac{v_{Ti}}{v_{Te}} \nu_e \simeq \sqrt{\frac{T_i}{T_e}} \sqrt{\frac{m_e}{m_i}} \nu_e .$$

For large $k \sim 1/\lambda_{De}$ we have $\gamma \sim \nu_e \sqrt{m_e/m_i} \gg \nu_i$, i.e., despite the quasilinear reduction of $\partial f/\partial v$, the growth rate is considerably larger than the collision frequency, and the oscillations continue to grow. For $k \sim \sqrt{\frac{T_i}{T_e}} \frac{1}{\lambda_{De}}$ the waves begin to decay.

Let us examine the nonlinear effects. The induced scattering of sound waves by ions, as already shown, leads to energy transfer into the region of small k, i.e., that of absorption. Consequently, induced scattering represents a mechanism resulting in the absorption of waves generated in the interval of unstable k. Now it is possible for a certain stationary or, more precisely, quasistationary oscillation spectrum to be created, which for each k corresponds to a balance between the inflow of oscillations due to instability from waves with large k and the outflow associated with the transfer of waves to small wave numbers. Taking account of nonlinear interaction, we find that the equation for the variation in the number of ion-acoustic waves becomes

$$\frac{\partial N^s_{k\Omega}}{\partial t} = \gamma N^s_{k\Omega} + N^s_{k\Omega} k^2 \frac{\partial}{\partial k} \int \frac{T_i \sin^2 2\theta}{T_e 16\pi^2 m_i n} N^s_{k\Omega_1} k^4 \, d\Omega_1, \qquad (9.27)$$

where $\gamma$ is the total growth rate (9.26) minus $\nu_i$.

Following the approach of Kadomtsev and Petviashvili [21], we can roughly estimate the shape of the spectrum, assuming small angles $\theta$ between the direction of the s-waves and the electron beam and averaging (9.27) over direction:

$$\frac{\partial \overline{N}^s_{\mathbf{k}}}{\partial t} = \beta k \overline{N}^s_{\mathbf{k}} + \overline{N}^s_{\mathbf{k}} k^2 \frac{\partial}{\partial k} k^4 \overline{N}^s_{\mathbf{k}} \alpha;$$

$$\alpha = \frac{T_i \theta_0^2}{4\pi^2 T_e n m_i}; \; \beta = \frac{\gamma}{k} = \frac{\nu}{\omega_{0e}} \frac{v_1}{v_s} \frac{m_e}{W^s} v_s^3 n; \; \gamma \gg \nu_i, \qquad (9.28)$$

where $\theta_0$ is a characteristic average angle. Equating $\frac{\partial N^s_{\mathbf{k}}}{\partial t}$, to zero, we have

$$\beta = -k\frac{\partial}{\partial k}k^4 N^s_{\mathbf{k}}\alpha,$$

i.e.,

$$N^s_{\mathbf{k}} = \frac{\beta}{\alpha}\frac{1}{k^4}\ln\frac{k_0}{k}, \tag{9.29}$$

where $k_0$ is an integration constant which depends on the boundary conditions for k-space.

If we assume that $N^s_{\mathbf{k}} = 0$ for $k = 1/\lambda_{De}$, then $k_0 = 1/\lambda_{De}$. The energy density of the oscillations, $\sim \hbar\omega N_{\mathbf{k}} = \hbar k v_s N_{\mathbf{k}}$, is proportional to $k^{-3}$ as a crude approximation, a result that has been qualitatively confirmed experimentally. The oscillation energy is concentrated in the region of small k through the transfer of waves there. Analysis shows that the oscillation intensity is also finite in the absorption interval over a "distance" $\Delta k$ of order $k_{min}$ from $k = k_{min}$.

It is a simple matter to estimate the oscillation energy by integrating (9.29) from $k_{min}$ to $k_{max}$. Since the dependence on $k_{min}$ and $k_{max}$ is logarithmic, by assuming the logarithm is of order unity and taking account of the dependence of $\beta$ on $W^s$, we obtain

$$(W^s)^2 \simeq \frac{\nu_i}{\omega_{0e}}\frac{T_e}{T_i}\frac{m_e}{m_i}(nm_e v_{Te})^2\frac{v_1}{v_s}\frac{1}{\theta_0^2}. \tag{9.30}$$

This value of $W^s$ is normally smaller than $nT_e$, i.e., the weak turbulence approximation is valid, but it can still be much larger than the quasilinear value of $W^s$:

$$\frac{W^s}{nm_e v_s^2} \simeq \sqrt{\frac{\nu_i}{\omega_{0e}}\frac{T_e}{T_i}\frac{m_i}{m_e}\frac{1}{\theta_0^2}}.$$

Notice that averaging over all directions can provide only a crude estimate of the effect. A more detailed analysis, carried out in [20], has shown that characteristic periodic contraction–expansion effects occur in the angular distribution of the oscillations.

Strictly speaking, we should also include the excitation of ion oscillations.

## 8. INDUCED SCATTERING EFFECTS AND PLASMA HEATING

It is most important to appreciate the fact that in induced scattering processes, unlike decay processes, the plasma particles play an important part. They always gain or lose some of the energy. This is well illustrated, for example, by the absorption of Langmuir waves with conversion into ion-acoustic modes in a Maxwellian plasma. In this process the Langmuir-wave energy $\hbar\omega_0$ greatly exceeds the ion-acoustic energy, i.e., most of the energy is acquired by the plasma particles (in this case electrons), which are thus heated. In the earlier example of the generation of Langmuir waves from ion-acoustic waves the energy of the Langmuir waves was drawn from the particle beam. In the excitation of acoustic oscillations by an electron current the oscillation energy was also greatly reduced by scattering, because it depended strongly on k, the scattering ions extracting energy. Finally, the transformation of oscillations from the generation interval to the absorption interval is another very effective plasma heating mechanism.

The heating of plasma by transfer into the absorption region needs a more detailed examination, particularly for magnetoactive plasma. In plasma in an external magnetic field new absorption bands appear, for example, near the ion and electron cyclotron frequencies. The transfer of oscillations into these bands might be a mechanism for cyclotron heating of plasma [18].

There are obvious difficulties attending cyclotron heating, connected, for example, with the poor penetration of electromagnetic waves into the plasma in a region of strong cyclotron absorption. If they are absorbed in the surface layer, they can heat only a relatively small volume. However, there are plasma transmission bands in which the absorption is small and in which various instabilities can easily excite oscillations. It should be possible, therefore, to generate waves in a transmission band to get efficient heating (throughout the entire volume). For a sufficiently large oscillation intensity nonlinear induced scattering can transfer the oscillations into a cyclotron absorption band.

## 9. EFFECTIVE COLLISION FREQUENCY FOR PARTICLES WITH WAVES IN INDUCED SCATTERING

For rough estimates of some of the effects associated with induced scattering it is convenient to introduce the idea of an effective collision frequency between particles and waves:

$$\nu_{\text{eff}} = \frac{D^t}{p^2};$$

$$D = \int \left((\mathbf{k} - \mathbf{k}_1)^2 - \frac{(\omega - \omega_1)^2}{v^2}\right) w_{\mathbf{p}}(\mathbf{k}\mathbf{k}_1)\, N^l_{\mathbf{k}_1} N^l_{\mathbf{k}_2} \frac{d\mathbf{k}_1\, d\mathbf{k}_2}{(2\pi)^6}.$$

Induced scattering leads to isotropization of the particle distribution within those velocity ranges for which they are most effective. These velocities are often much smaller than the mean thermal velocities of the particles.

## 10. ION ACCELERATION BY INDUCED WAVE SCATTERING

Let us consider the small group of ions whose velocities are greater than the mean ion thermal velocity but less than the mean electron thermal velocity. If $T_e \gg T_i$, this group can be accelerated by ion-acoustic oscillations. For $T_e \sim T_i$ ($T_e \lessapprox T_i$) ion-acoustic oscillations are impossible, leaving only interaction with Langmuir waves. The acceleration of ions in this case comes from the induced scattering of Langmuir waves. As an example let us consider isotropically distributed ions. The diffusion coefficient $D^l$, which governs the energy increase, is described by the relation

$$D^l = \frac{1}{v^2} \int \frac{N^l_{\mathbf{k}_1} N^l_{\mathbf{k}_2}}{(2\pi)^3}\, d\mathbf{k}_1\, d\mathbf{k}_2\, (\Omega^l(\mathbf{k}_1) - \Omega^l(\mathbf{k}_2))^2 \times$$

$$\times \frac{e^4 (\mathbf{k}_1\mathbf{k}_2)^2\, \delta(\Omega^l(\mathbf{k}_1) - \Omega^l(\mathbf{k}_2) - (\mathbf{k}_1 - \mathbf{k}_2)\mathbf{v})}{m_e^2 \omega_{0e}^2 k_1^2 k_2^2}. \tag{9.31}$$

The general formula expressing the rate of increase of the average particle energy $D^l$ was derived in Chap. VI:

$$\frac{\partial}{\partial t}\langle\varepsilon\rangle = \left\langle \frac{1}{p^2}\frac{\partial}{\partial p}p^2 vD \right\rangle. \tag{9.32}$$

Analysis shows that the most efficient acceleration occurs for

$$k \gg k_0 = \frac{\omega_{0e}}{3v_{Te}^2}v, \tag{9.33}$$

in which case [19]

$$\frac{\partial}{\partial t}\langle\varepsilon\rangle = \frac{62}{45}\frac{\omega_{0e}e^2}{m_i}\int\frac{(N_k^l)^2 k^5\, dk\hbar^2}{(2\pi)^3 nm_e v_{Te}^2}. \tag{9.34}$$

If the opposite inequality to (9.33) is satisfied the rate of acceleration falls off rapidly with increasing particle velocity:

$$\frac{\partial}{\partial t}\langle\varepsilon\rangle = \frac{189\pi\hbar^2}{4(2\pi)^3 m_i}\left(\frac{v_{Te}}{v}\right)^5\frac{v_{Te}^3 e^2}{\omega_{0e}^4 m_e n}\int_0^\infty N_{k_2}^l\frac{dk_2}{k_2}\int_0^{k_2}(k_1^2-k_2^2)^4 k_1^2 N_{k_1}^l dk_1. \tag{9.35}$$

However, the increase in particle energy caused by an increase in energy spread (fluctuation acceleration) is of the same order as (9.34) under these conditions. In the presence of an external magnetic field induced scattering can cause ion acceleration predominantly in a direction perpendicular to the magnetic field. The acceleration of ions with the development of beam instability has been observed, for example, by Nezlin and Solontsev [10]. The scattering of Langmuir waves by ions could also lead to the stabilization of beam instability and to beam "cutoff."

## 11. TRANSFER OF LANGMUIR WAVES IN INDUCED SCATTERING AND THE ACCELERATION OF HIGH-ENERGY PARTICLES

As shown in Chap. VI, the statistical acceleration of relativistic particles (cosmic rays in particular) is more effective the higher the phase velocity of the Langmuir waves causing their acceleration ($\sim v_\varphi^3$). The induced scattering of Langmuir waves excited in a cosmic plasma by various instabilities greatly enhances the acceleration efficiency. There is considerable interest in investigating possible quasistationary spectra of Langmuir waves caused by a balance between generation and transfer or between transfers in opposite directions [23-26].

## Literature Cited

1. I. S. Shklovskii, Cosmic Radio Waves. Harvard (1960).
2. V. L. Ginzburg and V. V. Zheleznyakov, "Mechanisms of the sporadic radio emission of the sun," Izv. Vuzov, Radiofizika, 1:9 (1958).
3. V. L. Ginzburg and V. V. Zheleznyakov, "Possible mechanisms of the sporadic radio emission of the sun (radiation in an isotropic plasma)," Astron. Zh., 35:694 (1958).
4. D. A. Tidman, "Radio emission by plasma oscillations in nonuniform plasmas," Phys. Rev., 117:366 (1960).
5. I. A. Akhiezer, A. G. Sitenko, and I. G. Prokhoda, "Theory of fluctuations in a plasma," Zh. Éksp. Teor. Fiz., 33:750 (1957).
6. V. L. Ginzburg and L. N. Ozernoi, "Role of coherent plasma radio emission for quasars and the remnants of supernova stars," Izv. Vuzov, Radiofizika, 9:221 (1966).
7. Ya. B. Fainberg, A. K. Berezin, G. N. Berezina, and L. I. Bolotin, "Interaction of pulsed high-current beams with a plasma in a magnetic field," Atomnaya Énergiya, 14:249 (1963).
8, E. K. Zavoiskii, et al., "New data on the turbulent heating of a plasma," Zh. Éksp. Teor. Fiz., 46:511 (1964).
9. I. Alexeff, R. Neidigh, W. Peed, and E. Shipley, "Hot-electron plasma by beam-plasma interaction," Phys. Rev. Lett., 10:273 (1963).
10. M. V. Nezlin and A. M. Solontsev, "Acceleration of ions in plasma beams," Zh. Éksp. Teor. Fiz., 45:840 (1963).
11. V. E. Golant, A. P. Zhilinskii, I. F. Liventsova, and I. E. Sakharov, "Electromagnetic radiation from a plasma permeated by an electron beam in a magnetic field," Zh. Éksp. Teor. Fiz., 35:2034 (1965).
12. I. F. Kharchenko, Ya. B. Fainberg, et al., "Interaction of an electron beam with a plasma in a magnetic field," Plasma Physics and the Problems of Controlled Thermonuclear Synthesis, Vol. 2. Izd. Akad. Nauk Ukr. SSR (1963).
13. V. L. Ginzburg and L. N. Ozernoi, "Temperature of the metagalactic and galactic gas," Astron. Zh., 42:943 (1965).
14. V. N. Tsytovich and V. D. Shapiro, "Nonlinear stabilization of the beam instabilities of a plasma," Yadernyi Sintez, 5:228 (1965).
15. Ya. B. Fainberg, "Particle acceleration in a plasma," Atomnaya Énergiya, 6:431 (1959).
16. V. N. Tsytovich, "Isotropization of cosmic rays," Astron. Zh., 43:528 (1966).
17. V. L. Ginzburg, "Cosmic rays and plasma effects in the galaxy and metagalaxy," Astron. Zh., 42:1129 (1965).
18. V. N. Tsytovich and A. B. Shvartsburg, "Nonlinear interaction of waves in a plasma in a powerful constant external magnetic field," Zh. Tekh. Fiz., 36 (1966).
19. V. N. Tsytovich, "Problems in the statistical acceleration of particles in a turbulent plasma," Paper at the Conference on Phenomena in Ionized Gases, Belgrade (August, 1965).

20. I. A. Akhiezer, "Theory of nonlinear motions of a nonequilibrium plasma," Zh. Éksp. Teor. Fiz., 47:952 (1964).
21. B. B. Kadomtsev and V. I. Petviashvili, "Weakly turbulent plasma in a magnetic field," Zh. Éksp. Teor. Fiz., 43:2234 (1962).
22. S. A. Kaplan and V. N. Tsytovich, "Radio emission from beams of charged particles under cosmic conditions," Astron. Zh., p. 44 (1967).
23. S. B. Pickel'nez and V. N. Tsytovich, "The spectram of plasma turbulence and acceleration of subeosmic rays," Zh. Éksp. Teor. Fiz., 55:977 (1968).
24. V. A. Lipezousky and V. N. Tsytovich, "Spectrum of the turbulence of hot plasmas," Zh. Éksp Teor. Fiz. 57:1252 (1969).
25. É. P. Jidkov, V. H. Machan'kov, V. N. Tsytovich and Ch'o Sai-han, "Numerical calculation of the spectra of stationary plasma turbulence," Joint Institute of Nuclear Research (Dubna) Preprint, P 9-4464.
26. V. N. Tsytovich, "Stochastic processes in plasma." Invited review paper on the 9th. International Conference on Phenomena in Ionized Gases, Bucharest, 1969, Preprint Lebedev Phys. Inst. (1969).

*Chapter X*

# Current Trends in Research on Nonlinear Effects in Plasma

In this final chapter we take a look at some of the most important trends in the development of nonlinear plasma theory. Current research may be grouped into: 1) investigations in which the theory of nonlinear interactions is elaborated on a broad scale, i.e., toward a more detailed analysis of nonlinear interactions within the theoretical context of weak nonlinearity; 2) investigations invoking new approaches. In the first category we include theories based on the kinetic equations for plasmons (Sec. 1 of the present chapter) and theories of the nonlinear plasma dispersive properties (Sec. 2). The new approaches attempt to transcend the weak nonlinearity concept, i.e., by and large to circumvent expansions in the amplitudes of the interacting fields. This category includes the investigation of nonlinear plasma instabilities (Sec. 3), the investigation of the dispersion characteristics of plasma in strong external fields (Sec. 4), the investigation of large-amplitude nonlinear waves (Sec. 5), etc.

## 1. DEVELOPMENT OF THEORY BASED ON THE KINETIC EQUATIONS FOR PLASMONS

There are some theories which start from the kinetic equation for plasmons [1] and those in which such kinetic equations are themselves derived [2, 3]. The only distinction between the two approaches lies in the method of calculating the probabilities of the processes. Investigations of this type have been made which attempt to include as completely as possible all the elementary interactions between the various oscillatory modes in the plasma. At present nonlinear interactions in a homogeneous magnetoactive plasma are being intensively studied [3, 4]. The effects of induced scattering by ions and electrons have been discussed in [1, 3, 4, 5], decay processes for electrostatic oscillations in both isothermal and nonisothermal plasmas in [6–8], and decay processes in magnetoactive plasma, including other oscillations, first in [8], and later in [1, 2, 4].

Another avenue of investigation aims at the analysis of nonlinear interactions in magnetoactive plasma in the presence of particle beams or of relative motion of the electrons and ions (see, e.g., [9]). Further attention needs to be devoted to the study of nonlinearities in inhomogeneous magnetoactive plasma [10], especially the interaction of the plasma drift waves with each other and with other waves. From the experimental point of view there is particular interest in the decay of transverse waves into drift waves, which causes the nonlinear scattering of electromagnetic waves by drift waves.

Finally, the kinetic equation for plasmons provides a method for describing effects bearing on the inverse influence of nonlinear wave interaction on the plasma particle distribution [13].

## 2. DEVELOPMENT OF RESEARCH ON THE NONLINEAR DISPERSION PROPERTIES OF PLASMA

The study of the nonlinear dispersion properties of plasma makes it possible to supersede the random-phase approximation for certain interacting waves, while retaining the concept of weak nonlinearity. An example of this is the "hydrodynamic" instability of longitudinal waves in the presence of intense transverse waves considered in Chap. III. The investigation of these instabilities reduces to the solution of a particular dispersion equation. The method used in Chap. III is capable of a simple extension, which leads to the concept of the nonlinear dielectric constant of a plasma. In broad outline the problem of the nonlinear variation of the dispersive properties of a plasma may be formulated as follows. In a plasma there are fairly strong random fields, which may be regarded more or less approximately as known and their amplitudes as invariant with time. The problem is to find the change in the dispersion of waves whose frequencies differ from and whose amplitudes are smaller than those of the strong waves. It is easy to carry out a general analysis of the nonlinear dielectric constant in the case of weak nonlinearity, which allows an expansion in the field amplitudes. This requires computation of the nonlinear plasma current to third-order terms in the field. Assuming that the total field consists of the self-consistent random field plus the weak field under investigation, the nonlinear current can be expanded in terms of the weak field. Recognizing that the self-consistent fields satisfy Maxwell's equations allowing for the nonlinearity of the plasma, and averaging them over the phases of the strong fields, we obtain the desired nonlinear equations for the weak fields, these equations also defining the nonlinear dielectric constant of the plasma. The appropriate analysis for an isotropic plasma is carried out in Appendix 4, in which a general formula is also given for the nonlinear dielectric constant.

Another method involves the dispersion equations. It is possible from an examination of nonlinear decay and induced scattering interactions to determine the nonlinear increments for the processes. They can be used to find the imaginary part of the nonlinear dielectric constant of the plasma, and this in turn can be used to determine the real part from the dispersion equations (see Appendix 4).

From the nonlinear dielectric constant it is a simple task to find the nonlinear corrections to the frequencies of the weak waves. For small intensity the imaginary parts of these corrections describe the rates for the spectral transfer of waves by decays and induced scattering. The real parts of the nonlinear corrections, on the other hand, describe the change in the wave dispersion. Note that for Langmuir waves with large phase velocities nonlinear changes in the dispersion can change the rate and direction of spectral transfer [11].

The nonlinear effects in the dispersion of Langmuir waves become significant for large phase velocities where the effect of thermal motion is very small:

$$\Omega^l(\mathbf{k}) \approx \omega_{0e}\left(1 + \frac{3}{2}\frac{v_{Te}^2}{v_\varphi^2}\right). \tag{10.1}$$

Let us now turn to the expression Eq. (A4.2) for the nonlinear permittivity derived in Appendix 4. For large phase velocities, when such corrections are especially important, the contribution of $\Sigma^{(2)}$ [see (A4.2)] becomes very small, which, as for induced scattering by electrons, is related to the part played by the ions in the screening effect. This may be seen from (A4.2) when one considers that the nonlinear current S contains only the electronic contribution, while $\varepsilon$ in (A4.4) [see (A3.11)] contains also the ionic contribution, and for $T_i \ll T_e$ and $v_\varphi/v_{Te} \gg \sqrt{m_i/m_e}$ we have $\varepsilon_i(\omega_-, k_-) \gg \varepsilon_e(\omega_-, k_-)$, which indicates the smallness of $\Sigma^{(2)}$. As for $\Sigma^{(3)}$, to first order with respect to the small parameter $v_{Te}/v_\varphi$ it yields the following expression for the nonlinear dielectric constant of the plasma (Appendix 4):

$$\varepsilon^l_{\text{nlr}} = 1 - \frac{e^2}{m_e\omega_{0e}^4}\int\frac{(\mathbf{k}\mathbf{k}_1)^2}{k^2k_1^2}\left(\varepsilon^l_e(\omega_-, \mathbf{k}_-) - 1\right)\mathbf{k}_-^2\,|E^l_{\mathbf{k}_1}|^2\,d\mathbf{k}_1. \tag{10.2}$$

Recognizing that

$$\varepsilon^l_e(\omega_-, \mathbf{k}_-) \approx 1 + \frac{\omega_{0e}^2}{k_-^2 v_{Te}^2},$$

we have*

$$\varepsilon^l_{\text{nlr}} \approx 1 - \int\frac{(\mathbf{k}\mathbf{k}_1)^2}{k^2k_1^2}\,|E^l_{\mathbf{k}_1}|^2\,\frac{d\mathbf{k}_1}{4\pi n m_e v_{Te}^2}. \tag{10.3}$$

Unlike the values for $\varepsilon^l_{\text{nlr}}$ discussed earlier in connection with decay processes, the values of $\varepsilon^l_{\text{nlr}}$ determined by (10.3) are related to scattering processes [in particular, by including the imaginary part of (10.2), as well as only the second term of (A4.18), we obtain the transfer effects (8.9) and (8.10)].

From the equation for longitudinal waves

$$\varepsilon^l = \varepsilon^l_e + \varepsilon^l_{\text{nlr}} - 1 = 0$$

we see that

$$\Omega^l(\mathbf{k}) \approx \omega_{0e}\left(1 + \frac{3}{2}\frac{v_{Te}^2}{v_\varphi^2} + \frac{1}{2}\int\frac{(\mathbf{k}\mathbf{k}_1)^2}{k^2k_1^2}\,\frac{|E^l_{\mathbf{k}_1}|^2\,d\mathbf{k}_1}{4\pi n m_e v_{Te}^2}\right). \tag{10.4}$$

* If the condition $\varepsilon_i(\omega_-, \mathbf{k}_-) \gg \varepsilon_e(\omega_-, \mathbf{k}_-)$ is not fulfilled, an additional factor $\varepsilon_i(\omega_-, \mathbf{k}_-)/\varepsilon(\omega_-, \mathbf{k}_-)$ appears in (10.3).

By comparing the nonlinear corrections with the thermal motion effects (10.1) we see that the nonlinear corrections are greater provided

$$v_{\varphi} \gg v_{Te} \left( \frac{W^l}{nT_e} \right)^{-1/2}. \qquad (10.5)$$

This leads to a very important statement concerning the direction of spectral transfer by induced scattering. When (10.5) holds, this direction is determined by the sign of the difference

$$\omega_- = \Omega^l(\mathbf{k}) - \Omega^l(\mathbf{k}') = \frac{\omega_{0e}}{2} \int \left[ \frac{(\mathbf{k}\mathbf{k}_1)^2}{k^2 k_1^2} - \frac{(\mathbf{k}'\mathbf{k}_1)^2}{k'^2 k_1^2} \right] \frac{|E^l_{\mathbf{k}_1}|^2 \, d\mathbf{k}_1}{4\pi n m_e v_{Te}^2}. \qquad (10.6)$$

For an isotropic distribution of the oscillations the sign of $\omega_-$ depends strongly on the direction of propagation of the interacting waves [11]. In the case of an isotropic distribution the nonlinear correction does not depend on **k** or on direction and, hence, it drops out of the difference $\omega_-$. This means that it may be essential to allow for particle collisions, resulting in nonlinear corrections of order $\nu/kv_{Te}$ for $\nu \ll kv_{Te}$ (where $\nu$ is the collision frequency) [12].

For large-intensity random waves the corrections to the imaginary part of the frequency become the growth rates for nonlinear "hydrodynamic" instabilities.

As an example consider the "hydrodynamic" instabilities of beams of Langmuir waves. Let all the $l$-waves have the same direction and close values of wave numbers $k_1 \simeq k_*$ and, hence, of group velocities $v_{gr} \simeq v_*$. Consider a low-frequency longitudinal wave propagating in the direction of the $l$-waves. Then, using Appendix 4, we obtain the following expression for the nonlinear dielectric constant:

$$\varepsilon^{nlr} = 1 - \frac{k^2 \omega_{0i}^2 v_\sim^2}{\omega^2 \left[ (\omega - kv_*)^2 - \frac{1}{4} \frac{k^4}{k_*^2} v_*^2 \right]}; \quad v_\sim^2 = \frac{3}{4} \frac{W^l}{nm_e} \qquad (10.7)$$

We have made no assumption here that k is small compared with $k_1 = k_1^l$.

Let us now investigate frequencies near $\omega = \Omega_s(\mathbf{k}) = \mathbf{k}\mathbf{v}_s$. We assume that the decay condition is approximately met, i.e.,

$$\text{i.e.,} \quad \left(\frac{v_s}{v_*} - 1\right)^2 \approx \frac{k^2}{4k_*^2}. \tag{10.8}$$

Setting $\omega = \Omega^s(\mathbf{k}) + \omega'$, for $\omega' \ll \Omega^s(\mathbf{k})$ we obtain

$$\varepsilon^{\text{nlr}} = 1 - \frac{v_\sim^2}{v_s^2} \frac{\omega_{0i}^2}{2\omega' v_* k \left(\frac{v_s}{v_*} - 1\right)}. \tag{10.9}$$

In the case of immediate concern the linear part of the dielectric constant is

$$\varepsilon^{\text{lr}} = \frac{2\omega_{0i}^2 \omega'}{k^3 v_s^3}. \tag{10.10}$$

Since $\varepsilon^{\text{lr}} + \varepsilon^{\text{nlr}} - 1 = 0$ we have

$$(\omega')^2 = \frac{1}{4} \frac{v_\sim^2}{v_s^2} \frac{k^2 v_s^2}{1 - v_*/v_s}. \tag{10.11}$$

Hence instability occurs only if $v_* > v_s$, i.e., if the group velocity of the Langmuir waves exceeds the velocity of sound. Then by (10.8)

$$\gamma^{\text{nlr}} = \frac{1}{2} \frac{v_\sim k}{|1 - v_*/v_s|^{1/2}} = \frac{k}{2} \sqrt{\frac{W^l}{nm_e}} \left(\frac{m_e}{m_i}\right)^{1/4} \left(\frac{\omega_{0e}}{kv_{Te}}\right)^{1/2} \frac{1}{2^{1/2}}. \tag{10.12}$$

It is important to remember that the hydrodynamic instabilities do not vanish with strong linear damping, and for $\gamma^{\text{lr}} \gg \gamma^{\text{nlr}}$

$$\gamma = \frac{(\gamma^{\text{nlr}})^2}{\gamma^{\text{lr}}}. \tag{10.13}$$

As the linear damping of sound is given by the familiar expression $\gamma^{\text{lr}} = \sqrt{\pi/8}\,\sqrt{m_e/m_i}\,kv_s$, for $\gamma^{\text{lr}} \gg \gamma^{\text{nlr}}$ we obtain the following growth rate:

$$\gamma = \frac{\omega_{0e}}{2\sqrt{2\pi}} \sqrt{\frac{m_i}{m_e}} \frac{W^l}{nT_e}. \tag{10.14}$$

The nonlinear instabilities resulting in the generation of low-frequency plasma waves can affect the quasilinear stage of beam relaxation. This requires that the nonlinear growth rate satisfy the condition

$$\gamma^{\text{nlr}} > k\Delta v^{l}_{\text{gr}} \approx 3k\frac{\Delta k v_{\text{T}e}^{2}}{\omega_{0e}}. \tag{10.15}$$

Assuming $k \simeq \omega_{0e}/v_1$ and $\Delta k/k = \Delta v/v_1$, we obtain

$$\gamma^{\text{nlr}} > 3\left(\frac{v_{\text{T}e}}{v_1}\right)^2 \omega_{0e}\frac{\Delta v}{v_1}. \tag{10.16}$$

However, according to (10.12),

$$\gamma^{\text{nlr}} \simeq \frac{k}{2}\sqrt{\frac{W^l}{n_0 m_e}}\left(\frac{m_e}{m_i}\right)^{1/4}\left(\frac{\omega_{0e}}{k v_{\text{T}e}}\right)^{1/2} \simeq \omega_{0e}\left(\frac{n_1}{n_0}\right)^{1/2}\left(\frac{\Delta v}{v_1}\right)^{1/2}\left(\frac{m_e}{m_i}\right)^{1/4}\left(\frac{v_1}{v_{\text{T}e}}\right)^{1/2} \tag{10.17}$$

This expression includes the substitution $W^l \simeq n_1 m_e v_1 \Delta v$. Thus, the condition for the onset of nonlinear instability in the quasilinear stage becomes

$$\frac{n_1}{n_0} > \sqrt{\frac{m_i}{m_e}}\, 9\left(\frac{v_{\text{T}e}}{v_1}\right)^5\left(\frac{\Delta v}{v_1}\right). \tag{10.18}$$

For low-velocity beams condition (10.18) is violated for $n_1/n_0 \ll 1$. For high-velocity beams, however, it can be fulfilled easily without being too severe.

Nonlinear instabilities can have an appreciable effect on the dynamics of beam–plasma interaction only if

$$\gamma^{\text{nlr}} > \gamma^{\text{quasil}} \simeq \omega_{0e}\frac{n_1}{n_0}\left(\frac{v_1}{\Delta v}\right)^2. \tag{10.19}$$

Substituting the nonlinear growth rate (10.17) into (10.19) yields

$$\frac{n_1}{n_0} < \left(\frac{\Delta v}{v_1}\right)^5\left(\frac{m_e}{m_i}\right)^{1/2}\left(\frac{v_1}{v_{\text{T}e}}\right). \tag{10.20}$$

Conditions (10.18) and (10.20) are compatible only if

$$\frac{\Delta v}{v} > \left(\frac{9m_i}{m_e}\right)^{1/4}\left(\frac{v_{\text{T}e}}{v_1}\right)^{3/2}. \tag{10.21}$$

Consequently, nonlinear instabilities can significantly affect the quasilinear relaxation of beams when (10.21) is satisfied.

In concluding our discussion on this line of investigation, notice that the nonlinear instabilities discussed cannot be described by the plasmon kinetic equation, hence the nonlinear dielectric constant method is in a certain sense more general. It also enables one to investigate important problems concerning the influence of particle collisions on nonlinear interaction [12]. In the presence of particle collisions the introduction of the plasmon concept becomes inapplicable. It is important to realize that collisions can affect the nonlinear interaction of waves whose frequencies are much greater than the collision frequency, because nonlinear scattering involves a virtual wave whose frequency is less than the collision frequency.* By using the method of investigating the nonlinear dispersion equations we can deduce considerably more information about the nonlinear properties of a plasma, since we avoid having to find self-consistent solutions in which allowance must be made for the influence of nonlinear growth on the distribution of the intense oscillations. This method therefore permits one to analyze only the initial stage of the process of nonlinear interaction of the oscillations. Another shortcoming lies in the assumption of weak nonlinearity, which implies that the nonlinear corrections must be very small compared to the oscillation frequency. This shortcoming becomes particularly obvious when one attempts to analyze the dispersion characteristics of a plasma at very low frequencies, much lower say, than the characteristic frequencies for nonlinear spectral transfer.

## 3. DEVELOPMENT OF THE THEORY OF THE LOW-FREQUENCY NONLINEAR CHARACTERISTICS OF PLASMA

This group of investigations is properly classified among the attempts to transcend theory that relies on the concept of weak nonlinearity of the plasma while retaining the notion of small amplitudes for the low-frequency oscillations in question. Mathematically, low-frequency oscillations must be described by a nonlinear dielectric constant which exhibits a more complex dependence on the intensity of the strong waves than a mere linear law. In the earliest papers following this approach [14, 15] several different methods were used (Lagrange formalism in [14] and the energy principle in [15]), but similar results were obtained on the existence of a new type of nonlinear instability. In this case the non-

* For further details see [12, 24, 25].

linear changes in the dispersion characteristics are not small. This lends support to the idea of a new type of collective motion in plasma caused by nonlinearity. The papers [14, 15] deal with the special problem of low-frequency instabilities in a plasma in which strongly excited Langmuir oscillations are present. We now present an analysis which departs from [14, 15] and is based on the correspondence principle, whereby we can generalize the results of [14, 15].

Let E be the average value of the weak field investigated. The field E can alter the concentration of electrons by an amount $\delta n_e$. The effective field acting on an electron must be different from E. Within the scope of the linear weak field approximation,* however, it will be linear in E and $\delta n_e$:

$$E_{\text{eff.}} = E + \beta\delta n_e. \tag{10.22}$$

The coefficient $\beta$ depends on the intensity W of the strong random waves, so that $\beta \to 0$ as $W \to 0$. We limit ourselves to an expression for $\beta$ linear in W, but without assuming that the additional field $\beta\delta n_e$ is small compared with E. In other words, the effective field can differ substantially from the average field. An example of the additional field $\beta\delta n_e$ is the Miller force exerted by strong high-frequency fields due to their weak nonuniformity created by E [14, 15].

We use the relation between $\delta n_e$ and E for longitudinal waves, replacing E with $E_{\text{eff}}$:

$$4\pi e\delta n_e = -(\varepsilon_e - 1)\, ikE_{\text{eff}}. \tag{10.23}$$

Thus

$$4\pi e\delta n_e = -(\varepsilon_e - 1)\left(1 + \frac{(\varepsilon_e - 1)\, ik\beta}{4\pi e}\right)^{-1} ikE \equiv -(\varepsilon_e^{*\text{nlr}} - 1)\, ikE. \tag{10.24}$$

To determine $\beta$ we use the correspondence principle. The analytic expression for $\beta$ does not depend on the actual value of the strong wave intensity. For small $\beta$ (small W), on the other hand, the result (10.24) can be expanded in $\beta$ and compared with the nonlinear dielectric constant $\varepsilon_e^{\text{nlr}}$ obtained by an expansion in the field amplitudes:

* In general $E_{\text{eff}}$ could depend on the current $\delta j$ generated by the field E. For electrostatic oscillations, however, to which the present discussion is limited, $\delta j$ is directly related to $\delta n$.

$$\varepsilon_e^{*\text{nlr}} \approx \varepsilon_e - \frac{(\varepsilon_e - 1)^2 ik\beta}{4\pi e} = \varepsilon_e + \varepsilon_e^{\text{nlr}} - 1. \tag{10.25}$$

Since their large mass exempts the ions from the action of the additional field $\beta\delta n_e$, they make only an additive contribution to the total $\varepsilon^{*\text{nlr}}$:

$$\varepsilon^{*\text{nlr}} = \varepsilon_i + \varepsilon_e^{*\text{nlr}} - 1 = \varepsilon_i + \frac{\varepsilon_e - 1}{1 - \dfrac{\varepsilon_e^{\text{nlr}} - 1}{\varepsilon_e - 1}}. \tag{10.26}$$

As an example consider a plasma in which there are intense Langmuir waves. In the low-frequency ($\omega \ll \omega_{0e}$ and $\omega \ll kv_{Te}$) and long-wavelength ($k \ll k_1$) limit the nonlinear dielectric constant $\varepsilon_e^{\text{nlr}}$ associated with the decay process $l \to l + s$ has the following form according to Appendix 4:*

$$\varepsilon_e^{\text{nlr}} = 1 + \frac{\pi e^2 \omega_{0e}^2}{m_e^2 k^2 v_{Te}^4} \int \frac{\mathbf{k}\dfrac{\partial N_{\mathbf{k}_1}^l}{\partial \mathbf{k}_1}\dfrac{d\mathbf{k}_1}{(2\pi)^3}}{\omega - \mathbf{k}\mathbf{v}_{\text{gr}}^l + i\delta} = 1 - \frac{3\pi e^2 \omega_{0e}}{m_e^2 v_{Te}^2} \int \frac{N_{\mathbf{k}_1}^l \dfrac{d\mathbf{k}_1}{(2\pi)^3}}{(\omega - \mathbf{k}\mathbf{v}_{\text{gr}}^l + i\delta)^2}. \tag{10.27}$$

Here $v_{\text{gr}}^l = (3/\omega_{0e})\, v_{Te}^2 \mathbf{k}_1$ is the Langmuir-wave group velocity. In (10.27) the retention of only linear terms in $W^l$, which correspond to the process of $ls$-decay, is justified by the presence in $\varepsilon_e^{\text{nlr}}$ of the resonance denominator $(\omega - \mathbf{k}\mathbf{v}_{\text{gr}}^l)^{-2}$, which does not occur for other processes of the same or higher order in $W^l$. In the frequency ranges $\omega \ll kv_{Te}$ and $\omega \ll \omega_{0i}$, $\omega \gg kv_{Ti}$ we have

$$\varepsilon_e - 1 \approx \frac{\omega_{0e}^2}{k^2 v_{Te}^2}\,; \quad \varepsilon_i \approx -\frac{\omega_{0i}^2}{\omega^2}$$

and the dispersion equation becomes

$$-\frac{\omega_{0i}^2}{\omega^2} + \frac{\omega_{0e}^2}{k^2 v_{Te}^2} + (\varepsilon_e^{\text{nlr}} - 1)\frac{\omega_{0i}^2 k^2 v_{Te}^2}{\omega^2 \omega_{0e}^2} = 0. \tag{10.28}$$

Consider now a plasma in which there are intense isotropic Langmuir waves. The dielectric constant of the plasma $\varepsilon_e^{\text{nlr}}$ can be derived by elementary integration over direction:

---

* It is readily deduced by comparing (A4.4) and (A4.16) that the denominator $\omega - \mathbf{k}\mathbf{v}_{\text{gr}}^l$ is obtained as the result of the approximate expansion $\varepsilon^l(\mathbf{k}_1 + \mathbf{k}) = \varepsilon^l(\omega_1 + \omega, \mathbf{k}_1 + \mathbf{k}) \simeq \varepsilon^l(\omega_1, \mathbf{k}_1) + (\partial\varepsilon^l/\partial\omega_1)(\omega - \mathbf{k}\mathbf{v}_{\text{gr}}^l)$, where $\varepsilon^l(\omega_1, \mathbf{k}_1) = 0$ if we can assume that $\omega_1 = \omega^l(\mathbf{k}_1)$ (see also the footnote after the next).

$$\varepsilon^{\text{nlr}} = 1 - \frac{3\pi e^2\omega_{0e}}{m_e^2 v_{Te}^2} \int \frac{N_{k_1}^l 4\pi k_1^2}{\omega^2 - k^2 (v_{\text{gr}}^l)^2} \frac{dk_1}{(2\pi)^3}. \tag{10.29}$$

Let $N_{k_1}^l$ represent a narrow spectrum of waves having large wave numbers $k_1$ near some $k_*$. We denote the Langmuir-wave group velocity for $k_1 = k_*$ by $v_*$. Then the dispersion equation (10.28) has the following form for $\varepsilon = 0$:

$$v^4 - v^2 (v_*^2 + v_s^2) - (v_\sim^2 - v_*^2) v_s^2 = 0. \tag{10.30}$$

Here $v = \omega/k$ is the phase velocity of the waves in question, $v_s = v_{Te}\sqrt{m_e/m_i}$ is the velocity of sound, and

$$v_\sim^2 = \frac{3}{4}\frac{W^l}{nm_e} \tag{10.31}$$

is a certain effective velocity for the oscillations of the plasma particles in the field of the plasma oscillations.* Consequently, the dispersion equation (10.30) reduces to one which determines the phase velocities of the waves:

$$v^2 = \frac{v_*^2 + v_s^2}{2} \pm \sqrt{\frac{(v_*^2 + v_s^2)^2}{4} + (v_\sim^2 - v_*^2) v_s^2}. \tag{10.32}$$

It is readily seen that expression under the square root (10.32) is always positive. Therefore, instability can occur only if $v^2 < 0$ (in which case $\omega$ is purely imaginary). This requires that the term with the minus sign in the solution (10.32) be larger than the first term, and this is possible only if

$$v_\sim^2 > v_*^2. \tag{10.33}$$

This means that the velocity of the electron oscillations in the field of the plasma oscillations must be greater than the group velocity of the plasma waves.

In the limit $v_*^2 \ll v_\sim^2 \ll v_s^2$ we obtain from (10.32)

$$v^2 = v_s^2; \quad v^2 = -v_\sim^2. \tag{10.34}$$

For the unstable solution $v = \omega/k = iv_\sim$, i.e.,

$$\operatorname{Im}\omega = \gamma = v_\sim k. \tag{10.35}$$

* Actually the velocity of the particle oscillations in the field of a wave of frequency $\omega$ is $v \sim eE/m\omega$; $v^2 = \frac{e^2E^2}{m^2\omega^2} = \frac{1}{4\pi}\frac{\omega_{0e}^2 E^2}{nm_e\omega^2} \sim \frac{W^l}{nm_e}$ for $\omega \sim \omega_{0e}$ $\left(W^l \sim \frac{E^2}{4\pi}\right)$.

The instability condition (10.33) was first obtained in [14] using the variational principle. We write this criterion in terms of the energy density $W^l$ of the plasma waves, remembering that $v_*^2 = 9v_{Te}^4/(v_\varphi^l)^2$,

$$\frac{W^l}{nT_e} > 12\,\frac{v_{Te}^2}{(v_\varphi^l)^2}\,. \qquad (10.36)$$

The stability condition derived in [15] by means of the variational principle differs from (10.36) in that $T_e$ on the left-hand side of the expression is replaced by $T_e + T_i$. It can be found similarly from the dispersion equation (10.28) as the criterion for the excitation of lower frequencies $\omega \ll kv_{Ti}$ $(\varepsilon_i \approx \omega_{0i}^2/k^2v_{Ti}^2)$.*

This line of research is interesting from the point of view of interpreting the mechanisms of transfer of oscillation energy from electrons to ions, the heating of ions, and the determination of additional plasma diffusion mechanisms resulting from the excitation of low-frequency oscillations [10]. Further investigations of the low-frequency characteristics of inhomogeneous plasma [10] might make it possible to consider the influence of intense high-frequency waves on the plasma drift instability. Finally, the generalization of the theory to arbitrary non-electrostatic oscillations should provide a general picture of the low-frequency nonlinear dispersion characteristics of a plasma [16] (skin effect, beams, etc.).†

## 4. NONLINEAR DISPERSION CHARACTERISTICS OF PLASMA IN THE FIELD OF A FIXED-PHASE WAVE

The dispersion of a plasma for waves having large amplitudes and large phase velocities has a special significance. Let us

* There is one more necessary condition for this instability; specifically, it must be allowable to assume $\varepsilon^l(\omega_1, k_1)$ nearly equal to zero (see the footnote before the last). But taking account of the nonlinear frequency corrections $\Delta\omega_{nlr}$ [see (10.4)], we find that $\varepsilon^l(\omega_1, k_1) \simeq \Delta\omega_{nlr}(\partial\varepsilon^l/\partial\omega_1)$, i.e., $\max(\omega, \mathbf{k}\mathbf{v}_{gr}) \gg \Delta\omega_{nlr}$. This condition imposes an upper limit on W. It may be inferred from this that instability is possible if $v_\varphi \ll 3v_{Te}^2/v_{Ti}$. Far less stringent conditions apply to the instability investigated in [16].

† The spectra of plasma drift waves in the presence of strong high-frequency oscillations have been investigated in [26].

consider the change in the dispersion characteristics caused by a monochromatic wave having $v_\varphi = \infty$ [17, 18]. For a cold plasma, as in Chap. II, we can use the fluid equations. Since we are not concerned exclusively with high-frequency oscillations, we must investigate the equations for both electrons and ions:

$$\frac{\partial \mathbf{v}_e}{\partial t} + (\mathbf{v}_e \nabla)\mathbf{v}_e = \frac{e}{m_e}\mathbf{E}_0 \cos \omega t; \quad \frac{\partial n_e}{\partial t} + \nabla(\mathbf{v}_e n_e) = 0; \tag{10.37}$$

$$\frac{\partial \mathbf{v}_i}{\partial t} + (\mathbf{v}_i \nabla)\mathbf{v}_i = -\frac{e}{m_i}\mathbf{E}_0 \cos \omega t; \quad \frac{\partial n_i}{\partial t} + \nabla(\mathbf{v}_i n_i) = 0. \tag{10.38}$$

In the zeroth approximation all particles oscillate with the field frequency:

$$\mathbf{v}_{e0} = \frac{e}{m_e}\frac{\mathbf{E}_0 \sin \omega t}{\omega}; \quad \mathbf{v}_{i0} = -\frac{e}{m_i}\frac{\mathbf{E}_0 \sin \omega t}{\omega}. \tag{10.39}$$

We wish to determine the behavior of perturbations which form plane waves in our coordinates r:

$$\mathbf{v}_e = \mathbf{v}_{e0} + \mathbf{v}_{1e}(t)\, e^{i\mathbf{kr}}; \quad \mathbf{v}_i = \mathbf{v}_{i0} + \mathbf{v}_{1i}(t)\, e^{i\mathbf{kr}}. \tag{10.40}$$

For $\mathbf{v}_1$ we at once obtain

$$\frac{\partial \mathbf{v}_{1e}}{\partial t} + i(\mathbf{k}\mathbf{v}_{e0})\mathbf{v}_{1e} = \frac{e}{m_e}\mathbf{E}_1; \quad \frac{\partial n_{1e}}{\partial t} + i\mathbf{k}(\mathbf{v}_{e0}n_{1e} + \mathbf{v}_{1e}n_0) = 0; \tag{10.41}$$

$$\frac{\partial \mathbf{v}_{1i}}{\partial t} + i(\mathbf{k}\mathbf{v}_{i0})\mathbf{v}_{1i} = -\frac{e}{m_i}\mathbf{E}_1; \quad \frac{\partial n_{1i}}{\partial t} + i\mathbf{k}(\mathbf{v}_{i0}n_{1i} + \mathbf{v}_{1i}n_0) = 0, \tag{10.42}$$

where $\mathbf{E}_1$ is the field of the perturbing wave.

Next we examine waves associated exclusively with the space charge. Their fields can be found from Poisson's equation

$$\operatorname{div} \mathbf{E}_1 = 4\pi e(n_e - n_i) = 4\pi e(n_{1e} - n_{1i}),$$

or

$$i(\mathbf{k}\mathbf{E}_1) = 4\pi e(n_{1e} - n_{1i}) \tag{10.43}$$

Together with the first equation of (10.41) this gives

$$\frac{\partial}{\partial t} i(\mathbf{k}\mathbf{v}_{1e}) + i(\mathbf{k}\mathbf{v}_{e0})\, i(\mathbf{k}\mathbf{v}_{1e}) = \frac{4\pi e^2}{m_e}(n_{1e} - n_{1i}). \tag{10.44}$$

Eliminating $(\mathbf{kv}_{1e})$ by means of the second equation of (10.41), we have

$$\left(\frac{\partial}{\partial t}+i\,(\mathbf{kv}_{e0})\right)^2 n_{1e}=-\,\omega_{0e}^2\,(n_{1e}-n_{1i}); \tag{10.45}$$

$$\left(\frac{\partial}{\partial t}+i\,(\mathbf{kv}_{i0})\right)^2 n_{1i}=\omega_{0i}^2\,(n_{1e}-n_{1i}). \tag{10.46}$$

We begin with the analysis of high-frequency oscillations. If we neglect the variation of the ion density ($n_{1i} \ll n_{1e}$), Eq. (10.45) becomes

$$\left(\frac{\partial}{\partial t}+i\,(\mathbf{kv}_{e0})\right)^2 n_{1e}=-\,\omega_{0e}^2 n_{1e} \tag{10.47}$$

and is reduced by the change of variable $n_{1e}=\nu_{1e}\exp\{-i\int\limits_{-\infty}^{t}(\mathbf{kv}_{e0})\,dt\}$ to

$$\frac{\partial^2}{\partial t^2}\nu_{1e}=-\,\omega_{0e}^2\nu_{1e}, \tag{10.48}$$

i.e., $\nu_{1e}=\nu_0\exp(\pm i\omega_{0e}t)$; $n_{1e}=n_{1e0}\exp(\pm i\omega_{0e}t+ia\cos\omega t)$, where $a=(e\mathbf{kE}_0)/(m_e\omega^2)$.

If $\omega \gg \omega_{0e}$, then when $a \ll 1$, a "small ripple" is superimposed on the Langmuir oscillations by the field of the monochromatic wave, but for $a \gg 1$ the density oscillations are mainly due to the external wave and tend to "contract and expand" with a frequency $\omega_{0e}$. A more detailed analysis for $\omega \gg \omega_{0e}$ and $\omega \gtrsim \omega_{0e}$ shows that the corrections associated with the motion of the ions, except in the case of parametric resonance, are very small, of order $m_e/m_i$. They need be included only when of the same order as thermal motion effects, i.e., when $v_\varphi/v_{Te} \gtrsim \sqrt{m_i/m_e}$. Then it is also required that $a \sim 1$, or that the fields be large:

$$\frac{E_0^2}{8\pi}\approx\left(\frac{\omega}{\omega_{0e}}\right)^4 nm_e v_\varphi^2 \gtrsim \left(\frac{\omega}{\omega_{0e}}\right)^4 nT_e\frac{m_i}{m_e}\,.$$

The most important feature is the change in the dispersion properties for low-frequency oscillations, which according to [17] become unstable with maximum growth rates of order $\omega_{0e}(m_e/m_i)^{1/3}$.

## 5. LARGE-AMPLITUDE NONLINEAR WAVES

Any external field is distorted by the plasma because of polarization. Consequently, a strong wave acting on the plasma can be influenced by this polarization. If a steady-state self-consistent field develops as a result, the implication is that there is a coordinate system in which the field is time-independent; then in the laboratory system the field is a function of one variable $x - v_\varphi t = \xi$. As shown by Akhiezer, Lyubarskii, and Fainberg [19, 20], the structure of these large-amplitude nonlinear waves is easily determined from fluid equations of the type (10.37) and (10.38), assuming all variables depend only on $\xi$. In the reference system in which $v_\varphi = 0$ we have for high-frequency electrostatic waves

$$v_e \frac{\partial v_e}{\partial \xi} = -\frac{e}{m_e}\frac{\partial}{\partial \xi}\Phi; \qquad \frac{\partial}{\partial \xi} v_e n_e = 0,$$

where $\Phi$ is the potential of the wave field. Hence

$$n_e v_e = \text{const} = n_0 v_0; \quad \frac{m_e v_e^2}{2} + e\Phi = \text{const} = \frac{m_e v_0^2}{2}, \tag{10.49}$$

where $n_0$ is the mean value of the electron density. The potential at a point where the density is equal to its mean value is taken as zero. The two equations (10.49) permit the potential to be expressed in terms of the electron concentration:

$$\frac{n_e}{n_0} = \frac{1}{\sqrt{1 - (2e\Phi/m_e v_0^2)}}\,. \tag{10.50}$$

In Poisson's equation the ion concentration may be regarded as constant for high-frequency fields:

$$\frac{\partial^2}{\partial \xi^2}\frac{e\Phi}{m_e} = \omega_{0e}^2\left(1 - \frac{1}{\sqrt{1 - (2e\Phi/m_e v_0^2)}}\right). \tag{10.51}$$

Letting $U = 1 - \frac{2e\Phi}{m_e v_0^2}$ and multiplying (10.51) by $dU/d\xi$, we have

$$\frac{d}{d\xi}\left(\frac{dU}{d\xi}\right)^2 = -\frac{4\omega_{0e}^2}{v_0^2}\frac{d}{d\xi}(U - 2\sqrt{U}). \tag{10.52}$$

Hence

$$\left(\frac{dU}{d\xi}\right)^2 + \frac{4\omega_{0e}^2}{v_0^2}(U - 2\sqrt{U}) = \text{const} = \frac{4\omega_{0e}^2}{v_0^2}(\beta^2 - 1), \tag{10.53}$$

where $\beta$ is a constant. Equation (10.53) is readily integrated:

$$\arccos\frac{\sqrt{U} - 1}{\beta} + \sqrt{\beta^2 - (\sqrt{U} - 1)^2} = \pm\frac{\omega_{0e}}{v_0}\xi. \tag{10.54}$$

We now clarify the meaning of the two parameters $\beta$ and $v_0$. Since the cosine of an angle cannot exceed unity, $\left|\frac{\sqrt{U}-1}{\beta}\right| < 1$, $\beta$ is the maximum value of the quantity $1-\sqrt{1-2e\Phi/mv_0^2}$, or $e\Phi_{max} = (1/2)mv_0^2(2\beta-\beta^2)$. Thus, $\beta$ characterizes the wave amplitude.

Notice that the velocity of the electrons in a reference frame in which the wave is not at rest is $v_e - v_{e0}$, where $v_{e0}$ is the velocity of the reference frame. We wish to find a frame of reference in which the average directional momentum of the electrons vanishes:

$$\overline{n_e(v_e - v_{e0})} = \overline{n_e v_e} - v_{e0}\bar{n}_e = n_0 v_0 - v_{e0} n_0 = 0,$$

i.e., $v_0 = v_{e0} = v_\varphi$. In other words, $v_0$ coincides with the phase velocity $v_\varphi$. By (10.50) the maximum potential $e\Phi$ cannot exceed $\frac{1}{2} m_e v_\varphi^2$. In the linear approximation $e\Phi \ll \frac{1}{2} m_e v_\varphi^2$. Consequently, unlike its weaker counterpart, strong nonlinearity merely imposes a mild, rather than a severe, limitation on the wave amplitude. If the opposite inequality holds, $e\Phi > \frac{1}{2} m_e v_\varphi^2$, the fluid description breaks down, because multistream motions occur which rapidly diffuse the wave.

Strong nonlinear waves have been studied in detail in connection with the problems of shock waves in plasma [21] and of the acceleration of particles by nonlinear waves [22, 23]. We have previously discussed the stability of strong nonlinear waves in the limit $v_\varphi = \infty$. For finite $v_\varphi$ the problem has not been satisfactorily investigated in detail.

One of the most fruitful avenues of research in the physics of nonlinear plasma processes should be the continued development of the theory of strong nonlinear waves.

## CONCLUSION

In summing up our study of the effects of nonlinear wave interaction in plasma we are drawn first of all to the possibility of extending the fundamental physical concepts to more complex situations, primarily to inhomogeneous plasma in a strong magnetic field, as well as to bounded plasma.

Second, the general method outlines for calculating the probabilities of nonlinear wave interaction can be used to study the interaction of waves in a so-called solid state plasma. Moreover, the results presented herein apply to any homogeneous medium and can also be used for a more detailed analysis of wave interaction in solids allowing for spatial dispersion effects.

Finally, one must be impressed with the tremendous progress that has been made recently in the interpretation of experimental data using the theory of nonlinear effects. They include, for instance, the effects of anomalous diffusion, anomalous resistance, beam–plasma interactions, interactions of a strong high-frequency field with plasma,* particle acceleration in plasma, radiation from turbulent plasma, plasma diagnostics, interactions of high-frequency and low-frequency oscillations in plasma, shock waves in plasma, etc. [2, 7, 13]. It is presumed that future extensive experimental and theoretical studies of nonlinear wave interaction will provide an increasing insight into the physical processes occurring in plasma and into the nature of this very interesting state of matter. This will aid in the interpretation of the many plasma phenomena observed in outer space.

## Literature Cited


1. V. N. Tsytovich and A. B. Shvartsburg, "Theory of nonlinear wave interaction in a magnetoactive anisotropic plasma," Zh. Éksp. Teor. Fiz., 49:797 (1965).
2. B. B. Kadomtsev, Plasma turbulence, Academic Press, New York (1965).
3. A. P. Kropotkin and V. V. Pustovalov, "Nonlinear wave interaction in a plasma in a magnetic field," Zh. Éksp. Teor. Fiz., 50:312 (1966).
4. V. N. Tsytovich and A. B. Shvartsburg, "Nonlinear wave interaction in a plasma in a strong constant external magnetic field," Zh. Tekh. Fiz., 36(10) (1966).
5. V. D. Shapiro and V. I. Shevchenko, "Induced scattering of Langmuir oscillations in a plasma in a strong magnetic field," Ukrainsk. Fiz. Zh., 10(9):960 (1965).
6. A. P. Kropotkin and V. V. Pustovalov, "Decays of longitudinal waves in a nonisothermal magnetoactive plasma," Zh. Éksp. Teor. Fiz., (in print).
7. A. A. Galeev, V. I. Karpman, and R. Z. Sagdeev, "Many-particle aspects of the theory of a turbulent plasma," Yadernyi Sintez, 5:20 (1965).
8. N. L. Tsintsadze, "Wave conversion in a magnetoactive plasma," Zh. Tekh. Fiz., 34:1807 (1964).
9. V. I. Karpman, "Theory of plasma turbulence in a magnetic field," Zh. Prikl, Mekhan. i Tekh. Fiz., No. 6, p. 34 (1963).
10. L. I. Rudakov, "Problems in the nonlinear theory of the oscillations of an inhomogeneous plasma," Zh. Éksp. Teor. Fiz., 48:1372 (1965).


* For a survey of this topic see [27].

11. P. A. Sturrock, "Nonlinear effects in an electron plasma," Proc. Roy. Soc., A242:277 (1957); L. M. Gorbunov, "Nonlinear variation of the spectra of Langmuir oscillations of a plasma," Zh. Éksp. Teor. Fiz., (in print).
12. V. G. Makhan'kov and V. N. Tsytovich, "Coulomb collisions of particles in a turbulent plasma," Zh. Éksp. Teor. Fiz., 53:1789 (1967).
13. V. N. Tsytovich, "Note added in proof to the article 'Nonlinear wave interactions in a plasma,' " Uspekhi Fiz. Nauk, 90 (1966).
14. A. A. Vedenov and L. I. Rudakov, "Wave interaction in continuous media," Dokl. Akad. Nauk SSSR, 159:767 (1964).
15. A. Gailitis, Problems in the Interaction of Radiation and Fast Particles with a Medium (author's abstract of dissertation). Moscow (1966).
16. V. N. Tsytovich, "Spontaneous excitation of magnetic fields in a turbulent plasma," Dokl. Akad. Nauk SSSR 1:181 (1968).
17. V. P. Silin, "Parametric resonance in a plasma," Zh. Éksp. Teor. Fiz., 40:1679 (1965).
18. Yu. M. Aliev and V. P. Silin, "Theory of the oscillations of a plasma in a high-frequency electric field," Zh. Éksp. Teor. Fiz., 48:901 (1965).
19. A. I. Akhiezer and G. Ya. Lyubarskii, "Nonlinear theory of the oscillations of an electron plasma," Dokl. Akad. Nauk SSSR, 80:193 (1951).
20. A. I. Akhiezer, G. Ya. Lyubarskii, and Ya. B. Fainberg, "Nonlinear theory of oscillations in a plasma," Uch. Zap. Kharkovsk. Gos. Univ., 6:73 (1955).
21. R. Z. Sagdeev, "Collective processes and shock waves in a rarefied plasma," Problems in Plasma Theory, Vol. 4. Atomizdat (1964), p. 20.
22. Ya. B. Fainberg, "Particle acceleration in a plasma," Atomnaya Énergiya, 6:431 (1959).
23. V. N. Tsytovich, "Structure of relativistic nonlinear waves in a plasma," Dokl. Akad. Nauk SSSR, 142:63 (1962).
24. V. G. Makhan'kov and V. N. Tsytovich, "Collisions in a turbulent plasma," Zh. Éksp. Teor. Fiz., 37:1381 (1967).
25. V. G. Makhan'kov and V. N. Tsytovich, "Ion-ion collisions and the nonlinear interaction of Langmuir waves in a turbulent plasma," Preprints of the Joint Institute for Nuclear Research (OIYaI), Dubna (1967), NP 3373-2.
26. É. N. Krivorutskii, V. G. Makhan'kov, and V. N. Tsytovich, "Drift oscillations of a magnetized turbulent plasma," Preprints of the Joint Institute for Nuclear Research (OIYaI), Dubna (1968), NP 3394-2.
27. V. P. Silin, "Interaction of a strong high-frequency electromagnetic field with a plasma," Paper at the Conference on Phenomena in Ionized Gases, Vienna (1967). FIAN Preprint No. 133 (1967).

*Appendix I*

# Derivation of the Nonlinear Equations for Slowly Varying Amplitudes of the Interacting Fields

Consider Eq. (2.47):

$$\omega^2 \varepsilon^l(\omega, \mathbf{k}) \mathbf{E}_k^l = \mathbf{j}_k^l. \tag{A1.1}$$

We write it in a form containing the longitudinal field as a function of the coordinates and time:

$$\mathbf{E}^l(r, t) = (2\pi)^{-4} \int \mathbf{E}_k^l \exp(-i\omega t + i\mathbf{k}\mathbf{r})\, dk.$$

For this it is sufficient to multiply (A1.1) by $\exp(-i\omega t + i\mathbf{kr})$ and integrate over k ($dk = d\mathbf{k}d\omega$):

$$\int \omega^2 \varepsilon^l(\omega, \mathbf{k}) \mathbf{E}_k^l \exp(-i\omega t + i\mathbf{k}\mathbf{r})\, dk = \\ = \int \mathbf{j}_k^l \exp(-i\omega t + i\mathbf{k}\mathbf{r})\, dk. \tag{A1.2}$$

Noting that $\omega \exp(-i\omega t) = i(\partial/\partial t) \exp(-i\omega t)$ and $\mathbf{k} \exp(i\mathbf{kr}) = i(\partial/\partial \mathbf{r}) \cdot \exp(i\mathbf{kr})$, we can replace $\omega$ by $i(\partial/\partial t)$ and $\mathbf{k}$ by $-i(\partial/\partial)\mathbf{r}$. The left-hand side of (A1.2) becomes

$$\left(i \frac{\partial}{\partial t}\right)^2 \varepsilon^l\left(i \frac{\partial}{\partial t},\ -i \frac{\partial}{\partial \mathbf{r}}\right) \int \mathbf{E}_k^l \exp(-i\omega t + i\mathbf{k}\mathbf{r})\, dk. \tag{A1.3}$$

The integral over k in (A1.3) gives the field $\mathbf{E}^l(\mathbf{r}, t)$.

Consequently, Eq. (A1.1) acquires the form

$$-\frac{\partial^2}{\partial t^2}\varepsilon^l\left(i\frac{\partial}{\partial t},\ -i\frac{\partial}{\partial \mathbf{r}}\right)\mathbf{E}^l(\mathbf{r},t)=\int \mathbf{j}_k^l\exp\left(-i\omega t+i\mathbf{k}\mathbf{r}\right)dk. \quad (A1.4)$$

The left-hand side of this equation becomes [with the insertion of (2.49)]

$$-\int\frac{\partial^2}{\partial t^2}\varepsilon^l\left(i\frac{\partial}{\partial t},\ -i\frac{\partial}{\partial \mathbf{r}}\right)\mathbf{E}_{\mathbf{k}}^l(\mathbf{r},t)\exp\left[-i\omega^l(\mathbf{k})t+i\mathbf{k}\mathbf{r}\right]d\mathbf{k}. \quad (A1.5)$$

Every differential operator now acts on two functions, the wave amplitude $\mathbf{E}_{\mathbf{k}}^l(\mathbf{r}, t)$ and $\exp[-i\omega^l(\mathbf{k})t+i\mathbf{k}\mathbf{r}]$. We observe that

$$i\frac{\partial}{\partial t}\exp\left[-i\omega^l(\mathbf{k})t+i\mathbf{k}\mathbf{r}\right]\mathbf{E}_{\mathbf{k}}^l(\mathbf{r},t)=$$
$$=\left(\omega^l(\mathbf{k})\mathbf{E}_{\mathbf{k}}^l+i\frac{\partial \mathbf{E}_{\mathbf{k}}^l}{\partial t}\right)\exp\left[-i\omega^l(\mathbf{k})t+i\mathbf{k}\mathbf{r}\right], \quad (A1.6)$$

i.e., $i(\partial/\partial t)$ goes over to $\omega^l(\mathbf{k})+i(\partial/\partial t)$, where the differential operator acts only on the wave amplitude. Similarly $-i(\partial/\partial\mathbf{r})$ goes over to $\mathbf{k}-i(\partial/\partial\mathbf{r})$. Thus, (A1.5) becomes

$$\int\exp\left[-i\omega^l(\mathbf{k})t+i\mathbf{k}\mathbf{r}\right]d\mathbf{k}\left(\omega^l(\mathbf{k})+i\frac{\partial}{\partial t}\right)^2\times$$
$$\times\,\varepsilon^l\left(\omega^l(\mathbf{k})+i\frac{\partial}{\partial t}\quad \mathbf{k}-i\frac{\partial}{\partial \mathbf{r}}\right)\mathbf{E}_{\mathbf{k}}^l(\mathbf{r},t). \quad (A1.7)$$

Now, recognizing that the amplitude $\mathbf{E}_{\mathbf{k}}^l(\mathbf{r}, t)$ is a slowly varying function, we need retain only its first derivatives, i.e., we can use the expansion

$$\left(\omega^l(\mathbf{k})+i\frac{\partial}{\partial t}\right)^2\cdot\varepsilon^l\left(\omega^l(\mathbf{k})+i\frac{\partial}{\partial t},\ \mathbf{k}-i\frac{\partial}{\partial \mathbf{r}}\right)=(\omega^l(\mathbf{k}))^2\varepsilon^l(\omega^l(\mathbf{k}),\mathbf{k})+$$
$$+\,i\frac{\partial}{\partial t}\frac{\partial}{\partial\omega}\omega^2\varepsilon^l\big|_{\omega=\omega^l(\mathbf{k})}-i\frac{\partial}{\partial\mathbf{r}}\frac{\partial}{\partial\mathbf{k}}\omega^2\varepsilon^l(\omega,\mathbf{k})_{\omega=\omega^l(\mathbf{k})}. \quad (A1.8)$$

The first term of (A1.8) vanishes, because $\omega_{\mathbf{k}}^l$ is by definition the solution of the equation

$$\varepsilon^l(\omega^l(\mathbf{k}),\mathbf{k})=0. \quad (A1.9)$$

Differentiating (A1.9) with respect to $\mathbf{k}$, we have

$$\frac{\partial\varepsilon^l}{\partial\omega}\frac{d\omega^l}{d\mathbf{k}}+\frac{\partial\varepsilon^l}{\partial\mathbf{k}}=0. \quad (A1.10)$$

We introduce the group velocity of the waves

$$\frac{d\omega^l}{d\mathbf{k}} = \mathbf{v}_{\mathrm{gr}}^l .$$

It follows from (A1.10) that the longitudinal-wave group velocity can be expressed in terms of the plasma dielectric constant:

$$\mathbf{v}_{\mathrm{gr}}^l = -\frac{\dfrac{\partial \varepsilon^l(\omega^l(\mathbf{k}),\, \mathbf{k})}{\partial \mathbf{k}}}{\left.\dfrac{\partial \varepsilon^l}{\partial \omega}\right|_{\omega=\omega^l(\mathbf{k})}} = -\left.\frac{\dfrac{\partial}{\partial \mathbf{k}}\, \omega^2 \varepsilon^l(\omega,\, \mathbf{k})}{\dfrac{\partial}{\partial \omega}(\varepsilon^l(\omega,\, \mathbf{k})\, \omega^2)}\right|_{\omega=\omega^l(\mathbf{k})} .$$

Here we have made use of the fact that $(\partial/\partial\omega)\omega^2\varepsilon^l = \omega^2(\partial\varepsilon^l/\partial\omega)$, since $\varepsilon^l = 0$. Hence (A1.8) may be written in the form

$$-\left(\frac{\partial}{\partial \omega}\, \omega^2 \varepsilon^l\right)_{\omega=\omega^l(\mathbf{k})} \frac{1}{i}\left(\frac{\partial}{\partial t} + \mathbf{v}_{\mathrm{gr}}^l \frac{\partial}{\partial \mathbf{r}}\right).$$

Thus, the left-hand side of (A1.4) may be written as follows:

$$\int \exp\left[-i\omega^l(\mathbf{k})\, t + i\mathbf{k}\mathbf{r}\right]\Big[-\left(\frac{\partial}{\partial \omega}\, \omega^2 \varepsilon^l\right)_{\omega=\omega^l(\mathbf{k})} \times \times \frac{1}{i}\left(\frac{\partial}{\partial t} + \mathbf{v}_{\mathrm{gr}}^l \frac{\partial}{\partial \mathbf{r}}\right) \mathbf{E}_{\mathbf{k}}^l(\mathbf{r},\, t)\Big]\, d\mathbf{k}. \qquad \text{(A1.11)}$$

We now consider a volume whose linear dimension is small in comparison with the characteristic scale length for the variation of $\mathbf{E}_{\mathbf{k}}^l(\mathbf{r},\, t)$, but large in comparison with the wavelength $\lambda = 2\pi/k$. We assume that $\partial \mathbf{E}_{\mathbf{k}}^l/\partial t$, as well as $\partial \mathbf{E}_{\mathbf{k}}^l/\partial \mathbf{r}$, are almost constant in this volume, i.e., the time variations of the amplitude are also small. This volume can be considered physically infinitesimal compared with changes in the field amplitude.

We multiply (A1.11) by $[1/(2\pi)^3] \exp[i\omega^l(\mathbf{k}')t - i\mathbf{k}'\mathbf{r}]$ and integrate the result over the physically infinitesimal volume. The expression

$$\frac{1}{i}\left(\frac{\partial}{\partial t} + \mathbf{v}_{\mathrm{gr}}^l \frac{\partial}{\partial \mathbf{r}}\right) \mathbf{E}_{\mathbf{k}}^l(\mathbf{r},\, t),$$

which is constant within this volume, can be taken outside the integral sign. The integral then consists of

$$\frac{1}{(2\pi)^3}\int e^{i(\mathbf{k}-\mathbf{k}')\mathbf{r}}\, d\mathbf{r} = \delta(\mathbf{k} - \mathbf{k}').$$

After integration over $\mathbf{k}$ only $\mathbf{k} = \mathbf{k}'$ remains. Thus, (A1.11) becomes

$$-\left(\frac{\partial}{\partial\omega}\,\omega^2\varepsilon^l(\omega,\mathbf{k}')\right)_{\omega=\omega^l(\mathbf{k}')}\frac{1}{i}\left(\frac{\partial}{\partial t}+\mathbf{v}^l_{\mathrm{gr}}(\mathbf{k}')\frac{\partial}{\partial\mathbf{r}}\right)\mathbf{E}^l_{\mathbf{k}'}(\mathbf{r},t).$$

We now see what form the right-hand side of (A1.4) takes. After multiplying by $\exp[-i\mathbf{k}'\mathbf{r}+i\omega^l(\mathbf{k}')t]$ and integrating over the above volume, the amplitudes $\mathbf{E}_{\mathbf{k}}(\mathbf{r},t)$ being assumed approximately constant on the right-hand side, i.e.,

$$\frac{1}{(2\pi)^3}\int\exp[-i\mathbf{k}'\mathbf{r}+i\omega^l(\mathbf{k}')t]\,\mathbf{j}_{\mathbf{k}\omega}\exp[-i\omega t+i\mathbf{k}\mathbf{r}]\,d\mathbf{k}\,d\omega\,d\mathbf{r}=$$
$$=\int\exp[i\omega^l(\mathbf{k}')t-i\omega t]\,\mathbf{j}_{\mathbf{k}\omega}\delta(\mathbf{k}-\mathbf{k}')\,d\mathbf{k}\,d\omega=$$
$$=\int\exp[i\omega^l(\mathbf{k}')t-i\omega t]\,\mathbf{j}_{\mathbf{k}'\omega}\,d\omega. \qquad \text{(A1.12)}$$

We then obtain the desired equation

$$\left(\frac{\partial}{\partial t}+\mathbf{v}^l_{\mathrm{gr}}\frac{\partial}{\partial\mathbf{r}}\right)\mathbf{E}^l_{\mathbf{k}}(\mathbf{r},t)=$$
$$=-i\,\frac{1}{\left(\frac{\partial}{\partial\omega}\,\omega^2\varepsilon^l\right)\Big|_{\omega=\omega^l(\mathbf{k})}}\int\exp[-i\omega t+i\omega^l(\mathbf{k})t]\,\mathbf{j}^l_{\mathbf{k}\omega}\,d\omega. \qquad \text{(A1.13)}$$

Similar calculations yield the following equation for the transverse-wave amplitudes $\mathbf{E}^t_{\mathbf{k}}$:

$$\left(\frac{\partial}{\partial t}+\mathbf{v}^t_{\mathrm{gr}}\frac{\partial}{\partial\mathbf{r}}\right)\mathbf{E}^t_{\mathbf{k}}(\mathbf{r},t)=$$
$$=i\,\frac{1}{\left(\frac{\partial}{\partial\omega}\,\omega^2\varepsilon^t\right)\Big|_{\omega=\omega^t(\mathbf{k})}}\int\exp[-i\omega t+i\omega^t(\mathbf{k})t]\,\mathbf{j}^t_{\mathbf{k}\omega}\,d\omega. \qquad \text{(A1.14)}$$

*Appendix 2*

# Averaging of the Nonlinear Equations over the Phases of the Interacting Waves

## Phases of All Waves Random

1. Consider Eq. (I) (Chap. 2, § 8). Multiplying it by $A_{\mathbf{k}}^{l*}$, we obtain the expression

$$-2\gamma_{\mathbf{k}}^{l} A_{\mathbf{k}}^{l} A_{\mathbf{k}'}^{l*} + \frac{\partial}{\partial t} A_{\mathbf{k}}^{l} A_{\mathbf{k}'}^{l*} + \mathbf{v}_{\mathrm{gr}}^{l} \frac{\partial}{\partial \mathbf{r}} A_{\mathbf{k}}^{l} A_{\mathbf{k}'}^{l*} = -2\gamma_{\mathbf{k}}^{l} A_{\mathbf{k}}^{l} A_{\mathbf{k}'}^{l*} +$$

$$+ A_{\mathbf{k}}^{l} \frac{\partial A_{\mathbf{k}'}^{l*}}{\partial t} + A_{\mathbf{k}'}^{l*} \frac{\partial}{\partial t} A_{\mathbf{k}}^{l} + \mathbf{v}_{\mathrm{gr}}^{l} A_{\mathbf{k}}^{l} \frac{\partial A_{\mathbf{k}'}^{l*}}{\partial \mathbf{r}} + \mathbf{v}_{\mathrm{gr}}^{l} A_{\mathbf{k}'}^{l*} \frac{\partial}{\partial \mathbf{r}} A_{\mathbf{k}}^{l} =$$

$$= -\int \alpha_{\mathbf{k},\,\mathbf{k}_1,\,\mathbf{k}_2}^{l} A_{\mathbf{k}'}^{l*} A_{\mathbf{k}_1}^{t} A_{\mathbf{k}_2}^{t*} \exp\left(-i\Delta\Omega_{\mathbf{k},\,\mathbf{k}_1,\,\mathbf{k}_2} t\right) d\mathbf{k}_1\, d\mathbf{k}_2 -$$

$$- \int \alpha_{\mathbf{k}',\,\mathbf{k}_1,\,\mathbf{k}_2}^{l*} A_{\mathbf{k}}^{l} A_{\mathbf{k}_1}^{t*} A_{\mathbf{k}_2}^{t} \exp\left(i\Delta\Omega_{\mathbf{k}',\,\mathbf{k}_1,\,\mathbf{k}_2} t\right) d\mathbf{k}_1\, d\mathbf{k}_2. \qquad \text{(A2.1)}$$

We average (A2.1) over the phases, incorporating (3.38) into the left-hand side of the equation and integrating over $\mathbf{k}'$:

$$\left(\frac{\partial}{\partial t} - 2\gamma_{\mathbf{k}}^{l} + \mathbf{v}_{\mathrm{gr}}^{l} \frac{\partial}{\partial \mathbf{r}}\right) |A_{\mathbf{k}}^{l}|^2 = -\int \alpha_{\mathbf{k},\,\mathbf{k}_1,\,\mathbf{k}_2}^{l} \langle A_{\mathbf{k}'}^{l*} A_{\mathbf{k}_1}^{t} A_{\mathbf{k}_2}^{t*} \rangle \times$$

$$\times \exp\left(-i\Delta\Omega_{\mathbf{k},\,\mathbf{k}_1,\,\mathbf{k}_2} t\right) d\mathbf{k}'\, d\mathbf{k}_1 d\mathbf{k}_2 -$$

$$- \int \alpha_{\mathbf{k}',\,\mathbf{k}_1,\,\mathbf{k}_2}^{l*} \langle A_{\mathbf{k}}^{l} A_{\mathbf{k}_1}^{t*} A_{\mathbf{k}_2}^{t} \rangle\, d\mathbf{k}'\, d\mathbf{k}_1\, d\mathbf{k}_2 \exp\left(i\Delta\Omega_{\mathbf{k}',\,\mathbf{k}_1,\,\mathbf{k}_2} t\right). \qquad \text{(A2.2)}$$

In order to express the three-field product in terms of four-field products, we use relations deduced from Eqs. (I), (II), and (III):

$$A_{\mathbf{k}}^{l} = \int \alpha_{\mathbf{k},\,\mathbf{k}_1,\,\mathbf{k}_2}^{l} e^{-i\Delta\Omega t} \frac{A_{\mathbf{k}_1}^{t} A_{\mathbf{k}_2}^{t*} d\mathbf{k}_1 \, d\mathbf{k}_2}{i(\Delta\Omega + i\delta)}; \tag{A2.3}$$

$$A_{\mathbf{k}_1}^{t} = \int \alpha_{1,\,\mathbf{k},\,\mathbf{k}_1,\,\mathbf{k}_2}^{t} e^{i\Delta\Omega t} \frac{A_{\mathbf{k}}^{l} A_{\mathbf{k}_2}^{t} d\mathbf{k} \, d\mathbf{k}_2}{i(\Delta\Omega - i\delta)}; \tag{A2.4}$$

$$A_{\mathbf{k}_2}^{t} = \int \alpha_{2,\,\mathbf{k},\,\mathbf{k}_1,\,\mathbf{k}_2}^{t} e^{-i\Delta\Omega t} \frac{A_{\mathbf{k}}^{l*} A_{\mathbf{k}_1}^{t} d\mathbf{k} \, d\mathbf{k}_1}{i(\Delta\Omega + i\delta)}. \tag{A2.5}$$

Here $\delta \to +0$.

The above solutions are only approximate, because their derivation rested on the assumption that, since $\Delta\Omega t_0 \gg 1$, the most rapidly varying quantity is $e^{i\Delta\Omega t}$, which passes through zero many times during the characteristic time $t_0$ for changes in $A_{\mathbf{k}_1}$ and $A_{\mathbf{k}_2}$, i.e., it executes many oscillations in that period. Consequently, in the solution of Eq. (I)

$$\left(\frac{\partial}{\partial t} - \gamma_{\mathbf{k}}^{l} + \mathbf{v}_{\mathrm{gr}} \frac{\partial}{\partial \mathbf{r}}\right) A_{\mathbf{k}}^{l} = \int \alpha_{\mathbf{k},\,\mathbf{k}_1,\,\mathbf{k}_2} e^{i\Delta\Omega t} A_{\mathbf{k}_1}^{t} A_{\mathbf{k}_2}^{t*} d\mathbf{k}_1 \, d\mathbf{k}_2 \tag{A2.6}$$

for $A_{\mathbf{k}}^{l}$, for example, $A_{\mathbf{k}_1}^{t}$ and $A_{\mathbf{k}_2}^{t}$ may be assumed roughly constant on the right-hand side of the equation, while on the left-hand side of (I) we can neglect $v_{gr}\, \partial/\partial\mathbf{r}$ and $\gamma_{\mathbf{k}}^{l}$, since $e^{i\Delta\Omega t}$ on the right-hand side does not depend on r, and $\gamma_{\mathbf{k}}^{l}$ is of the same order as the recip-characteristic time $1/t_0$ for change in $A_{\mathbf{k}}$. In short, perturbations of $A_{\mathbf{k}}^{l}$ due to oscillations with frequency $\Delta\Omega$ will also have a frequency on the order of $\Delta\Omega$, i.e., on the left-hand side of (I) $\partial/\partial t$ will be of order $\Delta\Omega$ and, for this reason, $\partial/\partial t \gg \gamma_{\mathbf{k}}$, $\mathbf{v}_{\mathrm{rp}} \cdot \partial/\partial\mathbf{r}$. The solution of an equation containing only $\partial/\partial t$ on the left-hand side is equal to the integral of the right-hand side over t. This integral is set between the limits $-\infty$ and t, and it is assumed that the integrand vanishes at $t = -\infty$, i.e., that $\Delta\Omega$ contains a negative infinitesimal imaginary part ($i\delta \to 0$). We thereby obtain (A2.3), where $i\delta$ indicates the proper path around the pole $\Delta\Omega = 0$.

If completely random fields are inserted into the right-hand sides of (A2.3)-(A2.5), the left-hand sides do not correspond to completely random fields; they can contain small corrections associated with phase correlation. It is possible to use (A2.3)-(A2.5) in order to express A in terms of $A_0$. For example, the first term on the right-hand side of (A2.2) assumes the form

$$-\int \alpha_{\mathbf{k},\,\mathbf{k}_1,\,\mathbf{k}_2}^{l} \langle A_{\mathbf{k}'}^{l*} A_{\mathbf{k}_1}^{t} A_{\mathbf{k}_2}^{t*} \rangle \, d\mathbf{k}' \, d\mathbf{k}_1 \, d\mathbf{k}_2 \exp(-i\Delta\Omega_{\mathbf{k},\,\mathbf{k}_1,\,\mathbf{k}_2} t) =$$

$$= \int \frac{\alpha^{l}_{\mathbf{k},\,\mathbf{k}_1,\,\mathbf{k}_2}\alpha^{l*}_{\mathbf{k}',\,\mathbf{k}_1',\,\mathbf{k}_2'}}{i\,(\Delta\Omega_{\mathbf{k}',\,\mathbf{k}_1',\,\mathbf{k}_2'} - i\delta)} \langle A^{t*}_{0\mathbf{k}_1'} A^{t}_{0\mathbf{k}_2'} A^{t}_{0\mathbf{k}_1} A^{t*}_{0\mathbf{k}_2}\rangle \times$$

$$\times \exp\left[-i\,(\Delta\Omega_{\mathbf{k},\,\mathbf{k}_1,\,\mathbf{k}_2} - \Delta\Omega_{\mathbf{k}',\,\mathbf{k}_1',\,\mathbf{k}_2'})\,t\right] d\mathbf{k}_1\, d\mathbf{k}_2\, d\mathbf{k}'\, d\mathbf{k}_1'\, d\mathbf{k}_2' -$$

$$- \int \frac{\alpha^{l}_{\mathbf{k},\,\mathbf{k}_1,\,\mathbf{k}_2}\alpha_{\mathbf{k}'',\,\mathbf{k}_1,\,\mathbf{k}_2'}}{i\,(\Delta\Omega_{\mathbf{k}_1'',\,\mathbf{k}_1,\,\mathbf{k}_2'} - i\delta)} \langle A^{l*}_{0\mathbf{k}'} A^{l}_{0\mathbf{k}''} A^{t}_{0\mathbf{k}_2'} A^{t*}_{0\mathbf{k}_2}\rangle \times$$

$$\times \exp\left[-i\,(\Delta\Omega_{\mathbf{k},\,\mathbf{k}_1,\,\mathbf{k}_2} - \Delta\Omega_{\mathbf{k}'',\mathbf{k}_1,\mathbf{k}_2'})\,t\right] d\mathbf{k}_1\, d\mathbf{k}_2\, d\mathbf{k}'\, d\mathbf{k}''\, d\mathbf{k}_2' +$$

$$+ \int \frac{\alpha^{l}_{\mathbf{k},\,\mathbf{k}_1,\,\mathbf{k}_2}\alpha^{t*}_{\mathbf{k}'',\,\mathbf{k}_1',\,\mathbf{k}_2}}{i\,(\Delta\Omega_{\mathbf{k}'',\,\mathbf{k}_1',\,\mathbf{k}_2} + i\delta)} \langle A^{l*}_{0\mathbf{k}'} A^{t}_{0\mathbf{k}_1} A^{l}_{0\mathbf{k}''} A^{t*}_{0\mathbf{k}_1'}\rangle \times$$

$$\times \exp\left[-i\,(\Delta\Omega_{\mathbf{k},\,\mathbf{k}_1,\,\mathbf{k}_2} - \Delta\Omega_{\mathbf{k}'',\,\mathbf{k}_1',\,\mathbf{k}_2})\,t\right] d\mathbf{k}_1\, d\mathbf{k}_2\, d\mathbf{k}'\, d\mathbf{k}''\, d\mathbf{k}_1'. \qquad \text{(A2.6)}$$

The average of the product of four random amplitudes is equal to a sum of terms in which the amplitudes are averaged in pairs as the different amplitudes $A_0$ are statistically independent. It is important to bear in mind that longitudinal waves do not correlate with transverse waves, and certain correlators give $\delta(\mathbf{k}_1 - \mathbf{k}_2)$, i.e., $\alpha \sim \delta(\mathbf{k})$. As $\mathbf{k} \neq 0$, they need not be included. Therefore, for example,

$$\langle A^{t*}_{0\mathbf{k}_1'} A^{t}_{0\mathbf{k}_2'} A^{t}_{0\mathbf{k}_1} A^{t*}_{0\mathbf{k}_2}\rangle = \langle A^{t*}_{0\mathbf{k}_1'} A^{t}_{0\mathbf{k}_1}\rangle \langle A^{t}_{0\mathbf{k}_2'} A^{t*}_{0\mathbf{k}_2}\rangle + \langle A^{t*}_{0\mathbf{k}_1'} A^{t*}_{0\mathbf{k}_2}\rangle \langle A^{t}_{0\mathbf{k}_2'} A^{t}_{0\mathbf{k}_1}\rangle. \qquad \text{(A2.7)}$$

The other pairing, $\langle A^{*}_{\mathbf{k}_1'}\, A_{\mathbf{k}_2'}\rangle \langle A_{\mathbf{k}_1}\;\; A^{*}_{\mathbf{k}_2}\rangle$, yields zero by virtue of the foregoing discussion.

Hence

$$\langle A^{t*}_{\mathbf{k}_1'} A^{t}_{\mathbf{k}_2'} A^{t}_{\mathbf{k}_1} A^{t*}_{\mathbf{k}_2}\rangle \simeq |A^{t}_{\mathbf{k}_1}|^2\, |A^{t}_{\mathbf{k}_2}|^2 \,(\delta(\mathbf{k}_1 - \mathbf{k}_1')\,\delta(\mathbf{k}_2 - \mathbf{k}_2') +$$

$$+ \delta(\mathbf{k}_1 + \mathbf{k}_2')\,\delta(\mathbf{k}_2 + \mathbf{k}_1')). \qquad \text{(A2.8)}$$

At once we see that $\mathbf{k}_1' \to \mathbf{k}_1$ and $\mathbf{k}_2' \to \mathbf{k}_2$ in the term on the right-hand side of the desired equation containing the first term of (A2.8) and that $\mathbf{k}' \to \mathbf{k}$ by virtue of the $\delta(\mathbf{k}' - \mathbf{k}_1' + \mathbf{k}_2')$ contained in $\alpha^{l}_{\mathbf{k}',\,\mathbf{k}_1',\,\mathbf{k}_2'}$. Consequently, $\Delta\Omega_{\mathbf{k}',\,\mathbf{k}_1',\,\mathbf{k}_2'} \to \Delta\Omega_{\mathbf{k},\,\mathbf{k}_1,\,\mathbf{k}_2}$. The second term of (A2.8) leads to the same result. This means that the first term of (A2.6) takes the form

$$\int d\mathbf{k}_1\, d\mathbf{k}_2 \frac{|A^{t}_{\mathbf{k}_1}|^2\, |A^{t}_{\mathbf{k}_2}|^2\, |(\mathbf{e}_{\mathbf{k}_1}\mathbf{e}^{*}_{\mathbf{k}_2})|^2 e^2 \omega^4_{0e} \mathbf{k}^2 \delta(\mathbf{k} - \mathbf{k}_1 + \mathbf{k}_2)}{4m_e^2\,(\Omega^t(\mathbf{k}_1))^2\,(\Omega^t(\mathbf{k}_2))^2 \left(\dfrac{\partial}{\partial\omega}\,\omega^2\varepsilon^l\right)^2_{\omega=\Omega^l(\mathbf{k})} i\,(\Delta\Omega_{\mathbf{k},\,\mathbf{k}_1,\,\mathbf{k}_2} - i\delta)}. \qquad \text{(A2.9)}$$

Equation (A2.9) depends on the polarization of the waves described by $\mathbf{e}_{\mathbf{k}_1}$ and $\mathbf{e}_{\mathbf{k}_2}$. Frequently one is concerned with the interaction of unpolarized waves. In this case (A2.9) must be averaged over the wave polarizations. It is helpful to choose the unit polarization vectors $\mathbf{e}_{k_1}$ and $\mathbf{e}_{k_2}$ such that one of them falls in the plane of the vectors $\mathbf{k}_1$ and $\mathbf{k}_2$, the second in the direction perpendicular to that plane. Then the terms $\alpha = \alpha'$ are not equal to zero in the scalar product $\sum_{\alpha\alpha'=1,2}(\mathbf{e}^{\alpha}_{\mathbf{k}_1}\mathbf{e}^{\alpha'}_{\mathbf{k}_2})^2$; the scalar product of two vectors perpendicular to the plane of $\mathbf{k}_1\mathbf{k}_2$ is equal to unity, and the scalar product of two vectors parallel to that plane are equal to $(\mathbf{k}_1\mathbf{k}_2)/k_1k_2$, i.e.,

$$\sum_{\alpha\alpha'=1,2}(\mathbf{e}^{\alpha}_{\mathbf{k}_1}\mathbf{e}^{\alpha'}_{\mathbf{k}_2})^2 = 1 + \frac{(\mathbf{k}_1\mathbf{k}_2)^2}{k_1^2k_2^2}. \tag{A2.10}$$

Henceforth for simplicity we analyze the effects of the interaction of unpolarized transverse waves, for which (A2.10) is valid.

The desired equation (A2.1) contains only the real part of (A2.9),* i.e., (A2.9) includes

$$\operatorname{Re}\frac{1}{i\Delta\Omega + \delta} = \pi\delta(\Delta\Omega). \tag{A2.11}$$

The other two terms are calculated similarly. The desired equation then takes the form

$$\left(\frac{\partial}{\partial t} - \gamma^l_{\mathbf{k}} + \mathbf{v}_{\text{gr}}\frac{\partial}{\partial \mathbf{r}}\right)|A^l_{\mathbf{k}}|^2 = \frac{e^2\omega_{0e}^4k^2}{2m_e^2}\int d\mathbf{k}_1\,d\mathbf{k}_2\left(1 + \frac{(\mathbf{k}_1\mathbf{k}_2)^2}{k_1^2k_2^2}\right)\times$$
$$\times\left\{\frac{|A^t_{\mathbf{k}_1}|^2|A^t_{\mathbf{k}_2}|^2}{4\omega_1^2\omega_2^2\left(\frac{\partial}{\partial\omega}\omega^2\varepsilon^l\right)^2_{\omega=\Omega^l(\mathbf{k})}} - \frac{|A^t_{\mathbf{k}_2}|^2|A^l_{\mathbf{k}}|^2}{2\omega_2^2\omega^2\left(\frac{\partial}{\partial\omega}\omega^2\varepsilon^t_1\right)_{\omega=\Omega^t(\mathbf{k}_1)}\left(\frac{\partial}{\partial\omega}\omega^2\varepsilon^l\right)_{\omega=\Omega^l(\mathbf{k})}} +\right.$$
$$\left. + \frac{|A^t_{\mathbf{k}_1}|^2|A^l_{\mathbf{k}}|^2}{2\omega_1^2\omega^2\left(\frac{\partial}{\partial\omega}\omega^2\varepsilon^t_2\right)_{\omega=\Omega^t(\mathbf{k}_2)}\left(\frac{\partial}{\partial\omega}\omega^2\varepsilon^l\right)_{\omega=\Omega^l(\mathbf{k})}}\right\}\pi\delta(\Delta\Omega)\,\delta(\mathbf{k} - \mathbf{k}_1 + \mathbf{k}_2); \tag{A2.12}$$
$$\omega_{1,2} = \Omega^t(\mathbf{k}_{1,2}).$$

The equations for $|A^t_{\mathbf{k}_1}|^2$ and $|A^t_{\mathbf{k}_2}|^2$ are obtained similarly.

**2.** Now consider the one-dimensional model, when the directions of all interacting waves coincide. We denote the projection

* The second term of (A2.1) is the complex conjugate of the first.

of the wave vector on to this direction by $\mathbf{k}_{\parallel}$, and that perpendicular to this direction by $\mathbf{k}_{\perp}$. Thus,

$$\mathbf{k}^t_{\perp_1} = \mathbf{k}^t_{\perp_2} = \mathbf{k}^l_{\perp} = 0; \; k_{\parallel} = k$$

With the above in mind, we must write the spectral energy density of the waves in the form

$$N_{\mathbf{k}} = N_k \delta(\mathbf{k}_{\perp}). \tag{A2.13}$$

The values of $N_k$ determined by (A2.13) are proportional to the one-dimensional spectral energy density divided by the interval dk in the direction of wave propagation. The equations for $N_k$ are easily derived by substituting (A2.13) into (3.50)-(3.52) and integrating both sides over $\mathbf{k}_{\perp}$. For example, let us write out the equation for $N_k$:

$$\mathbf{v}^l_{\text{gr}} \frac{\partial N^l_k}{\partial \mathbf{r}} + \frac{\partial N^l_k}{\partial t} - \gamma^l_k N^l_k = \int u(k_1, k_2)\, \delta(k - k_1 + k_2)\, \delta(\Omega^l - \Omega^t_1 + \Omega^t_2) \times$$
$$\times (N^t_{k_1} N^t_{k_2} - N^t_{k_2} N^l_k + N^t_{k_1} N^l_k)\, dk_1 dk_2;$$
$$u(\mathbf{k}_1, \mathbf{k}_2)\, \delta(\mathbf{k} - \mathbf{k}_1 + \mathbf{k}_2) = \widetilde{w}(\mathbf{k}_1, \mathbf{k}_2)\pi. \tag{A2.14}$$

The right-hand side of the equations thus obtained contain two differential intervals $dk_1 dk_2$ and two functions $\delta(k - k_1 + k_2)$ and $\delta[\omega^l(k) - \omega^t(k_1) + \omega^t(k_2)]$, which can be used to compute the integrals. Thus integration over $k_2$ causes $k_2$ to be replaced by $k_1 - k$ and, accordingly, $\delta[\omega^l(k) - \omega^t(k_1) + \omega^t(k_2)]$ by $\delta[\omega^l(k) - \omega^t(k_1) + \omega^t(k_1 - k)]$. For integration over $k_1$ by means of the latter delta function, it is convenient to use the following equation, which is easily checked:

$$\int f(k_1)\, \delta(\varphi(k_1))\, dk_1 = \frac{f(k_{10})}{\left| \dfrac{\partial \varphi(k_1)}{\partial k_1} \right|_{k_1 = k_{10}}}, \tag{A2.15}$$

where $k_{10}$ is the solution of the equation $\varphi(k_{10}) = 0$. Recognizing that $\Omega(k_1) = \sqrt{k^2c^2 + \omega^2_{0e}}$, we obtain

$$\mathbf{v}^l_{\text{gr}} \frac{\partial N^l}{\partial \mathbf{r}} + \left( \frac{\partial}{\partial t} - \gamma^l \right) N^l = \widetilde{\beta}^l (N^t_1 N^t_2 - N^t_2 N^l + N^t_1 N^l); \tag{A2.16}$$

$$\widetilde{\beta}^l = -\frac{e^2 \omega^3_{0e}}{8c^2 \pi m^2_e (\Omega^t)^2} \left( \frac{\Omega^t}{\omega_{0e}} \right)^3; \; \Omega^t \gg \omega_{0e}. \tag{A2.17}$$

We have not written here the values of the wave numbers k for which $N^t_1$ and $N^t_2$ are chosen, as they are uniquely determined for a fixed longtiudinal wave number k. In fact, from the conservation law

$$\Omega^l(k) - \Omega^t(k_1) + \Omega^t(k_1 - k) = 0$$

we see that $k_1$ bears a one-to-one relation with k, and from $k_2 = k_1 - k$ we infer that $k_2$ is also uniquely determined by k. The only assumption imposed was that for wave numbers $k_1$ and $k_2$ satisfying the conservation laws $N_1^t$ and $N_2^t$ do not become zero. In other words, the frequency difference between the two packets of transverse waves must be close to the longitudinal-wave frequency. As we assumed that the packets have no waves in common, the width $\delta\Omega^t$ of the transverse wave packets is smaller than $\omega_{0e}$. The one-dimensional equations for transverse waves have a form analogous to (A2.16):

$$v_{\mathrm{gr}}^t \frac{\partial N_1^t}{\partial \mathrm{r}} + \frac{\partial N_1^t}{\partial t} - \gamma_1 N_1^t = -\widetilde{\beta}^t (N_1^t N_2^t - N_2^t N^l + N_1^t N^l); \qquad \text{(A2.18)}$$

$$v_{\mathrm{gr}}^t \frac{\partial N_2^t}{\partial \mathrm{r}} + \frac{\partial N_2^t}{\partial t} - \gamma_2 N_2^t = \widetilde{\beta}^t (N_1^t N_2^t - N_2^t N^l + N_1^t N^l); \qquad \text{(A2.19)}$$

$$\widetilde{\beta}^t = \widetilde{\beta}^l \left(\frac{\omega_{0e}}{\Omega^t}\right)^3 \quad \text{for} \quad \Omega^t \gg \omega_{0e}. \qquad \text{(A2.20)}$$

## Phases of Intense Waves Random

Here we refer to the Fourier components $A_{\mathbf{k}\omega}$ in Eqs. (I), (II), and (III) (Chap. 2, § 8). Defining $\beta_{\mathbf{k},\mathbf{k}_1,\mathbf{k}_2}^l$ by the relation

$$\alpha_{\mathbf{k},\mathbf{k}_1,-\mathbf{k}_2}^l = \beta_{\mathbf{k},\mathbf{k}_1,\mathbf{k}_2}^l \delta(\mathbf{k} - \mathbf{k}_1 - \mathbf{k}_2),$$

we obtain

$$(-i\omega - \gamma_{\mathbf{k}}) A_{\mathbf{k}\omega}^l = -\int \beta_{\mathbf{k},\mathbf{k}_1,\mathbf{k}_2}^l A_{\mathbf{k}_1,\omega_1}^t A_{\mathbf{k}_2,\omega_2}^t \delta(\mathbf{k} - \mathbf{k}_1 - \mathbf{k}_2) \times$$
$$\times \delta(\omega - \omega_1 - \omega_2 - \Delta\Omega_{\mathbf{k},\mathbf{k}_1,-\mathbf{k}_2})\, d\mathbf{k}_1\, d\mathbf{k}_2\, d\omega_1\, d\omega_2. \qquad \text{(A2.21)}$$

We average this equation over the phases of the transverse waves. The average value of $\langle A_{\mathbf{k}_1,\omega_1}^t A_{\mathbf{k}_2,\omega_2}^t \rangle$ is proportional to $\delta(\mathbf{k}_1 + \mathbf{k}_2)$, i.e., by (A2.21) it yields a result proportional to $\delta(\mathbf{k})$. The fields with $\mathbf{k} = 0$ correspond to constant fields and do not concern us. Consequently, in the averaging process we must include corrections to $A_{\mathbf{k}_1,\omega_1}^t$ and $A_{\mathbf{k}_2,\omega_2}^t$ to account for nonlinear interaction. For this we write Eqs. (II) and (III) in a form analogous to (A2.21):

$$(-i\omega_1-\gamma^t_{\mathbf{k}_1})\,A^t_{\mathbf{k}_1\omega_1}=\int\beta^t_{\mathbf{k}',\,\mathbf{k}_1,\,\mathbf{k}_2'}A^l_{\mathbf{k}',\,\omega'}A^t_{\mathbf{k}_2',\,\omega_2'}\,\delta(-\mathbf{k}'+\mathbf{k}_1-\mathbf{k}_2')\times$$
$$\times\,\delta(-\omega'+\omega_1-\omega_2'+\Delta\Omega_{\mathbf{k}',\,\mathbf{k}_1,\mathbf{k}_2'})\,d\mathbf{k}'\,d\mathbf{k}_2'\,d\omega'\,d\omega_2'$$

and

$$(-i\omega_2-\gamma^t_{\mathbf{k}_2})\,A^t_{\mathbf{k}_2\omega_2}=\int\beta^t_{\mathbf{k}',\,\mathbf{k}_1',\,\mathbf{k}_2}A^l_{\mathbf{k}',\,\omega'}A^t_{\mathbf{k}_1',\,\omega_1'}\,\delta(-\mathbf{k}'+\mathbf{k}_2-\mathbf{k}_1')\times$$
$$\times\,\delta(-\omega'+\omega_2-\omega_1'+\Delta\Omega_{\mathbf{k}',\,\mathbf{k}_2,\mathbf{k}_1'})\,d\mathbf{k}'\,d\mathbf{k}_1'\,d\omega'\,d\omega_1', \qquad \text{(A2.23)}$$

where $\alpha^t_{\mathbf{k},\,\mathbf{k}_1,\,\mathbf{k}_2}=\beta^t_{\mathbf{k},\,\mathbf{k}_1,\,\mathbf{k}_2}\,\delta(\mathbf{k}_1-\mathbf{k}-\mathbf{k}_2)$.

Substituting the latter expressions into (A2.21) and averaging over the transverse-wave phases, we obtain the following expression for $A^l_{\mathbf{k}\omega}$ (here $\delta\to+0$):

$$(-i\omega-\gamma_{\mathbf{k}})\,A^l_{\mathbf{k}\omega}=-\int\beta^l_{\mathbf{k},\,\mathbf{k}_1,\,\mathbf{k}_2}\beta^t_{\mathbf{k}',\,\mathbf{k}_1,\,\mathbf{k}_2'}A^l_{\mathbf{k}',\,\omega'}\frac{\langle A^t_{\mathbf{k}_2',\,\omega_2'}A^t_{\mathbf{k}_2,\,\omega_2}\rangle}{-i(\omega_1+i\delta)}\times$$
$$\times\,\delta(\mathbf{k}-\mathbf{k}_1-\mathbf{k}_2)\,\delta(\omega-\omega_1-\omega_2-\Delta\Omega_{\mathbf{k},\,\mathbf{k}_1,-\mathbf{k}_2})\,\delta(-\mathbf{k}'+\mathbf{k}_1-\mathbf{k}_2')\times$$
$$\times\,\delta(-\omega'+\omega_1-\omega_2'+\Delta\Omega_{\mathbf{k}',\,\mathbf{k}_1,\,\mathbf{k}_2'})\,d\mathbf{k}'\,d\mathbf{k}_1\,d\mathbf{k}_2'\,d\mathbf{k}_2\,d\omega_1\,d\omega_2\,d\omega'\,d\omega_2'-$$
$$-\int\beta^l_{\mathbf{k},\,\mathbf{k}_1,\,\mathbf{k}_2}\beta^t_{\mathbf{k}',\,\mathbf{k}_1',\,\mathbf{k}_2}A^l_{\mathbf{k}',\,\omega'}\frac{\langle A^t_{\mathbf{k}_1,\,\omega_1}A^t_{\mathbf{k}_1',\,\omega_1'}\rangle}{-i(\omega_2+i\delta)}\,\delta(\mathbf{k}-\mathbf{k}_1-\mathbf{k}_2)\times$$
$$\times\,\delta(\omega-\omega_1-\omega_2-\Delta\Omega_{\mathbf{k},\mathbf{k}_1,-\mathbf{k}_2})\,\delta(-\mathbf{k}'+\mathbf{k}_2-\mathbf{k}_1')\,\delta(-\omega'+\omega_2-\omega_1'+$$
$$+\,\Delta\Omega_{\mathbf{k},\,\mathbf{k}_2,\,\mathbf{k}_1'})\,d\mathbf{k}'\,d\mathbf{k}_1'\,d\omega'\,d\omega_1'\,d\mathbf{k}_1\,d\mathbf{k}_2\,d\omega_1\,d\omega_2. \qquad \text{(A2.24)}$$

Assuming $\omega_{1,2}\gg\gamma^t_{\mathbf{k}}$, we must, in the solution of Eqs. (A2.22) and (A2.23) for $A^t_{\mathbf{k},\omega}$, replace $\gamma^t_{\mathbf{k}}$ by $\delta=+0$. This equation can be simplified by allowing for the fact that

$$A^t_{\mathbf{k},\omega}=A^t_{\mathbf{k}}\,\delta(\omega) \qquad \text{(A2.25)}$$

since the transverse waves are time-independent. Therefore

$$\langle A^t_{\mathbf{k}_2',\,\omega_2'}A^t_{\mathbf{k}_2,\,\omega_2}\rangle=\delta(\omega_2)\,\delta(\omega_2')\,|A^t_{\mathbf{k}_2}|^2\,\delta(\mathbf{k}_2+\mathbf{k}_2'). \qquad \text{(A2.26)}$$

Let us examine, for example, the first term of (A2.24). By virtue of (A2.26) integration over $\mathbf{k}_2'$ causes $\mathbf{k}_2'$ to be replaced by $-\mathbf{k}_2$, whereupon $\delta(-\mathbf{k}'+\mathbf{k}_1-\mathbf{k}_2')$ goes over to $\delta(-\mathbf{k}'+\mathbf{k}_1+\mathbf{k}_2)$. Then the sum $\mathbf{k}_1+\mathbf{k}_2$ in $\delta(\mathbf{k}-\mathbf{k}_1-\mathbf{k}_2)$ can be replaced by $\mathbf{k}'$, i.e., we obtain $\delta(\mathbf{k}-\mathbf{k}')$. Integration over $\mathbf{k}'$ causes $\mathbf{k}'$ to be replaced by $\mathbf{k}$. Consequently, $\Delta\Omega_{\mathbf{k}',\mathbf{k}_1,\mathbf{k}_2'}$ goes over to $\Delta\Omega_{\mathbf{k},\mathbf{k}_1,\mathbf{k}_2}$, and since $\omega_2'=\omega_2=0$ by (A2.26), $\delta(-\omega'+\omega_1-\omega_2'+\Delta\Omega_{\mathbf{k}',\mathbf{k}_1,\mathbf{k}_2'})$ goes over to $\delta(-\omega'+\omega_1+\Delta\Omega_{\mathbf{k},\,\mathbf{k}_1,-\mathbf{k}_2})$, which is caused by the other delta function $\delta(\omega-\omega_1-\Delta\Omega_{\mathbf{k},\,\mathbf{k}_1,-\mathbf{k}_2})$ to go over to $\delta(\omega-\omega')$, i.e., for integration over $\omega'$ it is necessary to replace $\omega'$ by $\omega$. Thus, the first term on the right-hand side of (A2.24) has the form

$$-A^l_{\mathbf{k},\omega}\int \beta^l_{\mathbf{k},\,\mathbf{k}_1,\,\mathbf{k}_2}\beta^t_{\mathbf{k},\,\mathbf{k}_1,-\mathbf{k}_2}\,|A^t_{\mathbf{k}_2}|^2\,d\mathbf{k}_1\,d\mathbf{k}_2\,\frac{1}{-i\,(\omega_1+i\delta)}\,\delta\,(\mathbf{k}-\mathbf{k}_1-\mathbf{k}_2)\,\times$$
$$\times\,\delta\,(\omega-\omega_1-\Delta\Omega_{\mathbf{k},\,\mathbf{k}_1,-\mathbf{k}_2})\,d\omega_1. \qquad (A2.27)$$

Integrating over $\omega_1$ and $\mathbf{k}_1$, we obtain instead of (A2.27)

$$-A^l_{\mathbf{k},\omega}\int \beta^l_{\mathbf{k},\,\mathbf{k}-\mathbf{k}_2,\,\mathbf{k}_2}\beta^t_{\mathbf{k},\,\mathbf{k}-\mathbf{k}_2,-\mathbf{k}_2}\,\frac{|A^t_{\mathbf{k}_2}|\,d\mathbf{k}_2}{-i\,(\omega-\Delta\Omega_{\mathbf{k},\,\mathbf{k}-\mathbf{k}_2,-\mathbf{k}_2}+i\delta)}\,. \qquad (A2.28)$$

The second term of (A2.24) undergoes a similar transformation, and Eq. (A2.21) assumes the form ($\delta \to 0$)

$$(-i\omega-\gamma^l_{\mathbf{k}})\,A^l_{\mathbf{k},\omega} = -A^l_{\mathbf{k},\omega}\int d\mathbf{k}_2\left\{\frac{|A^t_{\mathbf{k}_2}|^2\,\beta^l_{\mathbf{k},\,\mathbf{k}-\mathbf{k}_2,\,\mathbf{k}_2}\beta^t_{\mathbf{k},\,\mathbf{k}-\mathbf{k}_2,-\mathbf{k}_2}}{-i\,(\omega-\Delta\Omega_{\mathbf{k},\,\mathbf{k}-\mathbf{k}_2,-\mathbf{k}_2}+i\delta)}+\right.$$
$$\left.+\,\frac{|A^t_{\mathbf{k}_2-\mathbf{k}}|^2\,\beta^l_{\mathbf{k},\,\mathbf{k}_2,\,\mathbf{k}-\mathbf{k}_2}\beta^t_{\mathbf{k}_1,\,\mathbf{k}_2,\,\mathbf{k}_2-\mathbf{k}}}{-i\,(\omega-\Delta\Omega_{\mathbf{k},\,\mathbf{k}-\mathbf{k}_2,-\mathbf{k}_2}+i\delta)}\right\}. \qquad (A2.29)$$

Of vital consequence is the fact that $A^l_{\mathbf{k},\omega}$ drops out of (A2.29), i.e., the resulting equation is a dispersion relation, the solution of which should give the frequency $\omega$ of the excited longitudinal waves as a function of their wave number $\mathbf{k}$. If the longitudinal oscillations grow, $\omega$ will contain a positive imaginary part:

$$\omega = \mathrm{Re}\,\omega\,(\mathbf{k}) + i\Gamma_{\mathbf{k}}. \qquad (A2.30)$$

Here $\Gamma_{\mathbf{k}}$ is the growth rate (for $\Gamma_{\mathbf{k}} > 0$) for longitudinal waves. If we neglect the nonlinear effects described by the right-hand side of (A2.29), then

$$-i\omega-\gamma^l_{\mathbf{k}} = 0;\ \ \omega = i\gamma^l_{\mathbf{k}};\ \ \Gamma_{\mathbf{k}} = \gamma^l_{\mathbf{k}} \qquad (A2.31)$$

i.e., we obtain linear growth ($\gamma_{\mathbf{k}} > 0$) or damping ($\gamma_{\mathbf{k}} < 0$).

We now recall the condition whose fulfillment is required for the analysis of (A2.29). It was assumed in the derivation of Eq. (I) that the wave amplitude varied slowly compared with $1/\Omega$, i.e.,

$$|\omega| \ll \min\Omega. \qquad (A2.32)$$

If the lowest frequency is the Langmuir frequency $\omega_{0e}$, (A2.32) implies that $|\omega| \ll \omega_{0e}$. As apparent from (A2.29) including nonlinear effects, the dispersion characteristics of the longitudinal waves also depend on the intensity of the transverse waves. Substituting the values of the coefficients $\beta^l$ and $\beta^t$ [see (I) and (II) Chap. 2, § 8],

$$\beta^l_{\mathbf{k},\,\mathbf{k}_1,\,\mathbf{k}_2} = -\frac{e\omega_{0e}^2\,(\mathbf{e}_{\mathbf{k}_1}\mathbf{e}^*_{-\mathbf{k}_2})\,k}{2m_e\dfrac{\partial}{\partial\omega}\omega^2\varepsilon^l(\omega,\,\mathbf{k})\big|_{\omega=\Omega^l(\mathbf{k})}\,\Omega^t(\mathbf{k}_1)\,\Omega^t(\mathbf{k}_2)}\,; \tag{A2.33}$$

$$\beta^t_{\mathbf{k},\,\mathbf{k}_1,\,\mathbf{k}_2} = \frac{e\omega_{0e}^2 k\,(\Omega^l(\mathbf{k})+\Omega^t(\mathbf{k}_2))\,(\mathbf{e}^*_{\mathbf{k}_1}\mathbf{e}_{\mathbf{k}_2})}{m_e\dfrac{\partial}{\partial\omega}\omega^2\varepsilon^t(\omega,\,\mathbf{k}_1)\big|_{\omega=\Omega^t(\mathbf{k}_1)}\,\Omega^t(\mathbf{k}_2)\,(\Omega^l(\mathbf{k}))^2}\,, \tag{A2.34}$$

into the dispersion relation (A2.29), we obtain

$$-i\omega-\gamma^l_{\mathbf{k}} = -i\int\frac{e^2\omega_{0e}^4k^2\,|(\mathbf{e}_{\mathbf{k}-\mathbf{k}_2}\mathbf{e}^*_{-\mathbf{k}_2})|^2\,(N^t_{\mathbf{k}_2}-N^t_{\mathbf{k}_2-\mathbf{k}})}{(\omega-\Delta\Omega_{\mathbf{k},\,\mathbf{k}-\mathbf{k}_2,-\mathbf{k}_2}+i\delta)2\pi^2m_e^2\dfrac{\partial}{\partial\omega}\omega^2\varepsilon^l\big|_{\omega=\Omega^l(\mathbf{k})}(\Omega^l(\mathbf{k}))^2}\times$$

$$\times\frac{1}{\dfrac{\partial}{\partial\omega}\omega^2\varepsilon^t(\omega,\,\mathbf{k}-\mathbf{k}_2)\big|_{\omega=\Omega^t(\mathbf{k}-\mathbf{k}_2)}}\cdot\frac{1}{\dfrac{\partial}{\partial\omega}\omega^2\varepsilon^t(\omega,\,\mathbf{k}_2)\big|_{\omega=\Omega^t(\mathbf{k}_2)}}. \tag{A2.35}$$

Here we have neglected small terms of the order $\Delta\Omega/\Omega^t \ll 1$. The quantities $N^t_{\mathbf{k}_2}$ are consistent with the notation introduced above:

$$N^t_{\mathbf{k}} = \frac{\pi^2}{(\Omega^t(\mathbf{k}))^2}\left|\frac{\partial}{\partial\omega}\omega^2\varepsilon^t(\omega,\,k)\right|_{\omega=\Omega^t(\mathbf{k})}|A^t_{\mathbf{k}}|^2. \tag{A2.36}$$

We write the resulting dispersion relation, averaging over the polarizations:

$$\sum_{\alpha,\,\alpha'}(\mathbf{e}^{\alpha}_{\mathbf{k}-\mathbf{k}_2}\mathbf{e}^{\alpha*}_{-\mathbf{k}_2})^2 = 1+\frac{(\mathbf{k}_2(\mathbf{k}_2-\mathbf{k}))^2}{\mathbf{k}_2^2(\mathbf{k}_2-\mathbf{k})^2} \tag{A2.37}$$

and setting $\Omega^l(\mathbf{k}) \approx \omega_{0e}$:

$$-i\omega-\gamma^l_{\mathbf{k}} = i\int\frac{e^2\omega_{0e}k^2}{16\pi^2m_e^2\,|\,\Omega^t(\mathbf{k}_2)\,\Omega^t(\mathbf{k}_2-\mathbf{k})\,|}\left(1+\frac{(\mathbf{k}_2(\mathbf{k}_2-\mathbf{k}))^2}{\mathbf{k}_2^2(\mathbf{k}_2-\mathbf{k})^2}\right)\times$$

$$\times\frac{N^t_{\mathbf{k}_2}-N^t_{\mathbf{k}_2-\mathbf{k}}}{\omega-\Delta\Omega_{\mathbf{k},\,\mathbf{k}-\mathbf{k}_2,\,-\mathbf{k}_2}+i\delta}\,d\mathbf{k}_2. \tag{A2.38}$$

Introducing

$$w_{\mathbf{k},\,\mathbf{k}_2} = w_{\mathbf{k},\,\mathbf{k}_2-\mathbf{k},\,\mathbf{k}_2} = \frac{e^2\omega_{0e}k^2}{8\pi^2m_e^2\,|\,\Omega^t(\mathbf{k}_2)\,\Omega^t(\mathbf{k}_2-\mathbf{k})\,|}\left(1+\frac{(\mathbf{k}_2(\mathbf{k}_2-\mathbf{k}))^2}{\mathbf{k}_2^2(\mathbf{k}_2-\mathbf{k})^2}\right), \tag{A2.39}$$

we invest Eq. (A2.38) with the simple form

$$-i\omega-\gamma^l_{\mathbf{k}} = i\int\frac{w_{\mathbf{k},\,\mathbf{k}_2}}{2}\,\frac{N^t_{\mathbf{k}_2}-N^t_{\mathbf{k}_2-\mathbf{k}}}{\omega-\Delta\Omega_{\mathbf{k},\,\mathbf{k}-\mathbf{k}_2,-\mathbf{k}_2}+i\delta}\,d\mathbf{k}_2. \tag{A2.40}$$

Equation (A2.40) contains more information than (3.55) regarding the nonlinear properties of longitudinal waves.

*Appendix 3*

# General Method for Calculating the Probabilities of the Decay, Emission, and Scattering of Waves

The method described in this appendix is suitable for calculating the probabilities of the relevant processes, not only in a plasma, but also in any homogeneous and stationary medium.

We invoke the correspondence principle. We find the increase in energy of the $\sigma$-waves due to decay, emission, and scattering processes in the small-intensity limit ($N_{\mathbf{k}}^{\sigma} \to 0$):

$$I^{\sigma} = \frac{\partial W^{\sigma}}{\partial t} = \int \hbar\omega^{\sigma} \frac{\partial N_{\mathbf{k}}^{\sigma}}{\partial t} \frac{d\mathbf{k}}{(2\pi)^3} = \int w_{\sigma}^{\sigma', \sigma''} \hbar\omega^{\sigma} N_{\mathbf{k}_1}^{\sigma'} N_{\mathbf{k}_2}^{\sigma''} \frac{d\mathbf{k}\, d\mathbf{k}_1\, d\mathbf{k}_2}{(2\pi)^9} +$$

$$+ \int w_{\mathbf{p}}^{\sigma} f_{\mathbf{p}}\, \hbar\omega^{\sigma} \frac{d\mathbf{k}\, d\mathbf{p}}{(2\pi)^6} + \int w_{\mathbf{p}}^{\sigma\,\sigma'} \hbar\omega^{\sigma} N_{\mathbf{k}_1}^{\sigma'} \frac{d\mathbf{k}_1\, d\mathbf{k}\, d\mathbf{p}}{(2\pi)^9} f_{\mathbf{p}}\,. \tag{A3.1}$$

The first term of (A3.1) describes the rate of change of energy due to spontaneous decay processes, the second term for spontaneous emission, and the third term for spontaneous scattering.

Let us suppose that the current $\mathbf{j}_k$ inducing the emission of waves for each of these processes is known. Also let $e_{\mathbf{k}}^{\sigma}$ be the unit polarization vector for the wave $\sigma$. From Maxwell's equations

$$(k^2\delta_{i,j} - k_i k_j - \omega^2\varepsilon_{i,j}) E_{j,k} = 4\pi i\omega j_{i,k}. \tag{A3.2}$$

Assuming that $E_{j,k} = e_{\mathbf{k},j}^{\sigma} E_k^{\sigma}$, we obtain the field $\mathrm{E}_k^{\sigma}$ produced by the current $\mathbf{j}_k$:

$$(k^2 - \omega^2\varepsilon^\sigma) E_k^\sigma = 4\pi i\omega (\mathbf{e}_k^{\sigma *} \mathbf{j}_k),$$

where

$$\varepsilon^\sigma = e^*_{k,i}\varepsilon_{i,j}e_{kj} + \frac{1}{\omega^2}(\mathbf{k}\mathbf{e}_k^\sigma)(\mathbf{k}\mathbf{e}_k^{\sigma *})$$

is the dielectric constant for the wave $\sigma$.

The radiation power $I^\sigma$ is equal to the work done by the field $E_k^\sigma$ on the current $j_k$ producing it:

$$-I^\sigma = \int j_i E_i^\sigma \, d\mathbf{r} = \int (\mathbf{j}_{-\mathbf{k},-\omega}\, \mathbf{e}_\mathbf{k}^\sigma) E^\sigma_{\mathbf{k},\omega'} e^{i(\omega-\omega')t} d\omega\, d\omega' (2\pi)^3 =$$

$$= (2\pi)^3 \int (\mathbf{j}^*_{\mathbf{k},\omega} \mathbf{e}^\sigma_\mathbf{k}) \frac{4\pi i\omega}{k^2 - \omega^2\varepsilon^\sigma} (\mathbf{j}_{\mathbf{k},\omega'} \mathbf{e}^{\sigma *}_\mathbf{k}) e^{i(\omega-\omega')t} d\omega\, d\omega' d\mathbf{k}. \qquad (A3.3)$$

Having thus found $I^\sigma$, we express $\langle E_\mathrm{k}^\sigma E_{\mathrm{k}'}^\sigma \rangle$ in terms of $N_\mathbf{k}^\sigma$ and compare with (A3.1) to find the corresponding probabilities.

**1.** In order to find the probabilities of three-plasmon decay processes it is sufficient to know the nonlinear current to second order in the fields $E_{k_1}^{\sigma'}$ and $E_{k_2}^{\sigma''}$, generating the field $E_\mathrm{k}^\sigma$:

$$j_{k,i} = 2\int S_{i,j,l}(k, k_1, k_2) E^{\sigma'}_{j,k_1} E^{\sigma''}_{l,k_2} dk_1\, dk_2 \delta(k - k_1 - k_2). \qquad (A3.4)$$

Substituting (A3.4) into (A3.3), averaging over the phases of the waves (decomposing the four-field averages into the averages of the two-field products), and dividing by the volume of the plasma V ($V \to \infty$), we obtain*

$$w_\sigma^{\sigma',\sigma''}(\mathbf{k}, \mathbf{k}', \mathbf{k}'') = 32(2\pi)^7 \delta(\mathbf{k} - \mathbf{k}' - \mathbf{k}'') \delta[\Omega^\sigma(\mathbf{k}) - \Omega^{\sigma'}(\mathbf{k}') - \Omega^{\sigma''}(\mathbf{k}'')] \times$$

$$\times \frac{(\Omega^{\sigma'}(\mathbf{k}'))^2 (\Omega^{\sigma''}(\mathbf{k}''))^2 \,|\, S_{\sigma,\sigma',\sigma''}(\Omega^\sigma(\mathbf{k}), \mathbf{k}, \Omega^{\sigma'}(\mathbf{k}')\mathbf{k}', \Omega^{\sigma''}(\mathbf{k}''), \mathbf{k}'' \,|^2}{\left(\frac{\partial}{\partial\omega}\omega^2\varepsilon^\sigma\right)_{\omega=\Omega^\sigma(\mathbf{k})} \left(\frac{\partial}{\partial\omega}\omega^2\varepsilon^{\sigma'}\right)_{\omega=\Omega^\sigma(\mathbf{k}')} \left(\frac{\partial}{\partial\omega}\omega^2\varepsilon^{\sigma''}\right)_{\omega=\Omega^{\sigma''}(\mathbf{k}'')}};$$

$$S_{\sigma,\sigma',\sigma''} = e_i^{\sigma *}(\mathbf{k}) S_{i,j,l} e_j^{\sigma'}(\mathbf{k}') e_l^{\sigma''}(\mathbf{k}''). \qquad (A3.5)$$

---

* The coefficient 2 in (A3.4) arises from the assumption that $\sigma' \neq \sigma''$ and that $S_{ijl}(k, k_1, k_2)$ is symmetric, i.e., $S_{ijl}(k, k_1, k_2) = S_{ijl}(k, k_2, k_1)$, and from the inclusion of the two terms corresponding to $E_{k_1} = E_{k_1}^{\sigma'}$, $E_{k_2} = E_{k_2}^{\sigma''}$ and $E_{k_1} = E_{k_1}^{\sigma''}$, $E_{k_2} = E_{k_2}^{\sigma'}$. The insertion or omission of the 2 in (A3.4) depends on the definition of $S_{ijl}(k, k_1, k_2)$. The definition (A3.4) is somewhat more useful than that given in the original Soviet edition. We point out that if $\sigma' = \sigma''$, then the coefficient 2 does not occur in (A3.4), but it is still required to allow for the two possible decompositions of the average of four turbulent fields (coefficient 2). For this reason (A3.5) contains, in the case $\sigma' = \sigma''$ the coefficient 1/2.

2. For calculating the emission probabilities it is helpful to compare (A3.1) and the single-particle radiation intensity (without $f_p\, d\mathbf{p}/(2\pi)^3$).

The current due to a uniformly moving charge is

$$\mathbf{j}_{\mathbf{k},\,\omega} = e\mathbf{v}\int \delta(\mathbf{r} - \mathbf{v}t)\,\frac{d\mathbf{r}\,dt}{(2\pi)^4}\exp(i\omega t - i\mathbf{k}\mathbf{r}) = \frac{e\mathbf{v}}{(2\pi)^3}\,\delta(\omega - \mathbf{k}\mathbf{v}). \quad \text{(A3.6)}$$

Substituting (A3.6) into (A3.4) and comparing with the corresponding expression for $I^\sigma$, we obtain*

$$w_{\mathbf{p}}^{\sigma}(\mathbf{k}) = \frac{e^2}{\pi\hbar}\,\frac{(2\pi)^3\,|\mathbf{e}_{\mathbf{k}}^{\sigma}\mathbf{v}|^2}{\left.\frac{\partial}{\partial\omega}\,\omega^2\varepsilon^\sigma\right|_{\Omega=\Omega^\sigma(\mathbf{k})}}\,\delta(\Omega^\sigma(\mathbf{k}) - \mathbf{k}\mathbf{v}). \quad \text{(A3.7)}$$

3. For calculating the scattering probabilities it is necessary to include nonlinear and Compton scattering. The expression for the current associated with nonlinear scattering is obtained by replacing one of the fields in (A3.4) by the field generated by the charge current (A3.6). The current associated with Compton scattering can be found directly as the current of a charge oscillating in the field of the waves $\sigma$:

$$j_{i,\,k} = \int \Lambda_{i,\,j}(k, k_1)\,E_{j,\,k_1}^{\sigma'}\,dk_1\delta(\omega - \omega_1 - (\mathbf{k} - \mathbf{k}_1)\,\mathbf{v});$$

$$\Lambda_{i,j} = \Lambda_{i,j}^{\text{ns}} + \Lambda_{i,j}^{\text{cs}};$$

$$\delta(\omega - \omega_1 - (\mathbf{k} - \mathbf{k}_1)\mathbf{v})\,\Lambda_{i,\,j}^{\text{ns}} = [S_{i,\,l,\,j}(k, k - k_1, k_1) +$$

$$+ S_{i,\,j,\,l}(k, k_1, k - k_1)]\,\frac{4\pi i\,(\omega - \omega_1)\,e}{(2\pi)^3}\,\Pi_{l,\,m}(k - k_1)\,v_m, \quad \text{(A3.8)}$$

where

$$\Pi_{i,\,s}(k^2\delta_{s,\,j} - k_s k_j - \omega^2\varepsilon_{s,\,j}) = \delta_{i,\,j}, \quad \text{(A3.9)}$$

and $\Pi_{i,s}$ is the inverse Maxwell operator.

Inserting (A3.8) into (A3.4), we obtain

$$w_{\alpha,\,\sigma}^{\sigma'}(\mathbf{k}, \mathbf{k}', \mathbf{p}) = \frac{4\,[\Omega^{\sigma'}(\mathbf{k}')]^2\,|e_{\mathbf{k},\,i}^{\sigma*}\Lambda_{i,j}e_{\mathbf{k},j}^{\sigma'}|^2\,\delta(\Omega^\sigma(\mathbf{k}) - \Omega^{\sigma'}(\mathbf{k}') - (\mathbf{k} - \mathbf{k}')\,\mathbf{v})(2\pi)^9}{\left.\frac{\partial}{\partial\omega}(\omega^2\varepsilon^\sigma)\right|_{\omega=\Omega^\sigma(\mathbf{k})}\cdot\left.\frac{\partial}{\partial\omega}(\omega^2\varepsilon^{\sigma'})\right|_{\omega=\Omega^{\sigma'}(\mathbf{k}')}}$$

(A3.10)

* It is also appropriate to use the expression given for the probabilities when a magnetic field is present, provided the characteristic wavelengths are much smaller than the Larmor radius of the particles.

For an isotropic plasma

$$\Pi_{i,j} = -\frac{1}{k^2}\left(\frac{k_i k_j}{\omega^2 \varepsilon^l(\omega, \mathbf{k})} - \frac{k^2\delta_{i,j} - k_i k_j}{k^2c^2 - \omega^2\varepsilon^t(\omega, \mathbf{k})}\right), \qquad (A3.11)$$

where $\varepsilon^l$ and $\varepsilon^t$ are the longitudinal and transverse dielectric constants of the plasma. The first term, corresponding to the Green's function for the longitudinal field, describes the nonlinear scattering in terms of a virtual longitudinal wave, while the second term, containing the Green's function for the transverse field, describes it in terms of a virtual transverse wave. A general expression for the probability of scattering of transverse into longitudinal waves and of longitudinal into longitudinal waves in an isotropic plasma has been derived in [10], and a general expression for the probability of scattering of transverse into transverse waves in an isotropic plasma has been obtained in [11].

## Summary of the Decay Probabilities of Waves in an Isotropic Plasma

1. The probability of the decay of a Langmuir wave into a Langmuir and an ion-acoustic wave in an isotropic plasma [1, 2] has the form

$$w_l^{l,s}(\mathbf{k}_1, \mathbf{k}_2, \mathbf{k}_s) = \frac{\hbar e^2 \Omega_s^3 m_i (2\pi)^6}{8\pi m_e^3 v_{Te}^4 k_s^2}\left(\frac{\mathbf{k}_1\mathbf{k}_2}{k_1k_2}\right)^2 \delta(\mathbf{k}_1 - \mathbf{k}_2 - \mathbf{k}_s)\times$$
$$\times\,\delta(\Omega^l(\mathbf{k}_1) - \Omega^l(\mathbf{k}_2) - \Omega^s(\mathbf{k}_s)). \qquad (A3.12)$$

Here $\mathbf{k}_1$ and $\mathbf{k}_2$ are the Langmuir-wave vectors before and after decay.

2. The probability of the coalescence of two Langmuir waves into a transverse wave of frequency $\sim 2\omega_{0e}$ in an isotropic plasma [3] is described by the expression

$$w_t^{l,l}(\mathbf{k}_1, \mathbf{k}_2, \mathbf{k}_t) = \frac{\hbar e^2 (2\pi)^6}{32\pi m_e^2}\frac{(k_1^2 - k_2^2)^2}{k_t^2\,\omega_{0e}}\frac{[\mathbf{k}_1\mathbf{k}_2]^2}{k_1^2k_2^2}\delta(\mathbf{k}_t - \mathbf{k}_1 - \mathbf{k}_2)\times$$
$$\times\,\delta(\Omega^t - \Omega_1^l - \Omega_2^l). \qquad (A3.13)$$

3. The probability of the decay of a Langmuir wave $l$ into a transverse and an ion-acoustic wave in an isotropic plasma [4, 5] is equal to

$$w_l^{t,s}(\mathbf{k}_l, \mathbf{k}_t, \mathbf{k}_s) = \frac{\hbar e^2 (2\pi)^6}{8\pi m_e^2 v_{Te}^4} \frac{\omega_{0e}^3 \Omega_s^3}{\omega_{0i}^2 k_s^2 \Omega^t} \frac{[\mathbf{k}_t \mathbf{k}_l]^2}{(k_t k_l)^2} \delta(\mathbf{k}_l - \mathbf{k}_t - \mathbf{k}_s) \times$$
$$\times \delta(\Omega^l - \Omega^t - \Omega^s). \qquad \text{(A3.14)}$$

4. The probability of the decay of a transverse wave into a transverse and an ion-acoustic wave [6, 5] has the form

$$w_t^{t,s}(\mathbf{k}_1, \mathbf{k}_2, \mathbf{k}_s) = \frac{\hbar e^2 (2\pi)^6 \Omega_s^3 \omega_{0e}^4}{16\pi \omega_{0i}^2 m_e^2 v_{Te}^4 k_s^2 \Omega_1^t \Omega_2^t} \left(1 + \frac{(\mathbf{k}_1 \mathbf{k}_2)^2}{k_1^2 k_2^2}\right) \delta(\mathbf{k}_1 - \mathbf{k}_2 - \mathbf{k}_s) \times$$
$$\times \delta(\Omega_1^t - \Omega_2^t - \Omega^s). \qquad \text{(A3.15)}$$

5. The probability of the decay of a transverse wave into a transverse and a Langmuir wave [6] is equal to

$$w_t^{t,l}(\mathbf{k}_1, \mathbf{k}_2, \mathbf{k}_l) = \frac{\hbar e^2 (2\pi)^6 \omega_{0e} k^2}{16\pi m_e^2 \Omega^t(\mathbf{k}_1) \Omega^t(\mathbf{k}_2)} \left(1 + \frac{(\mathbf{k}_1 \mathbf{k}_2)^2}{k_1^2 k_2^2}\right) \delta(\mathbf{k}_1 - \mathbf{k}_2 - \mathbf{k}_l) \times$$
$$\times \delta(\Omega^t(\mathbf{k}_1) - \Omega^t(\mathbf{k}_2) - \Omega^l(\mathbf{k}_l)). \qquad \text{(A3.16)}$$

6. The probability of four-plasmon decay of Langmuir waves into Langmuir waves [7-9] ($\mathbf{k}_- = \mathbf{k}_2 - \mathbf{k}_3, \omega_- = \Omega_2 - \Omega_3$) is equal to

$$w_{l,l}^{l,l}(\mathbf{k}_1, \mathbf{k}_2, \mathbf{k}_3, \mathbf{k}_4) = \frac{(2\pi)^9 e^4 \hbar^2}{(4\pi)^3 m_e^4 v_{Te}^4} \left| \frac{\varepsilon_i^l(\omega_-, \mathbf{k}_-)}{\varepsilon^l(\omega_-, \mathbf{k}_-)} \frac{(\mathbf{k}_2\mathbf{k}_3)(\mathbf{k}_4\mathbf{k}_1)}{k_4 k_1 k_2 k_3} + \right.$$

+ terms in which

$$\left. 1 \rightleftarrows 2 \right|^2 \delta(\mathbf{k}_1 + \mathbf{k}_2 - \mathbf{k}_3 - \mathbf{k}_4) \delta(\Omega_1 + \Omega_2 - \Omega_3 - \Omega_4). \qquad \text{(A3.17)}$$

7. The probability of four-plasmon scattering of transverse waves by longitudinal waves, $l + t \rightleftarrows t + l$ [7], has the form

$$w_{l,t}^{l_2 t_2}(\mathbf{k}_{1t}, \mathbf{k}_{2t}, \mathbf{k}_{1l}, \mathbf{k}_{2l}) = \frac{\hbar^2 (2\pi)^9 e^4 (k_-^l)^4}{32 m_e^4 \omega_{0e}^2 (\Omega_1^t \Omega_2^t)} \frac{(\mathbf{k}_{1l}\mathbf{k}_{2l})^2}{k_{1l}^2 k_{2l}^2} \times$$
$$\times \left(1 + \frac{(\mathbf{k}_{1t}\mathbf{k}_{2t})^2}{k_{1t}^2 k_{2t}^2}\right) \left| \frac{\varepsilon_i^l(\mathbf{k}_-^l, \Omega_-^l)}{\varepsilon^l(\mathbf{k}_-^l, \Omega_-^l)} (1 - \varepsilon^l(\mathbf{k}_-^l, \Omega_-^l)) \right|^2 \times$$
$$\times \delta(\mathbf{k}_{1t} + \mathbf{k}_{1l} - \mathbf{k}_{2t} - \mathbf{k}_{2l}) \delta(\Omega_1^t + \Omega_1^l - \Omega_2^t - \Omega_2^l);$$
$$\mathbf{k}_-^l = \mathbf{k}_{1l} - \mathbf{k}_{2l}; \quad \Omega_-^l = \Omega_1^l - \Omega_2^l. \qquad \text{(A3.18)}$$

## Summary of Approximate Expressions for the Wave Scattering Probabilities in an Isotropic Plasma [12]

Here we denote Langmuir modes by the letter $l$, transverse modes by t, and ion-acoustic modes by s; $\boldsymbol{\omega}_- = \omega - \boldsymbol{\omega}_1$; $\mathbf{k}_- = \mathbf{k} - \mathbf{k}_1$.

Probability of $ll$-scattering by plasma electrons:

$$w_{\mathbf{p}}^{l,\,l} = \frac{(2\pi)^3 e^4}{m_e^2 \omega_{0e}^2} \left(\frac{\mathbf{k}\mathbf{k}_1}{kk_1}\right)^2 \left| \frac{2\,(\mathbf{k}\mathbf{v})}{\omega_{0e}} + \frac{\varepsilon_i^l(\omega_-,\,\mathbf{k}_-)}{\varepsilon^l(\omega_-,\,\mathbf{k}_-)} \right|^2 \delta(\omega_- - \mathbf{k}_-\mathbf{v}).$$

Probability of $ll$-scattering by plasma ions:

$$w_{\mathbf{p}}^{l,\,l} = \frac{(2\pi)^3 e^4}{m_e^2 \omega_{0e}^2} \left(\frac{\mathbf{k}\mathbf{k}_1}{kk_1}\right)^2 \left| \frac{\varepsilon_e^l(\omega_-,\,\mathbf{k}_-) - 1}{\varepsilon^l(\omega_-,\,\mathbf{k}_-)} \right|^2 \delta(\omega_- - \mathbf{k}_-\mathbf{v}).$$

Probability of ss-scattering by plasma ions:

$$w_{\mathbf{p}}^{s,\,s} = 4\,\frac{(2\pi)^3 e^4}{m_i^2 \omega_{0i}^4} (\mathbf{k}\mathbf{v})^2 \left(\frac{\mathbf{k}\mathbf{k}_1}{kk_1}\right)^2 \delta(\omega_- - \mathbf{k}_-\mathbf{v}).$$

Probability of $l$s-scattering by plasma electrons:

$$w_{\mathbf{p}}^{l,\,s} = \frac{(2\pi)^3 e^4 \omega_{0i}}{m_e^2 \omega_{0e}^3} \left(\frac{\mathbf{k}\mathbf{k}_1}{kk_1}\right)^2 \delta(\omega_- - \mathbf{k}_-\mathbf{v}).$$

Probability of t$l$-scattering by plasma thermal ions:

$$w_{\mathbf{p}}^{t,\,l} = \frac{(2\pi)^3 e^4}{m_e^2 \omega_{0e}^2} \frac{[\mathbf{k}\mathbf{k}_1]^2}{k^2 k_1^2} \left| \frac{\varepsilon_e^l(\omega_-,\,\mathbf{k}_-) - 1}{\varepsilon^l(\omega_-,\,\mathbf{k}_-)} \right|^2 \delta(\omega_- - \mathbf{k}_-\mathbf{v}).$$

Probability of t$l$-scattering by plasma thermal electrons:

$$w_{\mathbf{p}}^{t,\,l} = \frac{(2\pi)^3 e^4}{m_e^2 \omega_{0e}^2} \frac{[\mathbf{k}\mathbf{k}_1]^2}{k^2 k_1^2} \left| \frac{\varepsilon_i^l(\omega_-,\,\mathbf{k}_-)}{\varepsilon^l(\omega_-,\,\mathbf{k}_-)} \right|^2 \delta(\omega_- - \mathbf{k}_-\mathbf{v}).$$

Probability of tt-scattering by plasma electrons:

$$w_{\mathbf{p}}^{t,\,t} = \frac{(2\pi)^3 e^4}{m_e^2 2\Omega^t \Omega_1^t} \left(1 + \frac{(\mathbf{k}\mathbf{k}_1)^2}{k^2 k_1^2}\right) \left| \frac{\varepsilon_i^l(\omega_-,\,\mathbf{k}_-)}{\varepsilon^l(\omega_-,\,\mathbf{k}_-)} \right|^2 \delta(\omega_- - \mathbf{k}_-\mathbf{v}).$$

Probability of tt-scattering by plasma ions:

$$w_{\mathbf{p}}^{t,t} = \frac{(2\pi)^3 e^4}{m_e^2 2\Omega^t \Omega_1^t} \left(1 + \frac{(\mathbf{k}\mathbf{k}_1)^2}{k^2 k_1^2}\right) \left| \frac{\varepsilon_e^l(\omega_-, \mathbf{k}_-) - 1}{\varepsilon^l(\omega_-, \mathbf{k}_-)} \right|^2 \delta(\omega_- - \mathbf{k}_-\mathbf{v}).$$

Probability of tt-scattering by electrons in the high-frequency limit:

$$w_{\mathbf{p}}^{t,t} = \frac{(2\pi)^3 e^4}{m_e^2 2\Omega^t \Omega_1^t} \left(1 + \frac{(\mathbf{k}\mathbf{k}_1)^2}{k^2 k_1^2}\right) \delta(\omega_-).$$

*Appendix 4*

# Methods of Calculating the Nonlinear Dispersion Characteristics of a Plasma

## Calculation of the Nonlinear Dielectric Constant of a Plasma in the Presence of Strong Random Waves

From the nonlinear plasma current

$$j_{i,k} = \int S_{i,j,l}(k, k_1, k_2)\, E_{j,k_1} E_{l,k_2}\, dk_1\, dk_2\, \delta(k - k_1 - k_2) +$$
$$+ \int \Sigma_{i,j,l,s}(k, k_1, k_2, k_3) \cdot E_{j,k_1} E_{l,k_2} E_{s,k_3} \cdot dk_1\, dk_2\, dk_3\, \delta(k - k_1 - k_2 - k_3) \quad \text{(A4.1)}$$

we separate out linear terms in the weak field $E'_{i,k}$ : $E_{i,k} = E^{r}_{i,k} + E'_{i,k}$. We average the linear (in $E'_k$) current obtained from (A4.1) over the random field $E^{r}_k$, expressing the triple-product averages in terms of the products of pair-product averages (see Appendix 2). For the average value of the current linear in $E_k$ we obtain

$$\langle j'_{i,k} \rangle = \frac{\varepsilon^{\mathrm{nlr}}_{i,j}(k) - \delta_{ij}}{4\pi}\, i\omega E'_{j,k},$$

where $\varepsilon^{\mathrm{nlr}}_{i,j}$ is the nonlinear dielectric constant of the plasma and is equal to

$$\varepsilon^{\mathrm{nlr}}_{i,j} = \delta_{i,j} + \frac{4\pi}{i\omega} \int (\Sigma^{(1)}_{i,j}(k, k_1) + \Sigma^{(2)}_{i,j}(k, k_1))\, |E^{l}_{k}|^2\, dk. \quad \text{(A4.2)}$$

Here

$$\Sigma^{(1)}_{i,j}(k, k_1) = \frac{k_{1l}k_{1s}}{k_1^2}\{\Sigma_{i,j,l,s}(k, k, k_1, -k_1) + \Sigma_{i,l,j,s}(k, k_1, k, -k_1)\}; \tag{A4.3}$$

$$\Sigma^{(2)}_{i,j}(k, k_1) = \frac{k_{1l}k_{1p}}{k_1^2}\widetilde{S}_{i,n,l}(k, k_-, k_1)\, 16\pi i\omega_-\Pi_{n,m}(k_-)\widetilde{S}_{m,j,p}(k_-, k, k_1); \tag{A4.4}$$

$$\widetilde{S}_{i,l,m}(k, k_1, k_2) = \frac{1}{2}(S_{i,l,m}(k, k_1, k_2) + S_{i,m,l}(k, k_2, k_1)). \tag{A4.5}$$

In (A4.2) the random field is longitudinal, $\Pi_{n,m}$, and is determined by (A3.9).

## Calculation of the Nonlinear Dielectric Constant of a Plasma by the Dispersion Equation Method

In principle the nonlinear dielectric constant of a plasma can be calculated from general considerations when only the probabilities of the processes are known. Let us consider, for example, the elementary decay process corresponding to emission of a wave $l$ by a wave $\lambda$ (Fig. 34). Let the $\lambda$-waves be given and let the problem be to find the dielectric constant of the plasma for the $l$-wave. We assume that there is an arbitrary longitudinal wave whose frequency is much lower than the $\lambda$-wave frequency. We are concerned with the nonlinear dielectric constant of the plasma for electrostatic waves. We know that the general dispersion equation in this case is

$$\varepsilon(\omega, \mathrm{k}) = 0. \tag{A4.6}$$

We separate (A4.6) into the linear part $\varepsilon^{\mathrm{lr}}$ due to interaction of the $l$-waves with the plasma particles and the nonlinear part $\varepsilon^{\mathrm{nlr}}$ associated with interaction of the $l$-waves with the $\lambda$-waves:

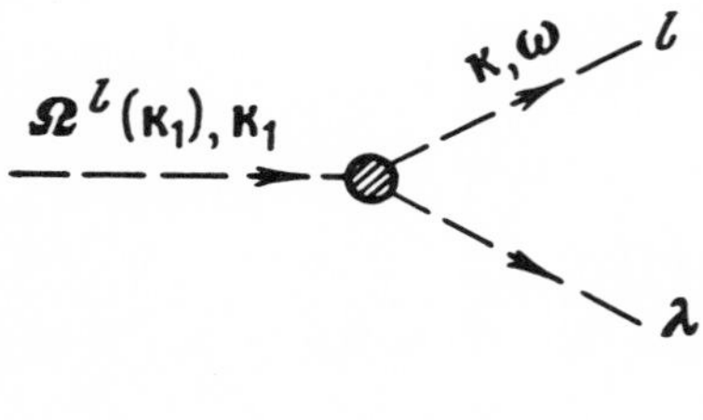


Fig. 34. Dispersion characteristics of $l$-waves in a field of $\lambda$-waves; $\mathbf{k}$, $\omega$ are the wave vector and frequency of the propagating longitudinal wave; $\mathbf{k}_1$, $\Omega_\lambda(\mathbf{k}_1)$ are the wave vector and frequency of strong electromagnetic waves in the plasma which alter the dispersion characteristics of the $l$-waves.

$$\varepsilon = \varepsilon^{lr} + \varepsilon^{nlr} - 1. \tag{A4.7}$$

The decomposition (A4.7) is always possible if the frequency $\omega$ is near one of the eigenfrequencies $\Omega^{\sigma}(k)$.

The process illustrated in Fig. 34 describes the absorption or emission of $l$-waves, i.e., it gives the contribution to the imaginary part of $\varepsilon^{nlr}$:

$$\varepsilon^{nlr} = \mathrm{Re}\,\varepsilon^{nlr} + i\,\mathrm{Im}\,\varepsilon^{nlr}. \tag{A4.8}$$

Since the conservation of energy and momentum must hold for these processes, Im $\varepsilon^{nlr}$ must have the form

$$\mathrm{Im}\,\varepsilon^{nlr}(\omega, \mathbf{k}) = \int \Phi_{\mathbf{k}\mathbf{k}_1} \delta(\omega - \Omega^{\lambda}(\mathbf{k}_1) + \Omega^{\lambda}(\mathbf{k}_1 - \mathbf{k}))\, d\mathbf{k}_1. \tag{A4.9}$$

We now calculate the weak damping of the $l$-waves caused by absorption and emission by $\lambda$-waves:

$$\gamma_{\mathbf{k}} = -\int \frac{\Phi_{\mathbf{k}\mathbf{k}_1}}{\left.\dfrac{\partial \varepsilon^{lr}(\omega, \mathbf{k})}{\partial \omega}\right|_{\omega = \Omega^{l}(\mathbf{k})}} \delta(\Omega^{l}(\mathbf{k}) - \Omega^{\lambda}(\mathbf{k}_1) + \Omega^{\lambda}(\mathbf{k}_1 - \mathbf{k}))\, d\mathbf{k}_1. \tag{A4.10}$$

Next we compare this result with that obtained for the rate of change in the number of $l$-waves from probability arguments:

$$\frac{\partial N_{\mathbf{k}}^{l}}{\partial t} = N_{\mathbf{k}}^{l} \int w_{l}^{\lambda}(\mathbf{k}\mathbf{k}_1)(N_{\mathbf{k}_1}^{\lambda} - N_{\mathbf{k}_1 - \mathbf{k}}^{\lambda}) \frac{d\mathbf{k}_1}{(2\pi)^3} \tag{A4.11}$$

Recognizing by virtue of the conservation law that

$$w_{l}^{\lambda}(\mathbf{k}\mathbf{k}_1) = u_{l}^{\lambda}(\mathbf{k}\mathbf{k}_1)\, \delta(\Omega^{l}(\mathbf{k}) - \Omega^{\lambda}(\mathbf{k}_1) + \Omega^{\lambda}(\mathbf{k}_1 - \mathbf{k})),$$

and making use of the dispersion relation

$$\mathrm{Re}\,\varepsilon(\omega, \mathbf{k}) - 1 = \frac{1}{\pi} \int \frac{\mathrm{Im}\,\varepsilon(\omega', \mathbf{k})\, d\omega'}{\omega' - \omega}, \tag{A4.12}$$

we obtain

$$\varepsilon^{nlr}(\omega, \mathbf{k}) - 1 = -\frac{1}{\pi} \int_{\delta \to +0} \frac{\Phi_{\mathbf{k}\mathbf{k}_1}}{\omega - \Omega^{\lambda}(\mathbf{k}_1) + \Omega^{\lambda}(\mathbf{k}_1 - \mathbf{k}) + i\delta}\, d\mathbf{k}_1, \tag{A4.13}$$

where

$$\Phi_{\mathbf{k}\mathbf{k}_1} = -\frac{\left.\dfrac{\partial \varepsilon^{l}(\omega, \mathbf{k})}{\partial \omega}\right|_{\omega = \Omega^{l}(\mathbf{k})} u_{l}^{\lambda}(\mathbf{k}\mathbf{k}_1)(N_{\mathbf{k}_1}^{\lambda} - N_{\mathbf{k}_1 - \mathbf{k}}^{\lambda})}{2(2\pi)^3}. \tag{A4.14}$$

This equation can be used only for frequencies near those which comply with the decay conditions.

## Summary of Formulas for the Nonlinear Dielectric Constants of an Isotropic Plasma

In the vicinity of different frequencies different processes contribute to the longitudinal nonlinear dielectric constant of a plasma.

Near the decay process $t \to t + l$ ($\omega \gg kv_{Te}$)

$$\varepsilon^{\mathrm{nlr}} = 1 + \int \frac{(N^t_{\mathbf{k}_1} - N^t_{\mathbf{k}_1-\mathbf{k}})\, e^2 k^2 \left(1 + \frac{(\mathbf{k}_1(\mathbf{k}_1-\mathbf{k}))^2}{\mathbf{k}_1^2(\mathbf{k}_1-\mathbf{k})^2}\right)}{8\pi^2 m_e^2 (\omega - \Omega^t(\mathbf{k}_1) + \Omega^t(\mathbf{k}_1-\mathbf{k}) + i\delta)\,\Omega^t(\mathbf{k}_1)\,\Omega^t(\mathbf{k}_1-\mathbf{k})}\, d\mathbf{k}_1. \quad (A4.15)$$

Near the process $l \to l + s$

$$\varepsilon^{\mathrm{nlr}} = 1 + \frac{e^2\omega_{0e}^2}{8\pi^2 m_e^2 k^2 v_{Te}^4} \int \frac{(\mathbf{k}_1(\mathbf{k}_1-\mathbf{k}))^2}{k_1^2(\mathbf{k}_1-\mathbf{k})^2} \frac{N^l_{\mathbf{k}_1} - N^l_{\mathbf{k}_1-\mathbf{k}}}{\omega - \frac{3}{2\omega_{0e}}[k_1^2 - (\mathbf{k}_1-\mathbf{k})^2]^2 v_{Te}^2 + i\delta}\, d\mathbf{k}_1. \quad (A4.16)$$

For the process $t \to t + s$ ($\omega \ll kv_{Te}$)

$$\varepsilon^{\mathrm{nlr}} = 1 + \frac{e^2\omega_{0e}^4}{8\pi^2 m_e^2 k^2 v_{Te}^4} \int \frac{1 + \frac{(\mathbf{k}_1(\mathbf{k}_1-\mathbf{k}))^2}{k_1^2(\mathbf{k}_1-\mathbf{k})^2}}{\Omega^t(\mathbf{k}_1)\,\Omega^t(\mathbf{k}_1-\mathbf{k})} \frac{(N^t_{\mathbf{k}_1} - N^t_{\mathbf{k}_1-\mathbf{k}})}{\omega - \Omega^t(\mathbf{k}_1) + \Omega^t(\mathbf{k}-\mathbf{k}_1) + i\delta}\, d\mathbf{k}_1. \quad (A4.17)$$

In order to relate the $|E^\sigma_{\mathbf{k}}|^2$ in (A4.2) to $N^\sigma_{\mathbf{k}}$ it is convenient to separate the parts with positive and negative frequencies:

$$|E^\sigma_{\mathbf{k}}|^2 = \frac{(\Omega^\sigma_{\mathbf{k}})^2}{2\pi^2 \frac{\partial}{\partial\omega}\omega^2\varepsilon^\sigma \Big|_{\omega=\Omega^\sigma_{\mathbf{k}}}} (N^\sigma_{\mathbf{k}}\delta(\omega - |\Omega^\sigma_{\mathbf{k}}|) + N^\sigma_{-\mathbf{k}}\delta(\omega + |\Omega^\sigma_{\mathbf{k}}|)). \quad (A4.18)$$

## Literature Cited in the Appendices


1. V. A. Liperovskii and V. N. Tsytovich, "Decay of longitudinal Langmuir oscillations into ion-acoustic oscillations," Zh. Prikl. Mekhan. i Tekh. Fiz., No. 5 (1965).
2. G. I. Suramlishvili, "Wave kinetics in a plasma," Dokl. Akad. Nauk SSSR, 153:317 (1963).
3. I. A. Akhiezer, N. L. Daneliya, and N. N. Tsintsadze, "Theory of the transformation and scattering of electromagnetic waves in a nonequilibrium plasma," Zh. Éksp. Teor. Fiz., 46:300 (1964).

4. I. A. Akhiezer, "Scattering and transformation of electromagnetic waves in a turbulent plasma," Zh. Éksp. Teor. Fiz., 48:1159 (1965).
5. L. M. Kovrizhnykh, "Theory of nonlinear wave interaction in a plasma," Trudy FIAN, 32:173 (1966).
6. L. M. Kovrizhnykh and V. N. Tsytovich, "Interaction of intense high-frequency radiation with a plasma," Dokl. Akad. Nauk SSSR, 158:1306 (1964).
7. V. A. Liperovskii and V. N. Tsytovich, "Nonlinear interaction of plasma waves in the presence of intense transverse waves," Izv. Vuzov, Radiofizika, 9:513 (1966).
8. A. A. Vedenov, "Introduction to the theory of a weakly turbulent plasma," Problems in Plasma Theory, Vol. 3. Atomizdat (1964).
9. A. A. Zakharov, "Spectra of turbulence in a plasma," Zh. Éksp. Teor. Fiz., 51: 310 (1966); L. M. Kovrizhnykh, "Correction to the article 'Effects of plasmon interaction,'" Zh. Éksp. Teor. Fiz., 49:1376 (1965).
10. A. Gailitis and V. N. Tsytovich, "Radiation during the scattering of charged particles by electromagnetic waves in an isotropic plasma," Zh. Éksp. Teor. Fiz., 46:1726 (1964).
11. A. Gailitis, L. M. Gorbunov, L. M. Kovrizhnykh, V. V. Pustovalov, V. P. Silin and V. N. Tsytovich, "Elementary processes of nonlinear interaction of charged particles with a plasma and the equation for a weakly turbulent plasma," Paper at the Conference on Phenomena in Ionized Gases, Belgrade (1965). FIAN Preprint No. A-136 (1965).
12. V. N. Tsytovich, "Nonlinear wave interaction in a plasma," Uspekhi Fiz. Nauk, 90 (1966).

# Index